Springer Optimization and Its Applications

VOLUME 86

Aims and Scope
Optimization has been expanding in all directions at an astonishing rate during the last few decades. New algorithmic and theoretical techniques have been developed, the diffusion into other disciplines has proceeded at a rapid pace, and our knowledge of all aspects of the field has grown even more profound. At the same time, one of the most striking trends in optimization is the constantly increasing emphasis on the interdisciplinary nature of the field. Optimization has been a basic tool in all areas of applied mathematics, engineering, medicine, economics, and other sciences.

The series *Springer Optimization and Its Applications* publishes undergraduate and graduate textbooks, monographs and state-of-the-art expository work that focus on algorithms for solving optimization problems and also study applications involving such problems. Some of the topics covered include nonlinear optimization (convex and nonconvex), network flow problems, stochastic optimization, optimal control, discrete optimization, multi-objective programming, description of software packages, approximation techniques and heuristic approaches.

For further volumes:
www.springer.com/series/7393

Yeol Je Cho • Themistocles M. Rassias •
Reza Saadati

Stability of Functional Equations in Random Normed Spaces

Springer

Yeol Je Cho
College of Education, Department of Mathematics Education
Gyeongsang National University
Chinju, Republic of South Korea

Reza Saadati
Department of Mathematics
Iran University of Science and Technology
Behshahr, Iran

Themistocles M. Rassias
Department of Mathematics
National Technical University of Athens
Athens, Greece

ISSN 1931-6828 Springer Optimization and Its Applications
ISBN 978-1-4939-0110-4 ISBN 978-1-4614-8477-6 (eBook)
DOI 10.1007/978-1-4614-8477-6
Springer New York Heidelberg Dordrecht London

Mathematics Subject Classification: 39B82, 39B52, 39B62, 39B72

Printed on acid-free paper

Springer is part of Springer Science+Business Media (www.springer.com)

To our beloved family:
In-Suk Kang (wife) and Eun-Hong Cho (granddaughter)
Ninetta (wife) and Michael, Stamatina (children)
Laleh (wife) and Shahriar, Shirin (children)

Preface

The study of functional equations has a long history. In 1791 and 1809, Legendre [152] and Gauss [90] attempted to provide a solution of the following functional equation:

$$f(x+y) = f(x) + f(y)$$

for all $x, y \in \mathbb{R}$, which is called the *Cauchy functional equation*. A function $f : \mathbb{R} \to \mathbb{R}$ is called an *additive function* if it satisfies the Cauchy functional equation. In 1821, Cauchy [30] first found the general solution of the Cauchy functional equation, that is, if $f : \mathbb{R} \to \mathbb{R}$ is a continuous additive function, then f is linear, that is, $f(x) = mx$, where m is a constant. Further, we can consider the biadditive function on $\mathbb{R} \times \mathbb{R}$ as follows:

A function $f : \mathbb{R} \times \mathbb{R} \to \mathbb{R}$ is called an *biadditive function* if it is additive in each variable, that is,

$$f(x+y, z) = f(x, z) + f(y, z)$$

and

$$f(x, y+z) = f(x, y) + f(x, z)$$

for all $x, y, z \in \mathbb{R}$. It is well known that every continuous biadditive function $f : \mathbb{R} \times \mathbb{R} \to \mathbb{R}$ is of the form

$$f(x, y) = mxy$$

for all $x, y \in \mathbb{R}$, where m is a constant.

Since the time of Legendre and Gauss, several mathematicians had dealt with additive functional equations in their books [2–4, 126, 145] and a number of them have studied Lagrange's mean value theorem and related functional equations, Pompeiu's mean value theorem and associated functional equations, two-dimensional mean value theorem and functional equations as well as several kinds of functional equations. We know that the mean value theorems have been motivated to study the functional equations (see the book "Mean Value Theorems and Functional Equations" by Sahoo and Riedel, 1998 [239]).

In 1940, S.M. Ulam [250] proposed the following stability problem of functional equations:

Given a group G_1, a metric group G_2 with the metric $d(\cdot,\cdot)$ and a positive number ε, does there exist $\delta > 0$ such that, if a mapping $f : G_1 \to G_2$ satisfies

$$d\big(f(xy), f(x)f(y)\big) \le \delta$$

for all $x, y \in G_1$, then a homomorphism $h : G_1 \to G_2$ exists with

$$d\big(f(x), h(x)\big) \le \varepsilon$$

for all $x \in G_1$?

Since then, several mathematicians have dealt with special cases as well as generalizations of Ulam's problem.

In fact, in 1941, D.H. Hyers [107] provided a partial solution to Ulam's problem for the case of approximately additive mappings in which G_1 and G_2 are Banach spaces with $\delta = \varepsilon$ as follows:

Let X and Y be Banach spaces and let $\varepsilon > 0$. Then, for all $g : X \to Y$ with

$$\sup_{x,y\in X} \big\| g(x+y) - g(x) - g(y) \big\| \le \varepsilon,$$

there exists a unique mapping $f : X \to Y$ such that

$$\sup_{x\in X} \big\| g(x) - f(x) \big\| \le \varepsilon,$$

$$f(x+y) = f(x) + f(y)$$

for all $x, y \in X$.

This proof remains unchanged if G_1 is an Abelian semigroup. Particularly, in 1968, it was proved by Forti (Proposition 1, [88]) that the following theorem can be proved.

Theorem F *Let $(S,+)$ be an arbitrary semigroup and E be a Banach space. Assume that $f : S \to E$ satisfies*

$$\big\| f(x+y) - f(x) - f(y) \big\| \le \varepsilon. \tag{A}$$

Then the limit

$$g(x) = \lim_{n\to\infty} \frac{f(2^n x)}{2^n} \tag{B}$$

exists for all $x \in S$ and $g : S \to E$ is the unique function satisfying

$$\big\| f(x) - g(x) \big\| \le \varepsilon, \quad g(2x) = 2g(x).$$

Finally, if the semigroup S is Abelian, then G is additive.

Here, the proof method which generates the solution g by the formula like (B) is called the *direct method*.

If f is a mapping of a group or a semigroup $(S, \cdot)$ into a vector space E, then we call the following expression:

$$Cf(x, y) = f(x \cdot y) - f(x) - f(y)$$

the *Cauchy difference* of f on $S \times S$. In the case that E is a topological vector space, we call the equation of homomorphism *stable* if, whenever the Cauchy difference Cf is bounded on $S \times S$, there exists a homomorphism $g : S \to E$ such that $f - g$ is bounded on S.

In 1980, Rätz [230] generalized Theorem F as follows: Let $(X, *)$ be a power-associative groupoid, that is, X is a nonempty set with a binary relation $x_1 * x_2 \in X$ such that the left powers satisfy $x^{m+n} = x^m * x^n$ for all $m, n \geq 1$ and $x \in X$. Let $(Y, |\cdot|)$ be a topological vector space over the field $\mathbb{Q}$ of rational numbers with $\mathbb{Q}$ topologized by its usual absolute value $|\cdot|$.

Theorem R *Let V be a nonempty bounded $\mathbb{Q}$-convex subset of Y containing the origin and assume that Y is sequentially complete. Let $f : X \to Y$ satisfy the following conditions: for all $x_1, x_2 \in X$, there exist $k \geq 2$ such that*

$$f\left((x_1 * x_2)^{k^n}\right) = f\left(x_1^{k^n} * x_2^{k^n}\right) \tag{C}$$

for all $n \geq 1$ and

$$f(x_1) + f(x_2) - f(x_1 * x_2) \in V. \tag{D}$$

Then there exists a function $g : X \to Y$ such that $g(x_1) + g(x_2) = g(x_1 * x_2)$ and $f(x) - g(x) \in \overline{V}$, where $\overline{V}$ is the sequential closure of V for all $x \in X$. When Y is a Hausdorff space, then g is uniquely determined.

Note that the condition (C) is satisfied when X is commutative and it takes the place of the commutativity in proving the additivity of g. However, as Rätz pointed out in his paper, the condition

$$(x_1 * x_2)^{k^n} = x_1^{k^n} * x_2^{k^n}$$

for all $x_1, x_2 \in X$, where X is a semigroup, and, for all $k \geq 1$, does not imply the commutativity.

In the proofs of Theorem F and Theorem R, the completeness of the image space E and the sequential completeness of Y, respectively, were essential in proving the existence of the limit which defined the additive function g. The question arises whether the completeness is necessary for the existence of an odd additive function g such that $f - g$ is uniformly bounded, given that the Cauchy difference is bounded.

For this problem, in 1988, Schwaiger [240] proved the following:

Theorem S *Let E be a normed space with the property that, for each function* $f:\mathbb{Z}\to E$*, whose Cauchy difference* $Cf = f(x+y)-f(x)-f(y)$ *is bounded for all* $x, y \in \mathbf{Z}$ *and there exists an additive mapping* $g:\mathbb{Z}\to E$ *such that* $f(x)-g(x)$ *is bounded for all* $x\in\mathbb{Z}$*. Then E is complete.*

Corollary 1 *The statement of theorem S remains true if* $\mathbb{Z}$ *is replaced by any vector space over* $\mathbb{Q}$*.*

In 1950, T. Aoki [14] generalized Hyers' theorem as follows:

Theorem A *Let* E_1 *and* E_2 *be two Banach spaces. If there exist* $K>0$ *and* $0\le p<1$ *such that*

$$\big\|f(x+y)-f(x)-f(y)\big\| \le K\big(\|x\|^p+\|y\|^p\big)$$

for all $x, y\in E_1$*, then there exists a unique additive mapping* $g: E_1\to E_2$ *such that*

$$\big\|f(x)-g(x)\big\| \le \frac{2K}{2-2^p}\|x\|^p$$

for all $x\in E_1$*.*

In 1978, Th.M. Rassias [216] formulated and proved the stability theorem for the linear mapping between Banach spaces E_1 and E_2 subject to the continuity of $f(tx)$ with respect to $t\in\mathbb{R}$ for each fixed $x\in E_1$. Thus, Rassias' theorem implies Aoki's theorem as a special case. Later, in 1990, Th.M. Rassias [218] observed that the proof of his stability theorem also holds true for $p<0$. In 1991, Gajda [89] showed that the proof of Rassias' theorem can be proved also for the case $p>1$ by just replacing n by $-n$ in (B). These results are stated in a generalized form as follows (see Rassias and Šemrl [228]):

Theorem RS *Let* $\beta(s,t)$ *be nonnegative for all nonnegative real numbers* s, t *and positive homogeneous of degree p, where p is real and* $p\ne 1$*, that is,* $\beta(\lambda s,\lambda t)=\lambda^p\beta(s,t)$ *for all nonnegative* λ, s, t*. Given a normed space* E_1 *and a Banach space* E_2*, assume that* $f: E_1\to E_2$ *satisfies the inequality*

$$\big\|f(x+y)-f(x)-f(y)\big\| \le \beta\big(\|x\|,\|y\|\big)$$

for all $x, y\in E_1$*. Then there exists a unique additive mapping* $g: E_1\to E_2$ *such that*

$$\big\|f(x)-g(x)\big\| \le \delta\|x\|^p$$

for all $x\in E_1$*, where*

$$\delta := \begin{cases} \frac{\beta(1,1)}{2-2^p}, & p<1,\\ \frac{\beta(1,1)}{2-2^p}, & p>1.\end{cases}$$

The proofs for the cases $p < 1$ and $p > 1$ were provided by applying the direct methods. For $p < 1$, the additive mapping g is given by (B), while in case $p > 1$ the formula is

$$g(x) = \lim_{n\to\infty} 2^n f\left(\frac{x}{2^n}\right).$$

Corollary 2 *Let $f : E_1 \to E_2$ be a mapping satisfying the hypotheses of Theorem RS and suppose that f is continuous at a single point $y \in E_1$, then the additive mapping g is continuous.*

Corollary 3 *If, under the hypotheses of Theorem RS, we assume that, for each fixed $x \in E_1$, the mapping $t \to f(tx)$ from $\mathbb{R}$ to E_2 is continuous, then the additive mapping g is linear.*

Remark 4 (1) For $p = 0$, Theorem RS, Corollaries 2 and 3 reduce to the results of Hyers in 1941. If we put $\beta(s,t) = \varepsilon(s^p + t^p)$, then we obtain the results of Rassias [216] in 1978 and Gajda [89] in 1991.

(2) The case $p = 1$ was excluded in Theorem RS. Simple counterexamples prove that one can not extend Rassias' Theorem when p takes the value one (see Z. Gajda [89], Rassias and Šemrl [228] and Hyers and Rassias [109] in 1992).

A further generalization of the Hyers-Ulam stability for a large class of mappings was obtained by Isac and Rassias [110] by introducing the following:

Definition 5 A mapping $f : E_1 \to E_2$ is said to be *ϕ-additive* if there exist $\Phi \geq 0$ and a function $\phi : \mathbb{R}_+ \to \mathbb{R}_+$ satisfying

$$\lim_{t\to+\infty} \frac{\phi(t)}{t} = 0$$

such that

$$\left\| f(x+y) - f(x) - f(y) \right\| \leq \Phi\left[\phi\left(\|x\|\right) + \phi\left(\|y\|\right)\right]$$

for all $x, y \in E_1$.

In [110], Isac and Rassias proved the following:

Theorem IR *Let E_1 be a real normed vector space and E_2 be a real Banach space. Let $f : E_1 \to E_2$ be a mapping such that $f(tx)$ is continuous in t for each fixed $x \in E_1$. If f is ϕ-additive and phi satisfies the following conditions:*

(a) $\phi(ts) \leq \phi(t)\phi(s)$ *for all $s, t \in \mathbb{R}$;*
(b) $\phi(t) < t$ *for all $t > 1$,*

then there exists a unique linear mapping $T : E_1 \to E_2$ *such that*

$$\| f(x) - T(x)\| \le \frac{2\theta}{2-\phi(2)}\phi(\|x\|)$$

for all $x \in E_1$.

Remark 4 (1) If $\phi(t) = t^p$ with $p < 1$, then, from Theorem IR, we obtain Rassias' theorem [216].

(2) If $p < 0$ and $\phi(t) = t^p$ with $t > 0$, then Theorem IR is implied by the result of Gajda in 1991.

Since the time the above stated results have been proven, several mathematicians (see [1, 5–13, 17, 18, 20–25, 28, 31–35, 37, 38, 42–44, 46, 47, 50–63, 67–99, 105, 108, 111–120, 124, 132–137, 144–159, 169–208, 212–215, 219–238, 243, 245–260, 262] and [263]) have extensively studied stability theorems for several kinds of functional equations in various spaces, for example, Banach spaces, 2-Banach spaces, Banach n-Lie algebras, quasi-Banach spaces, Banach ternary algebras, non-Archimedean normed and Banach spaces, metric and ultra metric spaces, Menger probabilistic normed spaces, probabilistic normed space, p-2-normed spaces, C^*-algebras, C^*-ternary algebras, Banach ternary algebras, Banach modules, inner product spaces, Heisenberg groups and others. Further, we have to pay attention to applications of the Hyers-Ulam-Rassias stability problems, for example, (partial) differential equations, Fréchet functional equations, Riccati differential equations, Volterra integral equations, group and ring theory and some kinds of equations (see [29, 114, 121–123, 128, 129, 142, 143, 153, 155, 157, 170–172, 209–211, 255, 257]). For more details on recent development in Ulam's type stability and its applications, see the papers of Brillouët-Belluot [19] and Ciepliński [41] in 2012.

The notion of random normed space goes back to Sherstnev [242] as well as the works published in [100, 101, 241] who were dulled from Menger [160], Schweizer and Sklar [241] works. After the pioneering works by several mathematicians including authors [9, 10, 148–150, 236] who focused at probabilistic functional analysis, Alsina [8] considered the stability of a functional equation in probabilistic normed spaces and, in 2008, Miheţ and Radu considered the stability of a Cauchy additive functional equation in random normed space via fixed point method [161].

The book provides a recent survey of both the latest and new results especially on the following topics:

(1) Basic theory of random normed spaces and related spaces;
(2) Stability theory for several new functional equations in random normed spaces via fixed point method, under the special t-norms as well as arbitrary t-norms;
(3) Stability theory of well known new functional equations in non-Archimedean random normed spaces;
(4) Applications in the class of fuzzy normed spaces.

We would like to express our thanks to Professors Claudi Alsina, Stefan Czerwik, Abbas Najati, Dorian Popa, Ioan A. Rus and G. Zamani Eskandani for reading the manuscript and providing valuable suggestions and comments which have helped to improve the presentation of the book.

Last but not least, it is our pleasure to acknowledge the superb assistance provided by the staff of Springer for the publication of the book.

Jinju, Republic of South Korea Yeol Je Cho
Athens, Greece Themistocles M. Rassias
Tehran, Iran Reza Saadati
June 2013

Contents

Acronyms

AQCQ	additive–quadratic–cubic–quartic
ACQ	additive–cubic–quartic
IRN-space	intuitionistic random normed space
IF-normed space	intuitionistic fuzzy normed space
LRN-space	lattice random normed space
RN-space	random normed space
QC	quadratic–cubic
Δ^+	space of all distribution functions
$\mathcal{L}$	complete lattice
$\mathcal{N}$	involutive negation
$(\mathbb{R}^n, \Phi, T)$	random Euclidean normed space
t-norm	triangular norm
T_L	Łukasiewicz t-norm
T_M	minimum t-norm
$(X, \mathcal{F}, T)$	Menger probabilistic metric space
(X, μ, T)	random normed space
τ	binary operation on Δ^+
$(X, M, *)$	fuzzy metric space
$(X, N, *)$	fuzzy normed space

Chapter 1
Preliminaries

In this chapter, we recall some definitions and results which will be used later on in the book.

1.1 Triangular Norms

Triangular norms first appeared in the framework of probabilistic metric spaces in the work of Menger [160]. It turns also out that this is a essential operation in several fields. Triangular norms are an indispensable tool for the interpretation of the conjunction in fuzzy logics [104] and, subsequently, for the intersection of fuzzy sets [261]. They are, however, interesting mathematical objects for themselves. We refer to some papers and books for further details (see [100, 138–141] and [241]).

Definition 1.1.1 A *triangular norm* (shortly, t-*norm*) is a binary operation on the unit interval $[0, 1]$, that is, a function $T : [0, 1] \times [0, 1] \to [0, 1]$ such that, for all $a, b, c \in [0, 1]$, the following four axioms are satisfied:

(T1) $T(a, b) = T(b, a)$ (commutativity);
(T2) $T(a, (T(b, c))) = T(T(a, b), c)$ (associativity);
(T3) $T(a, 1) = a$ (boundary condition);
(T4) $T(a, b) \leq T(a, c)$ whenever $b \leq c$ (monotonicity).

The commutativity of (T1), the boundary condition (T3) and the monotonicity (T4) imply that, for each t-norm T and $x \in [0, 1]$, the following boundary conditions are also satisfied:

$$T(x, 1) = T(1, x) = x,$$
$$T(x, 0) = T(0, x) = 0,$$

and so all the t-norms coincide with the boundary of the unit square $[0, 1]^2$.

Y.J. Cho et al., *Stability of Functional Equations in Random Normed Spaces*,
Springer Optimization and Its Applications 86, DOI 10.1007/978-1-4614-8477-6_1,

The monotonicity of a t-norm T in the second component (T4) is, together with the commutativity (T1), equivalent to the (joint) monotonicity in both components, that is, to

$$T(x_1, y_1) \leq T(x_2, y_2) \tag{1.1.1}$$

whenever $x_1 \leq x_2$ and $y_1 \leq y_2$.

Basic examples are the Łukasiewicz t-norm T_L:

$$T_L(a, b) = \max\{a + b - 1, 0\}$$

for all $a, b \in [0, 1]$ and the t-norms T_P, T_M, T_D defined as follows:

$$T_P(a, b) := ab,$$

$$T_M(a, b) := \min\{a, b\},$$

$$T_D(a, b) := \begin{cases} \min\{a, b\}, & \text{if } \max\{a, b\} = 1; \\ 0, & \text{otherwise.} \end{cases}$$

If, for any two t-norms T_1 and T_2, the inequality $T_1(x, y) \leq T_2(x, y)$ holds for all $(x, y) \in [0, 1]^2$, then we say that T_1 is *weaker* than T_2 or, equivalently, T_2 is *stronger* than T_2.

From (1.1.1), it follows that, for all $(x, y) \in [0, 1]^2$,

$$T(x, y) \leq T(x, 1) = x,$$

$$T(x, y) \leq T(1, y) = y.$$

Since $T(x, y) \geq 0 = T_D(x, y)$ for all $(x, y) \in (0, 1)^2$ holds trivially, for any t-norm T, we have

$$T_D \leq T \leq T_M,$$

that is, T_D is weaker and T_M is stronger than any others t-norms. Also, since $T_L < T_P$, we obtain the following ordering for four basic t-norms:

$$T_D < T_L < T_P < T_M.$$

Proposition 1.1.2 [100] (1) *The minimum T_M is the only t-norm satisfying $T(x, x) = x$ for all $x \in (0, 1)$.*

(2) *The weakest t-norm T_D is the only t-norm satisfying $T(x, x) = 0$ for all $x \in (0, 1)$.*

Proposition 1.1.3 [100] *A t-norm T is continuous if and only if it is continuous in its first component, i.e., for all $y \in [0, 1]$, if the one-place function*

$$T(\cdot, y) : [0, 1] \to [0, 1], \quad x \mapsto T(x, y),$$

is continuous.

For example, the minimum T_M and Łukasiewicz t-norm T_L are continuous, but the t-norm T^{Δ} defined by

$$T^{\Delta}(x, y) := \begin{cases} \frac{xy}{2}, & \text{if } \max\{x, y\} < 1; \\ xy, & \text{otherwise,} \end{cases}$$

for all $x, y \in [0, 1]$ is not continuous.

Definition 1.1.4 (1) A t-norm T is said to be *strictly monotone* if

$$T(x, y) < T(x, z)$$

whenever $x \in (0, 1)$ and $y < z$.

(2) A t-norm T is said to be *strict* if it is continuous and strictly monotone.

For example, the t-norm T^{Δ} is strictly monotone, but the minimum T_M and Łukasiewicz t-norm T_L are not strictly monotone.

Proposition 1.1.5 [100] *A t-norm T is strictly monotone if and only if*

$$T(x, y) = T(x, z), \quad x > 0 \quad \Longrightarrow \quad y = z.$$

If T is a t-norm, then $x_T^{(n)}$ for all $x \in [0, 1]$ and $n \geq 0$ is defined by 1 if $n = 0$ and $T(x_T^{(n-1)}, x)$ if $n \geq 1$.

Definition 1.1.6 A t-norm T is said to be *Archimedean* if, for all $(x, y) \in (0, 1)^2$, there exists an integer $n \geq 1$ such that

$$x_T^{(n)} < y.$$

Proposition 1.1.7 [100] *A t-norm T is Archimedean if and only if, for all $x \in (0, 1)$,*

$$\lim_{n\to\infty} x_T^{(n)} = 0.$$

Proposition 1.1.8 [100] *If t-norm T is Archimedean, then, for all $x \in (0, 1)$, we have*

$$T(x, x) < x.$$

For example, the product T_p, Łukasiewicz t-norm T_L and the weakest t-norm T_D are all Archimedean, but the minimum T_M is not an Archimedean t-norm.

A t-norm T is said to be *of Hadžić-type* (denoted by $T \in \mathcal{H}$) if the family $\{x_T^{(n)}\}$ is equicontinuous at $x = 1$, that is, for any $\varepsilon \in (0, 1)$, there exists $\delta \in (0, 1)$ such that

$$x > 1 - \delta \quad \Longrightarrow \quad x_T^{(n)} > 1 - \varepsilon \tag{1.1.2}$$

for all $n \geq 1$.

The t-norm T_M is a trivial example of Hadžić type, but T_P is not of Hadžić type.

Proposition 1.1.9 [100] *If a continuous t-norm T is Archimedean, then it can not be a t-norm of Hadžić-type.*

Other important t-norms are as follows (see [102]):

(1) The *Sugeno–Weber family* $\{T_\lambda^{SW}\}_{\lambda\in[-1,\infty]}$ is defined by $T_{-1}^{SW} = T_D$, $T_\infty^{SW} = T_P$ and

$$T_\lambda^{SW}(x, y) = \max\left\{0, \frac{x+y-1+\lambda xy}{1+\lambda}\right\}$$

if $\lambda \in (-1, \infty)$.

(2) The *Domby family* $\{T_\lambda^D\}_{\lambda\in[0,\infty]}$ is defined by T_D, if $\lambda = 0$, T_M if $\lambda = \infty$ and

$$T_\lambda^D(x, y) = \frac{1}{1+((\frac{1-x}{x})^\lambda + (\frac{1-y}{y})^\lambda)^{1/\lambda}}$$

if $\lambda \in (0, \infty)$.

(3) The *Aczel–Alsina family* $\{T_\lambda^{AA}\}_{\lambda\in[0,\infty]}$ is defined by T_D, if $\lambda = 0$, T_M if $\lambda = \infty$ and

$$T_\lambda^{AA}(x, y) = e^{-(|\log x|^\lambda + |\log y|^\lambda)^{1/\lambda}}$$

if $\lambda \in (0, \infty)$.

A t-norm T can be extended (by associativity) in a unique way to an n-array operation taking, for any $(x_1, \dots, x_n) \in [0, 1]^n$, the value $T(x_1, \dots, x_n)$ defined by

$$\mathrm{T}_{i=1}^0 x_i = 1, \qquad \mathrm{T}_{i=1}^n x_i = T\left(\mathrm{T}_{i=1}^{n-1} x_i, x_n\right) = T(x_1, \dots, x_n).$$

The t-norm T can also be extended to a countable operation taking, for any sequence $\{x_n\}$ in $[0, 1]$, the value

$$\mathrm{T}_{i=1}^\infty x_i = \lim_{n\to\infty} \mathrm{T}_{i=1}^n x_i. \tag{1.1.3}$$

The limit on the right side of (1.1.3) exists since the sequence $\{\mathrm{T}_{i=1}^n x_i\}$ is non-increasing and bounded from below.

Proposition 1.1.10 [102] (1) *For $T \ge T_L$ the following implication holds:*

$$\lim_{n\to\infty} \mathrm{T}_{i=1}^\infty x_{n+i} = 1 \iff \sum_{n=1}^\infty (1 - x_n) < \infty.$$

(2) *If T is of Hadžić-type, then we have*

$$\lim_{n\to\infty} \mathrm{T}_{i=1}^\infty x_{n+i} = 1$$

for any sequence $\{x_n\}_{n\ge1}$ in $[0, 1]$ such that $\lim_{n\to\infty} x_n = 1$.

(3) *If* $T \in \{T_\lambda^{AA}\}_{\lambda \in (0,\infty)} \cup \{T_\lambda^{D}\}_{\lambda \in (0,\infty)}$, *then we have*

$$\lim_{n\to\infty} \mathrm{T}_{i=1}^{\infty} x_{n+i} = 1 \quad \Longleftrightarrow \quad \sum_{n=1}^{\infty} (1 - x_n)^\alpha < \infty.$$

(4) *If* $T \in \{T_\lambda^{SW}\}_{\lambda \in [-1,\infty)}$, *then we have*

$$\lim_{n\to\infty} \mathrm{T}_{i=1}^{\infty} x_{n+i} = 1 \quad \Longleftrightarrow \quad \sum_{n=1}^{\infty} (1 - x_n) < \infty.$$

Definition 1.1.11 Let T and T' be two continuous t-norms. Then we say that T' *dominates* T (denoted by $T' \gg T$) if, for all $x_1, x_2, y_1, y_2 \in [0, 1]$,

$$T\big[T'(x_1, x_2), T'(y_1, y_2)\big] \le T'\big[T(x_1, y_1), T(x_2, y_2)\big].$$

1.2 Triangular Norms on Lattices

Now, we extend definitions and results on the t-norm to lattices.

Let $\mathcal{L} = (L, \ge_L)$ be a complete lattice, that is, a *partially ordered set* in which every nonempty subset admits supremum, infimum and $0_{\mathcal{L}} = \inf L$, $1_{\mathcal{L}} = \sup L$.

Definition 1.2.1 [49] A *t-norm* on L is a mapping $\mathcal{T} : L \times L \to L$ satisfying the following conditions:

(1) $\mathcal{T}(x, 1_{\mathcal{L}}) = x$ for all $x \in L$ (boundary condition);
(2) $\mathcal{T}(x, y) = \mathcal{T}(y, x)$ for all $x, y \in L$ (commutativity);
(3) $\mathcal{T}(x, \mathcal{T}(y, z)) = \mathcal{T}(\mathcal{T}(x, y), z)$ for all $x, y, z \in L$ (associativity);
(4) $x \le_L x'$ and $y \le_L y'$ implies that $\mathcal{T}(x, y) \le_L \mathcal{T}(x', y')$ for all $x, x', y, y' \in L$ (monotonicity).

Let $\{x_n\}$ be a sequence in L convergent to $x \in L$ (equipped order topology). The t-norm $\mathcal{T}$ is said to be a *continuous t-norm* if

$$\lim_{n\to\infty} \mathcal{T}(x_n, y) = \mathcal{T}(x, y)$$

for each $y \in L$.

Now, we put $\mathcal{T} = T$ whenever $L = [0, 1]$.

Definition 1.2.2 [49] A continuous t-norm $\mathcal{T}$ on $L = [0, 1]^2$ is said to be *continuous t-representable* if there exist a continuous t-norm $*$ and a continuous t-conorm $\diamond$ on $[0, 1]$ such that, for all $x = (x_1, x_2)$, $y = (y_1, y_2) \in L$,

$$\mathcal{T}(x, y) = (x_1 * y_1, x_2 \diamond y_2).$$

For example, the following t-norms

$$\mathcal{T}(a,b) = \big(a_1 b_1, \min\{a_2 + b_2, 1\}\big)$$

and

$$\mathbf{M}(a,b) = \big(\min\{a_1, b_1\}, \max\{a_2, b_2\}\big)$$

for all $a = (a_1, a_2)$, $b = (b_1, b_2) \in [0,1]^2$ are continuous t-representable.

Define the mapping $\mathcal{T}_\wedge : L^2 \to L$ by:

$$\mathcal{T}_\wedge(x,y) = \begin{cases} x, & \text{if } y \geq_L x, \\ y, & \text{if } x \geq_L y. \end{cases}$$

A *negation* on $\mathcal{L}$ is a decreasing mapping $\mathcal{N} : L \to L$ satisfying $\mathcal{N}(0_{\mathcal{L}}) = 1_{\mathcal{L}}$ and $\mathcal{N}(1_{\mathcal{L}}) = 0_{\mathcal{L}}$. If $\mathcal{N}(\mathcal{N}(x)) = x$ for all $x \in L$, then $\mathcal{N}$ is called an *involutive negation*. In the following, $\mathcal{L}$ is endowed with a (fixed) negation $\mathcal{N}$.

1.3 Distribution Functions

Let Δ^+ denote the space of all distribution functions, that is, the space of all mappings $F : \mathbb{R} \cup \{-\infty, +\infty\} \to [0,1]$ such that F is left-continuous, non-decreasing on $\mathbb{R}$, $F(0) = 0$ and $F(+\infty) = 1$. D^+ is a subset of Δ^+ consisting of all functions $F \in \Delta^+$ for which $l^- F(+\infty) = 1$, where $l^- f(x)$ denotes the left limit of the function f at the point x, that is, $l^- f(x) = \lim_{t \to x^-} f(t)$. The space Δ^+ is partially ordered by the usual point-wise ordering of functions, i.e., $F \leq G$ if and only if $F(t) \leq G(t)$ for all $t \in \mathbb{R}$. The maximal element for Δ^+ in this order is the distribution function ε_0 given by

$$\varepsilon_0(t) = \begin{cases} 0, & \text{if } t \leq 0, \\ 1, & \text{if } t > 0. \end{cases}$$

Example 1.3.1 The function $G(t)$ defined by

$$G(t) = \begin{cases} 0, & \text{if } t \leq 0, \\ 1 - e^{-t}, & \text{if } t > 0, \end{cases}$$

is a distribution function. Since $\lim_{t \to \infty} G(t) = 1$, $G \in D^+$. Note that $G(t+s) \geq T_p(G(t), G(s))$ for each $t, s > 0$.

Example 1.3.2 The function $F(t)$ defined by

$$F(t) = \begin{cases} 0, & \text{if } t \leq 0, \\ t, & \text{if } 0 \leq t \leq 1, \\ 1, & \text{if } 1 \leq t, \end{cases}$$

is a distribution function. Since $\lim_{t\to\infty} F(t) = 1$, $F \in D^+$. Note that $F(t+s) \geq T_M(F(t), F(s))$ for all $t, s > 0$.

Example 1.3.3 [9] The function $G_p(t)$ defined by

$$G_p(t) = \begin{cases} 0, & \text{if } t \leq 0, \\ \exp(-|p|^{1/2}), & \text{if } 0 < t < +\infty, \\ 1, & \text{if } t = +\infty, \end{cases}$$

is a distribution function. Since $\lim_{t\to\infty} G_p(t) \neq 1$, $G \in \Delta^+ \setminus D^+$. Note that $G_p(t+s) \geq T_M(G_p(t), G_p(s))$ for all $t, s > 0$.

Definition 1.3.4 A *non-measure distribution function* is a function $\nu : \mathbb{R} \to [0, 1]$ which is right continuous on $\mathbb{R}$, non-increasing and $\inf_{t\in\mathbb{R}} \nu(t) = 0$, $\sup_{t\in\mathbb{R}} \nu(t) = 1$.

We denote by B the family of all non-measure distribution functions and by G a special element of B defined by

$$G(t) = \begin{cases} 1, & \text{if } t \leq 0, \\ 0, & \text{if } t > 0. \end{cases}$$

If X is a nonempty set, then $\nu : X \to B$ is called a *probabilistic non-measure* on X and $\nu(x)$ is denoted by ν_x.

Let $\mathcal{L} = (L, \geq_L)$ be a complete lattice, that is, a partially ordered set in which every nonempty subset admits supremum, infimum and $0_{\mathcal{L}} = \inf L$, $1_{\mathcal{L}} = \sup L$. The space of latticetic random distribution functions, denoted by Δ_L^+, is defined as the set of all mappings $F : \mathbb{R} \cup \{-\infty, +\infty\} \to L$ such that F is left-continuous and non-decreasing on $\mathbb{R}$, $F(0) = 0_{\mathcal{L}}$ and $F(+\infty) = 1_{\mathcal{L}}$.

$D_L^+ \subseteq \Delta_L^+$ is defined as $D_L^+ = \{F \in \Delta_L^+ : l^- F(+\infty) = 1_{\mathcal{L}}\}$, where $l^- f(x)$ denotes the left limit of the function f at the point x. The space Δ_L^+ is partially ordered by the usual point-wise ordering of functions, that is, $F \geq G$ if and only if $F(t) \geq_L G(t)$ for all $t \in \mathbb{R}$. The maximal element for Δ_L^+ in this order is the distribution function given by

$$\varepsilon_0(t) = \begin{cases} 0_{\mathcal{L}}, & \text{if } t \leq 0, \\ 1_{\mathcal{L}}, & \text{if } t > 0. \end{cases}$$

1.4 Fuzzy Sets

In this section, we consider the definition of *fuzzy sets* and present some examples. For more details, see [264]. The first publication in fuzzy set theory by Zadeh [261] showed a generalization of the classical notation of a set. A classical (crisp) set is normally defined as a collection of elements or objects $x \in X$ which can be finite, countable or uncountable. Each single element can either belong to or not belong to a set A, $A \subseteq X$. In the former case, the statement "x belong to A" is true, whereas, in the latter case, this statement is false.

Such a classical set can be described in different ways. One way is defined the member element by using the characteristic function, in which 1 indicates membership and 0 non-membership. For a fuzzy set, the characteristic function allows various degrees of membership for the elements of a given set.

Definition 1.4.1 If W is a collection of objects denoted generically by w, then a fuzzy set A in W is a set of ordered pairs:

$$A = \{(w, \lambda_A(w)) : w \in W\},$$

where $\lambda_A(w)$ is called the *membership function* or *grade of membership* of w in A which maps W to the membership space M.

Note that, when M contains only the two points 0 and 1, A is non-fuzzy and $\lambda_A(w)$ is identical to the characteristic function of a non-fuzzy set. The range of the membership function is $[0, 1]$ or a complete lattice.

Example 1.4.2 Consider the following fuzzy set A which is real numbers considerably larger than 10:

$$A = \{(w, \lambda_A(w)) : w \in W\},$$

where

$$\lambda_A(w) = \begin{cases} 0, & \text{if } w < 10, \\ \frac{1}{1+(w-10)^{-2}}, & \text{if } w \geq 10. \end{cases}$$

Example 1.4.3 Consider the following fuzzy set A which is real numbers close to 10:

$$A = \{(w, \lambda_A(w)) : w \in W\},$$

where

$$\lambda_A(w) = \frac{1}{1 + (w - 10)^2}.$$

Note that, in this book, in short, we apply membership functions instead fuzzy sets.

Definition 1.4.4 [96] Let $\mathcal{L} = (L, \leq_L)$ be a complete lattice and U be a nonempty set called the *universe*. An *$\mathcal{L}$-fuzzy set* in U is defined as a mapping $\mathcal{A} : U \to L$. For each $u \in U$, $\mathcal{A}(u)$ represents the *degree* (in L) to which u is an element of $\mathcal{A}$.

Lemma 1.4.5 [49] *Consider the set L^* and the operation $\leq_{L^*}$ defined by*

$$L^* = \{(x_1, x_2) : (x_1, x_2) \in [0,1]^2,\ x_1 + x_2 \leq 1\},$$
$$(x_1, x_2) \leq_{L^*} (y_1, y_2) \quad \Longleftrightarrow \quad x_1 \leq y_1,\ x_2 \geq y_2$$

for all $(x_1, x_2), (y_1, y_2) \in L^*$. Then $(L^*, \leq_{L^*})$ is a complete lattice.

Definition 1.4.6 [16] An *intuitionistic fuzzy set* $\mathcal{A}_{\zeta,\eta}$ in the universe U is an object $\mathcal{A}_{\zeta,\eta} = \{(u, \zeta_{\mathcal{A}}(u), \eta_{\mathcal{A}}(u)) : u \in U\}$, where $\zeta_{\mathcal{A}}(u) \in [0,1]$ and $\eta_{\mathcal{A}}(u) \in [0,1]$ for all $u \in U$ are called the *membership degree* and the *non-membership degree*, respectively, of u in $\mathcal{A}_{\zeta,\eta}$ and, furthermore, satisfy $\zeta_{\mathcal{A}}(u) + \eta_{\mathcal{A}}(u) \leq 1$.

Example 1.4.7 Consider the following intuitionistic fuzzy set $\mathcal{A}_{\zeta,\eta}$ which is real numbers considerably larger than 10 for the first place and real numbers close to 10 in the second place:

$$\mathcal{A}_{\zeta,\eta} = \{(w, \zeta_A(w), \eta_A(w)) \ :\ w \in W\},$$

where

$$(\zeta_A(w), \eta_A(w)) = \begin{cases} (0, \frac{1}{1+(w-10)^2}), & \text{if } w < 10, \\ (\frac{1}{1+(w-10)^{-2}}, \frac{1}{1+(w-10)^2}), & \text{if } w \geq 10. \end{cases}$$

As we said in the above, forward, we will use $\mathcal{A}_{\zeta,\eta}(w) = (\zeta_A(w), \eta_A(w))$ in the next chapters.

Chapter 2
Generalized Spaces

In this chapter, we present some generalized spaces and their properties for the main results in this chapter.

2.1 Random Normed Spaces

Random (probabilistic) normed spaces were introduced by Šerstnev in 1962 [242] by means of a definition that was closely modelled on the theory of (classical) normed spaces, and used to study the problem of best approximation in statistics. In the sequel, we shall adopt usual terminology, notation and conventions of the theory of random normed spaces, as in [9, 10, 148, 241].

Definition 2.1.1 A *Menger probabilistic metric space* (or *random metric spaces*) is a triple $(X, \mathcal{F}, T)$, where X is a nonempty set, T is a continuous t-norm and $\mathcal{F}$ is a mapping from $X \times X$ into D^+ such that, if $F_{x,y}$ denotes the value of $\mathcal{F}$ at a point $(x, y) \in X \times X$, the following conditions hold: for all x, y, z in X,

(PM1) $F_{x,y}(t) = \varepsilon_0(t)$ for all $t > 0$ if and only if $x = y$;
(PM2) $F_{x,y}(t) = F_{y,x}(t)$;
(PM3) $F_{x,z}(t + s) \geq T(F_{x,y}(t), F_{y,z}(s))$ for all $x, y, z \in X$ and $t, s \geq 0$.

Definition 2.1.2 [242] A *random normed space* (briefly, a RN-space) or *a Šerstnev (Sherstnev) probabilistic normed space* (briefly, a Šerstnev PN-space) is a triple (X, μ, T), where X is a vector space, T is a continuous t-norm and μ is a mapping from X into D^+ such that the following conditions hold:

(RN1) $\mu_x(t) = \varepsilon_0(t)$ for all $t > 0$ if and only if $x = 0$ (0 is the null vector in X);
(RN2) $\mu_{\alpha x}(t) = \mu_x(\frac{t}{|\alpha|})$ for all $x \in X$ and $\alpha \neq 0$;
(RN3) $\mu_{x+y}(t + s) \geq T(\mu_x(t), \mu_y(s))$ for all $x, y \in X$ and $t, s \geq 0$, where μ_x denotes the value of μ at a point $x \in X$.

Y.J. Cho et al., *Stability of Functional Equations in Random Normed Spaces*, Springer Optimization and Its Applications 86, DOI 10.1007/978-1-4614-8477-6_2,

Note that a *triangular function* $\tau : \Delta^+ \times \Delta^+ \to \Delta^+$ is a binary operation on Δ^+ which is associative, commutative and nondecreasing in each argument and has ε_0 as the unit, that is, for all $F, G, H \in \Delta^+$,

$$\begin{aligned}\tau\big(\tau(F,G),H\big) &= \tau\big(F,\tau(G,H)\big),\\ \tau(F,G) &= \tau(G,F),\\ \tau(F,\varepsilon_0) &= F,\\ F \le G &\implies \tau(F,H) \le \tau(G,H).\end{aligned}$$

The continuity of a triangular function means the continuity with respect to the topology of weak convergence in Δ^+. Triangular functions are recursively defined by $\tau^1 = \tau$ and

$$\tau^n(F_1,\ldots,F_{n+1}) = \tau\big(\tau^{n-1}(F_1,\ldots,F_n),F_{n+1}\big)$$

for each $n \ge 2$.

Typical continuous triangular functions are as follows:

$$\tau_T(F,G)(x) = \sup_{s+t=x} T\big(F(s),G(t)\big),$$

and

$$\tau_{T^*}(F,G) = \inf_{s+t=x} T^*\big(F(s),G(t)\big),$$

where T is a continuous t-norm, that is, a continuous binary operation on $[0,1]$ that is commutative, associative, nondecreasing in each variable and has 1 as the identity element and T^* is a continuous t-conorm, that is, a continuous binary operation on $[0,1]$ which is related to the continuous t-norm T through $T^*(x,y) = 1 - T(1-x, 1-y)$.

Examples of such t-norms and t-conorms are M and M^*, respectively, defined by

$$M(x,y) = \min(x,y)$$

and

$$M^*(x,y) = \max(x,y).$$

Let τ_1 and τ_2 be two triangular functions. Then τ_1 dominates τ_2 (which is denoted by $\tau_1 \gg \tau_2$) if, for all $F_1, F_2, G_1, G_2 \in \Delta^+$,

$$\tau_1\big(\tau_2(F_1,G_1),\tau_2(F_2,G_2)\big) \ge \tau_2\big(\tau_1(F_1,F_2),\tau_1(G_1,G_2)\big).$$

In 1993, Alsina, Schweizer and Sklar gave a new definition of a probabilistic normed space [9] as follows:

A *probabilistic normed space* (briefly, PN-space) is a quadruple (V,ν,τ,τ^*), where V is a real vector space, τ, τ^* are continuous triangulares functions and ν

is a mapping from $V \to \Delta^+$ such that, for all $p, q \in V$, the following conditions hold:

(PN1) $\nu_p = \varepsilon_0$ if and only if $p = \theta$, where θ is the null vector in V;
(PN2) $\nu_{-p} = \nu_p$ for all $p \in V$;
(PN3) $\nu_{p+q} \geq \tau(\nu_p, \nu_q)$ for all $p, q \in V$;
(PN4) $\nu_p \leq \tau^*(\nu_{\alpha p}, \nu_{(1-\alpha)p})$ for all $\alpha \in [0, 1]$.

If the inequality (PN4) is replaced by the equality $\nu_p = \tau_M(\nu_{\alpha p}, \nu_{(1-\alpha)p})$, then the PN-space (V, ν, τ, τ^*) is called a *Šerstnev probabilistic normed space* or a *random normed space* (see Definition 2.1.2) and, as a consequence, we have the following condition stronger than (N2):

$$\nu_{\lambda p}(x) = \nu_p\left(\frac{x}{|\lambda|}\right)$$

for all $p \in V$, $\lambda \neq 0$ and $x \in \mathbb{R}$.

Example 2.1.3 Let $(X, \|\cdot\|)$ be a linear normed spaces. Define a mapping

$$\mu_x(t) = \begin{cases} 0, & \text{if } t \leq 0, \\ \frac{t}{t+\|x\|}, & \text{if } t > 0. \end{cases}$$

Then (X, μ, T_p) is a random normed space. In fact, (RN1) and (RN2) are obvious. Now, we show (RN3).

$$\begin{aligned} T_p\big(\mu_x(t), \mu_y(s)\big) &= \frac{t}{t+\|x\|} \cdot \frac{s}{s+\|y\|} \\ &= \frac{1}{1+\frac{\|x\|}{t}} \cdot \frac{1}{1+\frac{\|y\|}{s}} \\ &\leq \frac{1}{1+\frac{\|x\|}{t+s}} \cdot \frac{1}{1+\frac{\|y\|}{t+s}} \\ &\leq \frac{1}{1+\frac{\|x\|+\|y\|}{t+s}} \\ &\leq \frac{1}{1+\frac{\|x+y\|}{t+s}} \\ &= \frac{t+s}{t+s+\|x+y\|} \\ &= \mu_{x+y}(t+s) \end{aligned}$$

for all $x, y \in X$ and $t, s \geq 0$. Also, (X, μ, T_M) is a random normed space.

Example 2.1.4 Let $(X, \|\cdot\|)$ be a linear normed spaces. Define a mapping

$$\mu_x(t) = \begin{cases} 0, & \text{if } t \le 0, \\ e^{-(\frac{\|x\|}{t})}, & \text{if } t > 0. \end{cases}$$

Then (X, μ, T_p) is a random normed space. In fact, (RN1) and (RN2) are obvious and so, now, we show (RN3).

$$\begin{aligned} T_p\big(\mu_x(t), \mu_y(s)\big) &= e^{-(\frac{\|x\|}{t})} \cdot e^{-(\frac{\|y\|}{s})} \\ &\le e^{-(\frac{\|x\|}{t+s})} \cdot e^{-(\frac{\|y\|}{t+s})} \\ &= e^{-(\frac{\|x\|+\|y\|}{t+s})} \\ &\le e^{-(\frac{\|x+y\|}{t+s})} \\ &= \mu_{x+y}(t+s) \end{aligned}$$

for all $x, y \in X$ and $t, s \ge 0$. Also, (X, μ, T_M) is a random normed space.

Example 2.1.5 [164] Let $(X, \|\cdot\|)$ be a linear normed space. For all $x \in X$, define a mapping

$$\mu_x(t) = \begin{cases} \max\{1 - \frac{\|x\|}{t}, 0\}, & \text{if } t > 0, \\ 0, & \text{if } t \le 0. \end{cases}$$

Then (X, μ, T_L) is a RN-space (this was essentially proved by Musthari in [179], see also [213]). Indeed, we have

$$\mu_x(t) = 1 \quad \Longrightarrow \quad \frac{\|x\|}{t} = 0 \quad \Longrightarrow \quad x = 0$$

for all $t > 0$ and, obviously,

$$\mu_{\lambda x}(t) = \mu_x\left(\frac{t}{\lambda}\right)$$

for all $x \in X$ and $t > 0$. Next, for any $x, y \in X$ and $t, s > 0$, we have

$$\begin{aligned} \mu_{x+y}(t+s) &= \max\left\{1 - \frac{\|x+y\|}{t+s}, 0\right\} \\ &= \max\left\{1 - \|\frac{x+y}{t+s}\|, 0\right\} \\ &= \max\left\{1 - \|\frac{x}{t+s} + \frac{y}{t+s}\|, 0\right\} \\ &\ge \max\left\{1 - \|\frac{x}{t}\| - \|\frac{y}{s}\|, 0\right\} \end{aligned}$$

$$= T_L\big(\mu_x(t), \mu_y(s)\big).$$

Let φ be a function defined on the real field $\mathbb{R}$ into itself with the following properties:

(a) $\varphi(-t) = \varphi(t)$ for all $t \in \mathbb{R}$;
(b) $\varphi(1) = 1$;
(c) φ is strictly increasing and continuous on $[0, \infty)$, $\varphi(0) = 0$ and $\lim_{\alpha \to \infty} \varphi(\alpha) = \infty$.

Examples of such functions are as follows:

$$\varphi(t) = |t|, \qquad \varphi(t) = |t|^p \quad \big(p \in (0, \infty)\big), \qquad \varphi(t) = \frac{2t^{2n}}{|t| + 1}$$

for all $t \in \mathbb{R}$ and $n \geq 1$.

Definition 2.1.6 [97] A *random φ-normed space* is a triple (X, ν, T), where X is a real vector space, T is a continuous t-norm and ν is a mapping from X into D^+ such that the following conditions hold:

(φ-RN1) $\nu_x(t) = \varepsilon_0(t)$ for all $t > 0$ if and only if $x = 0$;
(φ-RN2) $\nu_{\alpha x}(t) = \nu_x(\frac{t}{\varphi(\alpha)})$ for all x in X, $\alpha \neq 0$ and $t > 0$;
(φ-RN3) $\nu_{x+y}(t + s) \geq T(\nu_x(t), \nu_y(s))$ for all $x, y \in X$ and $t, s \geq 0$.

Example 2.1.7 [165] An important example is the space (X, ν, T_M), where $(X, \|\cdot\|^p)$ is a p-normed space and

$$\nu_x(t) = \begin{cases} 0, & \text{if } t \leq 0, \\ \frac{t}{t + \|x\|^p}, \quad p \in (0, 1], & \text{if } t > 0. \end{cases}$$

(φ-RN1) and (φ-RN2) are obvious and so we show (φ-RN3). In fact, let $\nu_x(t) \leq \nu_y(s)$. Then we have

$$\frac{\|y\|^p}{s} \leq \frac{\|x\|^p}{t}$$

for any $x, y \in X$

Now, if $x = y$, we have $t \leq s$. Thus, otherwise, we have

$$\begin{aligned} \frac{\|x\|^p}{t} + \frac{\|x\|^p}{t} &\geq \frac{\|x\|^p}{t} + \frac{\|y\|^p}{s} \\ &\geq 2\frac{\|x\|^p}{t + s} + 2\frac{\|y\|^p}{t + s} \\ &\geq 2\frac{\|x + y\|^p}{t + s} \end{aligned}$$

and so

$$1+\frac{\|x\|^p}{t} \geq 1+\frac{\|x+y\|^p}{t+s},$$

which implies that $\nu_x(t) \leq \nu_{x+y}(t+s)$. Hence, $\nu_{x+y}(t+s) \geq T_M(\nu_x(t), \nu_y(s))$ for all $x, y \in X$ and $t, s \geq 0$.

Definition 2.1.8 Let μ and ν be measure and non-measure distribution function from $X \times (0, +\infty)$ to $[0, 1]$, respectively, such that $\mu_x(t) + \nu_x(t) \leq 1$ for all $x \in X$ and $t > 0$, where X is a real vector space. The triple $(X, \mathcal{P}_{\mu,\nu}, \mathcal{T})$ is said to be an *intuitionistic random normed space* (briefly, IRN-space) if X is a real vector space, $\mathcal{T}$ is a continuous t-representable and $\mathcal{P}_{\mu,\nu}$ is a mapping $X \times (0, +\infty) \to L^*$ satisfying the following conditions: for all $x, y \in X$ and $t, s > 0$,

(IRN1) $\mathcal{P}_{\mu,\nu}(x, 0) = 0_{L^*}$;
(IRN2) $\mathcal{P}_{\mu,\nu}(x, t) = 1_{L^*}$ if and only if $x = 0$;
(IRN3) $\mathcal{P}_{\mu,\nu}(\alpha x, t) = \mathcal{P}_{\mu,\nu}(x, \frac{t}{|\alpha|})$ for all $\alpha \neq 0$;
(IRN4) $\mathcal{P}_{\mu,\nu}(x + y, t + s) \geq_{L^*} \mathcal{T}(\mathcal{P}_{\mu,\nu}(x, t), \mathcal{P}_{\mu,\nu}(y, s))$.

In this case, $\mathcal{P}_{\mu,\nu}$ is called an *intuitionistic random norm*, where

$$\mathcal{P}_{\mu,\nu}(x, t) = \big(\mu_x(t), \nu_x(t)\big).$$

Definition 2.1.9 A *lattice random normed space* (LRN-space shortly) is a triple $(X, \mu, \mathcal{T}_\wedge)$, where X is a vector space and μ is a mapping from X into D_L^+ such that the following conditions hold:

(LRN1) $\mu_x(t) = 1_{\mathcal{L}}$ for all $t > 0$ if and only if $x = 0$;
(LRN2) $\mu_{\alpha x}(t) = \mu_x(\frac{t}{|\alpha|})$ for all x in X, $\alpha \neq 0$ and $t \geq 0$;
(LRN3) $\mu_{x+y}(t + s) \geq_L \mathcal{T}_\wedge(\mu_x(t), \mu_y(s))$ for all $x, y \in X$ and $t, s \geq 0$.

We note that, from (LRN2), $\mu_{-x}(t) = \mu_x(t)$ for all $x \in X$ and $t \geq 0$.

Example 2.1.10 Let $L = [0, 1] \times [0, 1]$ and the operation $\leq_L$ be defined by:

$$L = \{(a_1, a_2) : (a_1, a_2) \in [0, 1] \times [0, 1]\ a_1 + a_2 \leq 1\},$$
$$(a_1, a_2) \leq_L (b_1, b_2) \iff a_1 \leq b_1,\ a_2 \geq b_2$$

for all $a = (a_1, a_2), b = (b_1, b_2) \in L$. Then $(L, \leq_L)$ is a complete lattice (see [49]). In this complete lattice, we denote its units by $0_L = (0, 1)$ and $1_L = (1, 0)$.

Let $(X, \|\cdot\|)$ be a normed linear space. Let $\mathcal{T}(a, b) = (\min\{a_1, b_1\}, \max\{a_2, b_2\})$ for all $a = (a_1, a_2), b = (b_1, b_2) \in [0, 1] \times [0, 1]$ and μ be a mapping defined by

$$\mu_x(t) = \left(\frac{t}{t + \|x\|}, \frac{\|x\|}{t + \|x\|}\right)$$

for all $t \in \mathbb{R}^+$. Then $(X, \mu, \mathcal{T})$ is a lattice random normed space.

2.2 Random Topological Structures

In this section, we give some topological structures of random normed spaces.

Definition 2.2.1 Let (X, μ, T) be an RN-space. We define the *open ball* $B_x(r, t)$ and the closed ball $B_x[r, t]$ with center $x \in X$ and radius $0 < r < 1$ for all $t > 0$ as follows:

$$B_x(r,t) = \{y \in X : \mu_{x-y}(t) > 1 - r\},$$

$$B_x[r,t] = \{y \in X \ : \ \mu_{x-y}(t) \geq 1 - r\},$$

respectively.

Theorem 2.2.2 *Let (X, μ, T) be an RN-space. Every open ball $B_x(r, t)$ is open set.*

Proof Let $B_x(r, t)$ an open ball with center x and radius r for all $t > 0$. Let $y \in B_x(r, t)$. Then $\mu_{x-y}(t) > 1 - r$. Since $\mu_{x-y}(t) > 1 - r$, there exists $t_0 \in (0, t)$ such that $\mu_{x-y}(t_0) > 1 - r$. Put $r_0 = \mu_{x,y}(t_0)$. Since $r_0 > 1 - r$, there exists $s \in (0, 1)$ such that $r_0 > 1 - s > 1 - r$. Now, for any r_0 and s such that $r_0 > 1 - s$, there exists $r_1 \in (0, 1)$ such that $T(r_0, r_1) > 1 - s$. Consider the open ball $B_y(1 - r_1, t - t_0)$.

Now, we claim that $B_y(1 - r_1, t - t_0) \subset B_x(r, t)$. In fact, let $z \in B_y(1 - r_1, t - t_0)$. Then $\mu_{y-z}(t - t_0) > r_1$ and so

$$\begin{aligned}\mu_{x-z}(t) &\geq T\big(\mu_{x-y}(t_0), \mu_{y-z}(t - t_0)\big)\\ &\geq T(r_0, r_1)\\ &\geq 1 - s\\ &> 1 - r.\end{aligned}$$

Thus, $z \in B_x(r, t)$ and hence $B_y(1 - r_1, t - t_0) \subset B_x(r, t)$. This completes the proof. □

Now, different kinds of topologies can be introduced in a random normed space [241]. The (r, t)-*topology* is introduced by a family of neighborhoods

$$\{B_x(r,t)\}_{x \in X, t>0, r\in(0,1)}.$$

In fact, every random norm μ on X generates a topology ((r, t)-topology) on X which has as a base the family of open sets of the form

$$\{B_x(r,t)\}_{x \in X, t>0, r\in(0,1)}.$$

Remark 2.2.3 Since $\{B_x(\frac{1}{n}, \frac{1}{n}) : n \geq 1\}$ is a local base at x, the (r, t)-topology is first countable.

Theorem 2.2.4 *Every RN-space* (X, μ, T) *is a* Hausdorff space.

Proof Let (X, μ, T) be an RN-space. Let x and y be two distinct points in X and $t > 0$. Then $0 < \mu_{x-y}(t) < 1$. Put $r = \mu_{x-y}(t)$. For each $r_0 \in (r, 1)$, there exists r_1 such that $T(r_1, r_1) \geq r_0$. Consider the open balls $B_x(1 - r_1, \frac{t}{2})$ and $B_y(1 - r_1, \frac{t}{2})$. Then, clearly, $B_x(1 - r_1, \frac{t}{2}) \cap B_y(1 - r_1, \frac{t}{2}) = \emptyset$. In fact, if there exists

$$z \in B_x\left(1 - r_1, \frac{t}{2}\right) \cap B_y\left(1 - r_1, \frac{t}{2}\right),$$

then we have

$$\begin{aligned} r &= \mu_{x-y}(t) \\ &\geq T\left(\mu_{x-z}\left(\frac{t}{2}\right), \mu_{y-z}\left(\frac{t}{2}\right)\right) \\ &\geq T(r_1, r_1) \\ &\geq r_0 \\ &> r, \end{aligned}$$

which is a contradiction. Hence (X, μ, T) is a Hausdorff space. This completes the proof. □

Definition 2.2.5 Let (X, μ, T) be an RN-space. A subset A of X is said to be *R-bounded* if there exist $t > 0$ and $r \in (0, 1)$ such that $\mu_{x-y}(t) > 1 - r$ for all $x, y \in A$.

Theorem 2.2.6 *Every compact subset A of an RN-space* (X, μ, T) *is R-bounded.*

Proof Let A be a *compact* subset of an RN-space (X, μ, T). Fix $t > 0$, $0 < r < 1$ and consider an open cover $\{B_x(r, t) : x \in A\}$. Since A is compact, there exist $x_1, x_2, \ldots, x_n \in A$ such that

$$A \subseteq \bigcup_{i=1}^{n} B_{x_i}(r, t).$$

Let $x, y \in A$. Then $x \in B_{x_i}(r, t)$ and $y \in B_{x_j}(r, t)$ for some $i, j \geq 1$. Thus we have $\mu_{x-x_i}(t) > 1 - r$ and $\mu_{y-x_j}(t) > 1 - r$. Now, let

$$\alpha = \min\{\mu_{x_i, x_j}(t) : 1 \leq i, j \leq n\}.$$

Then we have $\alpha > 0$ and

$$\begin{aligned} \mu_{x-y}(3t) &\geq T^2\left(\mu_{x-x_i}(t), \mu_{x_i, x_j}(t), \mu_{y-x_j}(t)\right) \\ &\geq T^2(1 - r, 1 - r, \alpha) \\ &> 1 - s. \end{aligned}$$

Taking $t' = 3t$, it follows that $\mu_{x-y}(t') > 1 - s$ for all $x, y \in A$. Hence A is R-bounded. This completes the proof. □

Remark 2.2.7 In an RN-space (X, μ, T), every compact set is closed and R-bounded.

Definition 2.2.8 Let (X, μ, T) be an RN-space.

(1) A sequence $\{x_n\}$ in X is said to be *convergent* to a point $x \in X$ if, for any $\epsilon > 0$ and $\lambda > 0$, there exists a positive integer N such that

$$\mu_{x_n - x}(\epsilon) > 1 - \lambda$$

whenever $n \geq N$.

(2) A sequence $\{x_n\}$ in X is called a *Cauchy sequence* if, for any $\epsilon > 0$ and $\lambda > 0$, there exists a positive integer N such that

$$\mu_{x_n - x_m}(\epsilon) > 1 - \lambda$$

whenever $n \geq m \geq N$.

(3) An RN-space (X, μ, T) is said to be *complete* if every Cauchy sequence in X is convergent to a point in X.

Theorem 2.2.9 [241] *If (X, μ, T) is an RN-space and $\{x_n\}$ is a sequence such that $x_n \to x$, then $\lim_{n\to\infty} \mu_{x_n}(t) = \mu_x(t)$ almost everywhere.*

Theorem 2.2.10 *Let (X, μ, T) be an RN-space such that every Cauchy sequence in X has a convergent subsequence. Then (X, μ, T) is complete.*

Proof Let $\{x_n\}$ be a Cauchy sequence in X and $\{x_{i_n}\}$ be a subsequence of $\{x_n\}$ which converges to a point $x \in X$.

Now, we prove that $x_n \to x$. Let $t > 0$ and $\epsilon \in (0, 1)$ such that

$$T(1 - r, 1 - r) \geq 1 - \epsilon.$$

Since $\{x_n\}$ is a Cauchy sequence, there exists $n_0 \geq 1$ such that

$$\mu_{x_m - x_n}(t) > 1 - r$$

for all $m, n \geq n_0$. Since $x_{i_n} \to x$, there exists a positive integer i_p such that $i_p > n_0$ and

$$\mu_{x_{i_p} - x}\left(\frac{t}{2}\right) > 1 - r.$$

Then, if $n \geq n_0$, we have

$$\mu_{x_n - x}(t) \geq T\left(\mu_{x_n - x_{i_p}}\left(\frac{t}{2}\right), \mu_{x_{i_p} - x}\left(\frac{t}{2}\right)\right)$$

$$> T(1-r, 1-r)$$
$$\geq 1-\epsilon.$$

Therefore, $x_n \to x$ and hence (X, μ, T) is complete. This completes the proof. □

Lemma 2.2.11 *Let (X, μ, T) be an RN-space. If we define*

$$F_{x,y}(t) = \mu_{x-y}(t)$$

for all $x, y \in X$ and $t > 0$, then F is a random (probabilistic) metric on X, which is called the random (probabilistic) metric induced by the random norm μ.

Lemma 2.2.12 *A random (probabilistic) metric F which is induced by a random norm on a RN-space (X, μ, T) has the following properties: for all $x, y, z \in X$ and scalar $\alpha \neq 0$,*

(1) $F_{x+z,y+z}(t) = F_{x,y}(t)$;
(2) $F_{\alpha x,\alpha y}(t) = F_{x,y}(\frac{t}{|\alpha|})$.

Proof We have the following:

$$F_{x+z,y+z}(t) = \mu_{(x+z)-(y+z)}(t) = \mu_{x-y}(t) = F_{x,y}(t)$$

and, also,

$$F_{\alpha x,\alpha y}(t) = \mu_{\alpha x-\alpha y}(t) = \mu_{x-y}\left(\frac{t}{|\alpha|}\right) = F_{x,y}\left(\frac{t}{|\alpha|}\right).$$

Therefore, we have (1) and (2). This completes the proof. □

Lemma 2.2.13 *If (X, μ, T) is an RN-space, then we have*

(1) *The function $(x, y) \to x + y$ is continuous;*
(2) *The function $(\alpha, x) \to \alpha x$ is continuous.*

Proof If $x_n \to x$ and $y_n \to y$ as $n \to \infty$, then we have

$$\mu_{(x_n+y_n)-(x+y)}(t) \geq T\left(\mu_{x_n-x}\left(\frac{t}{2}\right), \mu_{y_n-y}\left(\frac{t}{2}\right)\right) \to 1$$

as $n \to \infty$. This proves (1).

Now, if $x_n \to x$ and $\alpha_n \to \alpha$ as $n \to \infty$, where $\alpha_n \neq 0$, then we have

$$\mu_{\alpha_n x_n-\alpha x}(t) = \mu_{\alpha_n(x_n-x)+x(\alpha_n-\alpha)}(t)$$
$$\geq T\left(\mu_{\alpha_n(x_n-x)}\left(\frac{t}{2}\right)\mu_{x(\alpha_n-\alpha)}\left(\frac{t}{2}\right)\right)$$

$$= T\left(\mu_{x_n - x}\left(\frac{t}{2\alpha_n}\right), \mu_x\left(\frac{t}{2(\alpha_n - \alpha)}\right)\right) \to 1$$

as $n \to \infty$. This proves (2). This completes the proof. □

Definition 2.2.14 An RN-space (X, μ, T) is called a *random Banach space* whenever X is complete with respect to the random metric induced by random norm.

Lemma 2.2.15 *Let (X, μ, T) be an RN-space and define*

$$E_{\lambda,\mu} : X \to \mathbb{R}^+ \cup \{0\}$$

by

$$E_{\lambda,\mu}(x) = \inf\{t > 0 : \mu_x(t) > 1 - \lambda\}$$

for all $\lambda \in (0, 1)$ and $x \in X$. Then we have

(1) $E_{\lambda,\mu}(\alpha x) = |\alpha| E_{\lambda,\mu}(x)$ *for all $x \in X$ and $\alpha \in \mathbb{R}$;*
(2) *If T satisfies* (1.1.2), *then, for any $\alpha \in (0, 1)$, there exists $\beta \in (0, 1)$ such that*

$$E_{\gamma,\mu}(x_1 + \cdots + x_n) \le E_{\lambda,\mu}(x_1) + \cdots + E_{\lambda,\mu}(x_n)$$

for all $x, y \in X$;
(3) *A sequence $\{x_n\}$ is convergent with respect to the random norm μ if and only if $E_{\lambda,\mu}(x_n - x) \to 0$. Also, the sequence $\{x_n\}$ is a Cauchy sequence with respect to the random norm μ if and only if it is a Cauchy sequence with $E_{\lambda,\mu}$.*

Proof For (1), we have

$$\begin{aligned}
E_{\lambda,\mu}(\alpha x) &= \inf\{t > 0 : \mu_{\alpha x}(t) > 1 - \lambda\} \\
&= \inf\left\{t > 0 : \mu_x\left(\frac{t}{|\alpha|}\right) > 1 - \lambda\right\} \\
&= |\alpha| \inf\{t > 0 : \mu_x(t) > 1 - \lambda\} \\
&= |\alpha| E_{\lambda,\mu}(x).
\end{aligned}$$

For (2), by (1.1.2), for all $\alpha \in (0, 1)$, we can find $\lambda \in (0, 1)$ such that

$$T^{n-1}(1 - \lambda, \ldots, 1 - \lambda) \ge 1 - \alpha.$$

Thus, we have

$$\begin{aligned}
&\mu_{x_1 + \cdots + x_n}\big(E_{\lambda,\mu}(x_1) + \cdots + E_{\lambda,\mu}(x_n) + n\delta\big) \\
&\quad \ge_L T^{n-1}\big(\mu_{x_1}\big(E_{\lambda,\mathcal{M}}(x_1) + \delta\big), \ldots, \mu_{x_n}\big(E_{\lambda,\mathcal{P}}(x_n) + \delta\big)\big)
\end{aligned}$$

$$\geq T(1-\lambda, \dots, 1-\lambda)$$
$$\geq 1-\alpha$$

for all $\delta > 0$, which implies that

$$E_{\alpha,\mu}(x_1 + \dots + x_n) \leq E_{\lambda,\mu}(x_1) + \dots + E_{\lambda,\mu}(x_n) + n\delta.$$

Since $\delta > 0$ is arbitrary, we have

$$E_{\alpha,\mu}(x_1 + \dots + x_n) \leq E_{\lambda,\mu}(x_1) + \dots + E_{\lambda,\mu}(x_n).$$

For (3), since μ is continuous, $E_{\lambda,\mu}(x)$ is not an element of the set $\{t > 0 : \mu_x(t) > 1-\lambda\}$ for all $x \in X$ with $x \neq 0$. Hence, we have

$$\mu_{x_n - x}(\eta) > 1-\lambda \quad \Longleftrightarrow \quad E_{\lambda,\mu}(x_n - x) < \eta$$

for all $\eta > 0$. This completes the proof. □

Definition 2.2.16 A function f from a RN-space (X, μ, T) to a RN-space (Y, ν, T') is said to be *uniformly continuous* if, for all $r \in (0,1)$ and $t > 0$, there exist $r_0 \in (0,1)$ and $t_0 > 0$ such that

$$\mu_{x-y}(t_0) > 1 - r_0 \quad \Longrightarrow \quad \nu_{f(x),f(y)}(t) > 1 - r.$$

Theorem 2.2.17 (Uniform Continuity Theorem) *If f is continuous function from a compact RN-space (X, μ, T) to an RN-space (Y, ν, T'), then f is uniformly continuous.*

Proof Let $s \in (0,1)$ and $t > 0$ be given. Then we can find $r \in (0,1)$ such that

$$T'(1-r, 1-r) > 1-s.$$

Since $f : X \to Y$ is continuous, for any $x \in X$, we can find $r_x \in (0,1)$ and $t_x > 0$ such that

$$\mu_{x-y}(t_x) > 1 - r_x \quad \Longrightarrow \quad \nu_{f(x)-f(y)}\left(\frac{t}{2}\right) > 1 - r.$$

But $r_x \in (0,1)$ and then we can find $s_x < r_x$ such that

$$T(1-s_x, 1-s_x) > 1-r_x.$$

Since X is compact and

$$\left\{B_x\left(s_x, \frac{t_x}{2}\right) : x \in X\right\}$$

is an open covering of X, there exist $x_1, x_2, \ldots, x_k$ in X such that

$$X = \bigcup_{i=1}^{k} B_{x_i}\left(s_{x_i}, \frac{t_{x_i}}{2}\right).$$

Put $s_0 = \min s_{x_i}$ and $t_0 = \min \frac{t_{x_i}}{2}$, $i = 1, 2, \ldots, k$. For any $x, y \in X$, if $\mu_{x-y}(t_0) > 1 - s_0$, then $\mu_{x-y}(\frac{t_{x_i}}{2}) > 1 - s_{x_i}$. Since $x \in X$, there exists $x_i \in X$ such that

$$\mu_{x-x_i}\left(\frac{t_{x_i}}{2}\right) > 1 - s_{x_i}.$$

Hence, we have

$$\nu_{f(x),f(x_i)}\left(\frac{t}{2}\right) > 1 - r.$$

Now, note that

$$\begin{aligned}\mu_{y-x_i}(t_{x_i}) &\geq T\left(\mu_{x-y}\left(\frac{t_{x_i}}{2}\right), \mu_{x-x_i}\left(\frac{t_{x_i}}{2}\right)\right)\\ &\geq T(1 - s_{x_i}, 1 - s_{x_i})\\ &> 1 - r_{x_i}.\end{aligned}$$

Therefore, we have

$$\nu_{f(y)-f(x_i)}\left(\frac{t}{2}\right) > 1 - r$$

and so

$$\begin{aligned}\nu_{f(x)-f(y)}(t) &\geq T\left(\nu_{f(x)-f(x_i)}\left(\frac{t}{2}\right), \nu_{f(y)-f(x_i)}\left(\frac{t}{2}\right)\right)\\ &\geq T(1 - r, 1 - r)\\ &> 1 - s.\end{aligned}$$

Therefore, f is uniformly continuous. This completes the proof. □

Remark 2.2.18 Let f be an uniformly continuous function from an RN-space (X, μ, T) to an RN-space (Y, ν, T'). If $\{x_n\}$ is a Cauchy sequence in X, then $\{f(x_n)\}$ is also a Cauchy sequence in Y.

Theorem 2.2.19 *Every compact RN-space is separable.*

Proof Let (X, μ, T) be a compact RN-space. Let $r \in (0, 1)$ and $t > 0$. Since X is compact, there exist $x_1, x_2, \dots, x_n$ in X such that

$$X = \bigcup_{i=1}^{n} B_{x_i}(r, t).$$

In particular, for each $n \geq 1$, we can choose a finite subset A_n of X such that

$$X = \bigcup_{a \in A_n} B_a\left(r_n, \frac{1}{n}\right)$$

in which $r_n \in (0, 1)$. Let

$$A = \bigcup_{n \geq 1} A_n.$$

Then A is countable.

Now, we claim that $X \subset \overline{A}$. Let $x \in X$. Then, for each $n \geq 1$, there exists $a_n \in A_n$ such that $x \in B_{a_n}(r_n, \frac{1}{n})$. Thus, $\{a_n\}$ converges to the point $x \in X$. But, since $a_n \in A$ for all $n \geq 1$, $x \in \overline{A}$ and so A is dense in X. Therefore, X is separable. This completes the proof. □

Definition 2.2.20 Let X be a nonempty set and (Y, ν, T') be an RN-space. Then a sequence $\{f_n\}$ of functions from X to Y is said to be *converge uniformly* to a function f from X to Y if, for any $r \in (0, 1)$ and $t > 0$, there exists $n_0 \geq 1$ such that

$$\nu_{f_n(x)-f(x)}(t) > 1 - r$$

for all $n \geq n_0$ and $x \in X$.

Definition 2.2.21 A family $\mathcal{F}$ of functions from an RN-space (X, μ, T) to a complete RN-space (Y, ν, T') is said to be *equicontinuous* if, for any $r \in (0, 1)$ and $t > 0$, there exist $r_0 \in (0, 1)$ and $t_0 > 0$ such that

$$\mu_{x-y}(t_0) > 1 - r_0 \quad \Longrightarrow \quad \nu_{f(x)-f(y)}(t) > 1 - r$$

for all $f \in \mathcal{F}$.

Lemma 2.2.22 *Let $\{f_n\}$ be an equicontinuous sequence of functions from an RN-space (X, μ, T) to a complete RN-space (Y, ν, T'). If $\{f_n\}$ converges for each point of a dense subset D of X, then $\{f_n\}$ converges for each point of X and the limit function is continuous.*

Proof Let $s \in (0, 1)$ and $t > 0$ be given. Then we can find $r \in (0, 1)$ such that

$$T'^2(1 - r, 1 - r, 1 - r) > 1 - s.$$

Since $\mathcal{F} = \{f_n\}$ is an equicontinuous family, for any $r \in (0,1)$ and $t > 0$, there exist $r_1 \in (0,1)$ and $t_1 > 1$ such that, for each $x, y \in X$,

$$\mu_{x-y}(t_1) > 1 - r_1 \quad \Longrightarrow \quad \nu_{f_n(x)-f_n(y)}\left(\frac{t}{3}\right) > 1 - r$$

for all $f_n \in \mathcal{F}$. Since D is dense in X, there exists

$$y \in B_x(r_1, t_1) \cap D$$

and $\{f_n(y)\}$ converges for the point y. Since $\{f_n(y)\}$ is a Cauchy sequence, for any $r \in (0,1)$ and $t > 0$, there exists $n_0 \geq 1$ such that

$$\nu_{f_n(y)-f_m(y)}\left(\frac{t}{3}\right) > 1 - r$$

for all $m, n \geq n_0$. Now, for any $x \in X$, we have

$$\begin{aligned}
&\nu_{f_n(x)-f_m(x)}(t) \\
&\geq T'^2\left(\nu_{f_n(x)-f_n(y)}\left(\frac{t}{3}\right), \nu_{f_n(y)-f_m(y)}\left(\frac{t}{3}\right), \nu_{f_m(x)-f_m(y)}\left(\frac{t}{3}\right)\right) \\
&\geq T'^2(1-r, 1-r, 1-r) \\
&> 1 - s.
\end{aligned}$$

Hence, $\{f_n(x)\}$ is a Cauchy sequence in Y. Since Y is complete, $f_n(x)$ converges and so let $f(x) = \lim f_n(x)$.

Now, we claim that f is continuous. Let $s_o \in 1 - r$ and $t_0 > 0$ be given. Then we can find $r_0 \in 1 - r$ such that

$$T'^2(1 - r_0, 1 - r_0, 1 - r_0) > 1 - s_0.$$

Since $\mathcal{F}$ is equicontinuous, for any $r_0 \in (0,1)$ and $t_0 > 0$, there exist $r_2 \in (0,1)$ and $t_2 > 0$ such that

$$\mu_{x-y}(t_2) > 1 - r_2 \quad \Longrightarrow \quad \nu_{f_n(x)-f_n(y)}\left(\frac{t_0}{3}\right) > 1 - r_0$$

for all $f_n \in \mathcal{F}$. Since $f_n(x)$ converges to $f(x)$, for any $r_0 \in (0,1)$ and $t_0 > 0$, there exists $n_1 \geq 1$ such that

$$\nu_{f_n(x)-f(x)}\left(\frac{t_0}{3}\right) > 1 - r_0.$$

Also, since $f_n(y)$ converges to $f(y)$, for any $r_0 \in (0,1)$ and $t_0 > 0$, there exists $n_2 \geq 1$ such that

$$\nu_{f_n(y)-f(y)}\left(\frac{t_0}{3}\right) > 1 - r_0$$

for all $n \geq n_2$. Now, for all $n \geq \max\{n_1, n_2\}$, we have

$$\begin{aligned}
&\nu_{f(x)-f(y)}(t_0) \\
&\geq T'^2\left(\nu_{f(x)-f_n(x)}\left(\frac{t_0}{3}\right), \nu_{f_n(x)-f_n(y)}\left(\frac{t_0}{3}\right), \nu_{f_n(y)-f(y)}\left(\frac{t_0}{3}\right)\right) \\
&\geq T'^2(1-r_0, 1-r_0, 1-r_0) \\
&> 1-s_0.
\end{aligned}$$

Therefore, f is continuous. This completes the proof. □

Theorem 2.2.23 (Ascoli–Arzela Theorem) *Let (X, μ, T) be a compact RN-space and (Y, ν, T') be a complete RN-space. Let $\mathcal{F}$ be an equicontinuous family of functions from X to Y. If $\{f_n\}$ is a sequence in $\mathcal{F}$ such that*

$$\overline{\{f_n(x) : n \in \mathbb{N}\}}$$

is a compact subset of Y for any $x \in X$, then there exists a continuous function f from X to Y and a subsequence $\{g_n\}$ of $\{f_n\}$ such that $\{g_n\}$ converges uniformly to f on X.

Proof Since (X, μ, T) be a compact RN-space, by Theorem 2.2.19, X is separable. Let

$$D = \{x_i : i = 1, 2, \ldots\}$$

be a countable dense subset of X. By hypothesis, for each $i \geq 1$,

$$\overline{\{f_n(x_i) : n \geq 1\}}$$

is compact subset of Y. Since every $\mathcal{L}$-fuzzy metric space is first countable space, every compact subset of Y is sequentially compact. Thus, by standard argument, we have a subsequence $\{g_n\}$ of $\{f_n\}$ such that $\{g_n(x_i)\}$ converges for each $i \geq 1$. Thus, by Lemma 2.2.22, there exists a continuous function f from X to Y such that $\{g_n(x)\}$ converges to $f(x)$ for all $x \in X$.

Now, we claim that $\{g_n\}$ converges uniformly to a functions f on X. Let $s \in (0, 1)$ and $t > 0$ be given. Then we can find $r \in (0, 1)$ such that

$$T'^2(1-r, 1-r, 1-r) > 1-s.$$

Since $\mathcal{F}$ is equicontinuous, there exist $r_1 \in (0, 1)$ and $t_1 > 0$ such that

$$\mu_{x-y}(t_1) > 1-r_1 \quad \Longrightarrow \quad \nu_{g_n(x), g_n(y)}\left(\frac{t}{3}\right) > 1-r$$

for all $n \geq 1$. Since X is compact, by Theorem 2.2.17, f is uniformly continuous. Hence, for any $r \in (0, 1)$ and $t > 0$, there exist $r_2 \in (0, 1)$ and $t_2 > 0$ such that

$$\mu_{x-y}(t_2) > 1 - r_2 \quad \Longrightarrow \quad \nu_{f(x)-f(y)}\left(\frac{t}{3}\right) > 1 - r$$

for all $x, y \in X$. Let $r_0 = \min\{r_1, r_2\}$ and $t_0 = \min\{t_1, t_2\}$. Since X is compact and D is dense in X, we have

$$X = \bigcup_{i=1}^{k} B_{x_i}(r_0, t_0)$$

for some $k \geq 1$. Thus, for any $x \in X$, there exists i, $i \leq i \leq k$, such that

$$\mu_{x-x_i}(t_0) > 1 - r_0.$$

But, since $r_0 = \min\{r_1, r_2\}$ and $t_0 = \min\{t_1, t_2\}$, we have, by the equicontinuity of $\mathcal{F}$,

$$\nu_{g_n(x)-g_n(x_i)}\left(\frac{t}{3}\right) > 1 - r$$

and we also have, by the uniform continuity of f,

$$\nu_{f(x)-f(x_i)}\left(\frac{t}{3}\right) > 1 - r.$$

Since $\{g_n(x_j)\}$ converges to $f(x_j)$, for any $r \in (0, 1)$ and $t > 0$, there exists $n_0 \geq 1$ such that

$$\nu_{g_n(x_j)-f(x_j)}\left(\frac{t}{3}\right) > 1 - r$$

for all $n \geq n_0$. Now, for all $x \in X$, we have

$$\begin{aligned}
&\nu_{g_n(x)-f(x)}(t) \\
&\quad \geq T'^2\left(\nu_{g_n(x)-g_n(x_i)}\left(\frac{t}{3}\right), \nu_{g_n(x_i)-f(x_i)}\left(\frac{t}{3}\right), \nu_{f(x_i)-f(x)}\left(\frac{t}{3}\right)\right) \\
&\quad \geq T'^2(1-r, 1-r, 1-r) \\
&\quad > 1 - s.
\end{aligned}$$

Therefore, $\{g_n\}$ converges uniformly to a function f on X. This completes the proof. □

We recall that a subset A is said R-bounded in (X, μ, T), if there exist $t_0 > 0$ and $r_0 \in (0, 1)$ such that $\mu_x(t_0) > 1 - r_0$ for all $x \in A$.

Lemma 2.2.24 *A subset A of $\mathbb{R}$ is R-bounded in $(\mathbb{R}, \mu, T)$ if and only if it is bounded in $\mathbb{R}$.*

Proof Let A be a subset in $\mathbb{R}$ which is R-bounded in $(\mathbb{R}, \mu, T)$. Then there exist $t_0 > 0$ and $r_0 \in (0, 1)$ such that $\mu_x(t_0) > 1 - r_0$ for all $x \in A$. Thus, we have

$$t_0 \geq E_{r_0,\mu}(x) = |x| E_{r_0,\mu}(1).$$

Now, $E_{r_0,\mu}(1) \neq 0$. If we put $k = \frac{t_0}{E_{r_0,\mu}(1)}$, then we have $|x| \leq k$ for all $x \in A$, that is, A is bounded in $\mathbb{R}$.

The converse is easy to see. This completes the proof. □

Lemma 2.2.25 *A sequence* $\{\beta_n\}$ *is convergent in an RN-space* $(\mathbb{R}, \mu, T)$ *if and only if it is convergent in* $(\mathbb{R}, |\cdot|)$.

Proof Let $\beta_n \to \beta$ in $\mathbb{R}$. Then, by Lemma 2.2.15(1), we have

$$E_{\lambda,\mu}(\beta_n - \beta) = |\beta_n - \beta| E_{\lambda,\mu}(1) \to 0.$$

Thus, by Lemma 2.2.15 (3), $\beta_n \stackrel{\mu}{\to} \beta$.

Conversely, let $\beta_n \stackrel{\mu}{\to} \beta$. Then, by Lemma 2.2.15,

$$\lim_{n\to+\infty} |\beta_n - \beta| E_{\lambda,\mu}(1) = \lim_{n\to+\infty} E_{\lambda,\mu}(\beta_n - \beta) = 0.$$

Now, $E_{\lambda,\mu}(1) \neq 0$ and so $\beta_n \to \beta$ in $\mathbb{R}$. This completes the proof. □

Corollary 2.2.26 *If a real sequence* $\{\beta_n\}$ *is R-bounded, then it has at least one limit point.*

Lemma 2.2.27 *A subset* A *of* $\mathbb{R}$ *is R-bounded in* $(\mathbb{R}, \mu, T)$ *if and only if it is bounded in* $\mathbb{R}$.

Proof Let the subset A is R-bounded in $(\mathbb{R}, \mu, T)$. Then there exist $t_0 > 0$ and $r_0 \in (0, 1)$ such that

$$\mu_x(t_0) > 1 - r_0$$

for all $x \in A$ and so

$$t_0 \geq E_{r_0,\mu}(x) = |x| E_{r_0,\mu}(1).$$

Now, $E_{r_0,\mu}(1) \neq 0$. If we put $k = \frac{t_0}{E_{r_0,\mu}(1)}$, then we have $|x| \leq k$ for all $x \in A$, i.e., A is bounded in $\mathbb{R}$.

The converse is easy. This completes the proof. □

Definition 2.2.28 A triple $(\mathbb{R}^n, \Phi, T)$ is called an *random Euclidean normed space* if T is a continuous t-norm and $\Phi_x(t)$ is a random Euclidean norm defined by

$$\Phi_x(t) = \prod_{j=1}^{n} \mu_{x_j}(t),$$

where $\prod_{j=1}^{n} a_j = T'^{n-1}(a_1, \ldots, a_n)$, $T' \gg T$, $x = (x_1, \ldots, x_n)$, $t > 0$ and μ is a random norm.

For example, let $\Phi_x(t) = \exp(\frac{\|x\|}{t})^{-1}$, $\mu_{x_j}(t) = \exp(\frac{|x_j|}{t})^{-1}$ and $T = \min$. Then we have $\Phi_x(t) = \min_j \mu_{x_j}(t)$ or, equivalently, $\|x\| = \max_j |x_j|$.

Lemma 2.2.29 *Suppose that the hypotheses of Definition* 2.2.28 *are satisfied. Then* $(\mathbb{R}^n, \Phi, T)$ *is an RN-space.*

Proof The properties of (RN1) and (RN2) follow immediately from the definition. For the triangle inequality (RN3) suppose that $x, y \in X$ and $t, s > 0$. Then we have

$$\begin{aligned}
T\big(\Phi_x(t), \Phi_y(s)\big) &= T\left(\prod_{j=1}^{n} \mathcal{P}_{x_j}(t), \prod_{j=1}^{n} \mathcal{P}_{y_j}(s)\right) \\
&= T\big(T'^{n-1}\big(\mathcal{P}_{x_1}(t), \ldots, \mathcal{P}_{x_n}(t)\big), T'^{n-1}\big(\mathcal{P}_{y_1}(t), \ldots, \mathcal{P}_{y_n}(t)\big)\big) \\
&\le T'^{n-1}\big(T\big(\mathcal{P}_{x_1}(t), \mathcal{P}_{y_1}(t)\big), \ldots, T\big(\mathcal{P}_{x_n}(t), \mathcal{P}_{y_n}(t)\big)\big) \\
&\le T'^{n-1}\big(\mathcal{P}_{x_1+y_1}(t+s), \ldots, \mathcal{P}_{x_n+y_n}(t+s)\big) \\
&= \prod_{j=1}^{n} \mathcal{P}_{x_j+y_j}(t+s) \\
&= \Phi_{x+y}(t+s).
\end{aligned}$$

This completes the proof. □

Lemma 2.2.30 *Suppose that* $(\mathbb{R}^n, \Phi, T)$ *is a random Euclidean normed space and* A *is an infinite and* R*-bounded subset of* $\mathbb{R}^n$. *Then* A *has at least one limit point.*

Proof Let $\{x^{(m)}\}$ be an infinite sequence in A. Since A is R-bounded, so is $\{x^{(m)}\}_{m\ge1}$. Therefore, there exist $t_0 > 0$ and $r_0 \in (0, 1)$ such that

$$1 - r_0 < \Phi_x(t_0)$$

for all $x \in A$, which implies that $E_{r_0,\Phi}(x) \le t_0$. However, we have

$$\begin{aligned}
E_{r_0,\Phi}(x) &= \inf\big\{t > 0 : 1 - r_0 < \Phi_x(t)\big\} \\
&= \inf\left\{t > 0 : 1 - r_0 < \prod_{j=1}^{n} \mu_{x_j}(t)\right\} \\
&\ge \inf\big\{t > 0 : 1 - r_0 < \mu_{x_j}(t)\big\} \\
&= E_{r_0,\mu}(x_j)
\end{aligned}$$

for each $1 \le j \le n$. Therefore, $|x_j| \le k$ in which $k = \frac{t_0}{E_{r_0,\mu}(1)}$, that is, the real sequences $\{x_j^{(m)}\}$ for each $j \in \{1, \dots, n\}$ are bounded. Hence, there exists a subsequence $\{x_1^{(m_{k_1})}\}$ which converges to x_1 in A with respect to the random norm μ. The corresponding sequence $\{x_2^{(m_{k_1})}\}$ is bounded and so there exists a subsequence $\{x_2^{(m_{k_2})}\}$ of $\{x_2^{(m_{k_1})}\}$ which converges to x_2 with respect to the random norm μ.

Continuing like this, we find a subsequence $\{x^{(m_k)}\}$ converging to $x = (x_1, \dots, x_n) \in \mathbb{R}^n$. This completes the proof. □

Lemma 2.2.31 *Let $(\mathbb{R}^n, \Phi, T)$ be a random Euclidean normed space. Let $\{Q_1, Q_2, \dots\}$ be a countable collection of nonempty subsets in $\mathbb{R}^n$ such that $Q_{k+1} \subseteq Q_k$, each Q_k is closed and Q_1 is R-bounded. Then $\bigcap_{k=1}^{\infty} Q_k$ is nonempty and closed.*

Proof Using the above lemma, the proof proceeds as in the classical case (see Theorem 3.25 in [15]). □

We call an n-dimensional ball $B_x(r, t)$ a *rational ball* if $x \in \mathbb{Q}^n$, $r_0 \in (0, 1)$ and $t \in \mathbb{Q}^+$.

Theorem 2.2.32 *Let $(\mathbb{R}^n, \Phi, T)$ be a random Euclidean normed space in which T satisfies* (1.1.2). *Let $G = \{A_1, A_2, \dots\}$ be a countable collection of n-dimensional rational open balls. If $x \in \mathbb{R}^n$ and S is an open subset of $\mathbb{R}^n$ containing x, then there exists $A_k \in G$ such that $x \in A_k \subseteq S$ for some $k \ge 1$.*

Proof Since $x \in S$ and S is open, there exist $r \in (0, 1)$ and $t > 0$ such that $B_x(r, t) \subseteq S$. By (1.1.2), we can find $\eta \in\in (0, 1)$ such that $1 - r < T(1 - \eta, 1 - \eta)$. Let $\{\xi_k\}_{k=1}^n$ be a finite sequence such that $1 - \eta < \prod_{k=1}^n (1 - \xi_k)$ and $x = (x_1, \dots, x_n)$. Then we can find $y = (y_1, \dots, y_n) \in \mathbb{Q}^n$ such that $(1 - \xi_k) < \mu_{x_k - y_k}(\frac{t}{2})$. Therefore, we have

$$1 - \eta < \prod_{k=1}^{n} (1 - \xi_k) \le \Phi_{x-y}\left(\frac{t}{2}\right) = \prod_{k=1}^{n} \mu_{x_k - y_k}\left(\frac{t}{2}\right)$$

and so $x \in B_y(\eta, \frac{t}{2})$.

Now, we prove that $B_y(\eta, \frac{t}{2}) \subseteq B_x(r, t)$. Let $z \in B_y(\eta, \frac{t}{2})$. Then $\Phi_{y-z}(\frac{t}{2}) > 1 - \eta$ and hence

$$1 - r < T(1 - \eta, 1 - \eta) \le T\left(\Phi_{x-y}\left(\frac{t}{2}\right), \Phi_{y-z}\left(\frac{t}{2}\right)\right) \le \Phi_{x-z}(t).$$

On the other hand, there exists $t_0 \in \mathbb{Q}$ such that $t_0 < \frac{t}{2}$ and $x \in B_y(\eta, t_0) \subseteq B_y(\eta, \frac{t}{2}) \subseteq B_x(r, t) \subseteq S$. Now, $B_y(\eta, t_0) \in G$. This completes the proof. □

Corollary 2.2.33 *In a random Euclidean normed space $(\mathbb{R}^n, \Phi, T)$ in which T satisfies* (1.1.2), *every closed and R-bounded set is compact.*

Proof The proof is similar to the proof of Theorem 3.29 in [15]. □

Corollary 2.2.34 *Let* $(\mathbb{R}^n, \Phi, T)$ *be a random Euclidean normed space in which* T *satisfies* (1.1.2) *and* $S \subseteq \mathbb{R}^n$. *Then* S *is compact set if and only if it is* R*-bounded and closed.*

Corollary 2.2.35 *The random Euclidean normed space* $(\mathbb{R}^n, \Phi, T)$ *is complete.*

Proof Let $\{x_m\}$ be a Cauchy sequence in the random Euclidean normed space $(\mathbb{R}^n, \Phi, T)$. Since

$$\begin{aligned} E_{\lambda,\Phi}(x_n - x_m) &= \inf\{t > 0 : \Phi_{x_n - x_m}(t) > 1 - \lambda\} \\ &= \inf\left\{t > 0 : \prod_{j=1}^{n} \mathcal{P}_{x_{m,j} - x_{n,j}}(t) > 1 - \lambda\right\} \\ &\geq \inf\{t > 0 : \mathcal{P}_{x_{m,j} - x_{n,j}}(t) > 1 - \lambda\} \\ &= E_{\lambda,\mathcal{P}}(x_{m,j} - x_{n,j}) = |x_{m,j} - x_{n,j}| E_{\lambda,\mathcal{P}}(1), \end{aligned}$$

the sequence $\{x_{m,j}\}$ for each $j = 1, \dots, n$ is a Cauchy sequence in $\mathbb{R}$ and so it convergent to $x_j \in \mathbb{R}$. Then, by Lemma 2.2.15, the sequence $\{x_{m,j}\}$ is convergent in RN-space $(\mathbb{R}, \mu, T)$.

Now, we prove that $\{x_m\}$ convergent to $x = (x_1, \dots, x_n)$. In fact, we have

$$\lim_m \Phi_{x_m - x}(t) = \lim_m \prod_{j=1}^{n} \mathcal{P}_{x_{m,j} - x_j}(t) = T'^{n-1}(1, \dots, 1) = 1.$$

This completes the proof. □

2.3 Random Functional Analysis

In this section, we discuss some important results dealing with topological isomorphisms and also give the proofs of Open Mapping Theorem, Closed Graph Theorem and some other fundamental theorems in the framework of Random Functional Analysis.

Theorem 2.3.1 *Let* $\{x_1, \dots, x_n\}$ *be a linearly independent set of vectors in vector space* X *and* (X, μ, T) *be an RN-space. Then there exist* $c \neq 0$ *and an RN-space* $(\mathbb{R}, \mu', T)$ *such that, for every choice of the* n *real scalars* $\alpha_1, \dots, \alpha_n$,

$$\mu_{\alpha_1 x_1 + \cdots + \alpha_n x_n}(t) \leq \mu'_{c \sum_{j=1}^{n} |\alpha_j|}(t). \tag{2.3.1}$$

Proof Put $s = |\alpha_1| + \cdots + |\alpha_n|$. If $s = 0$, all α_j's must be zero and so (2.3.1) holds for any c. Let $s > 0$. Then (2.3.1) is equivalent to the inequality that we obtain from (2.3.1) by dividing by s and putting $\beta_j = \frac{\alpha_j}{s}$, that is,

$$\mu_{\beta_1 x_1 + \cdots + \beta_n x_n}(t') \leq \mu'_c(t'), \tag{2.3.2}$$

where $t' = \frac{t}{s}$ and $\sum_{j=1}^{n} |\beta_j| = 1$. Hence, it suffices to prove the existence of $c \neq 0$ and the random norm μ' such that (2.3.2) holds. Suppose that this is not true. Then there exists a sequence $\{y_m\}$ of vectors

$$y_m = \beta_{1,m} x_1 + \cdots + \beta_{n,m} x_n, \qquad \sum_{j=1}^{n} |\beta_{j,m}| = 1,$$

such that

$$\mu_{y_m}(t) \to 1$$

as $m \to \infty$ for any $t > 0$. Since $\sum_{j=1}^{n} |\beta_{j,m}| = 1$, we have $|\beta_{j,m}| \leq 1$ and so, by the Lemma 2.2.24, the sequence of $\{\beta_{j,m}\}$ is R-bounded. According to Corollary 2.2.26, $\{\beta_{1,m}\}$ has a convergent subsequence. Let β_1 denote the limit of the subsequence and let $\{y_{1,m}\}$ denote the corresponding subsequence of $\{y_m\}$. By the same argument, $\{y_{1,m}\}$ has a subsequence $\{y_{2,m}\}$ for which the corresponding of real scalars $\beta_2^{(m)}$ convergence. Let β_2 denote the limit. Continuing this process, after n steps, we obtain a subsequence $\{y_{n,m}\}_{m \geq 1}$ of $\{y_m\}$ such that

$$y_{n,m} = \sum_{j=1}^{n} \gamma_{j,m} x_j,$$

where $\sum_{j=1}^{n} |\gamma_{j,m}| = 1$, and $\gamma_{j,m} \to \beta_j$ as $m \to \infty$. By the Lemma 2.2.15 (2), for any $\alpha \in (0, 1)$, there exists $\lambda \in (0, 1)$ such that

$$E_{\alpha,\mu}\left(y_{n,m} - \sum_{j=1}^{n} \beta_j x_j\right) = E_{\alpha,\mu}\left(\sum_{j=1}^{n} (\gamma_{j,m} - \beta_j) x_j\right)$$
$$\leq \sum_{j=1}^{n} |\gamma_{j,m} - \beta_j| E_{\lambda,\mu}(x_j) \to 0$$

as $m \to \infty$. By Lemma 2.2.15 (3), we conclude

$$\lim_{m \to \infty} y_{n,m} = \sum_{j=1}^{n} \beta_j x_j,$$

where $\sum_{j=1}^{n} |\beta_j| = 1$, and so all β_j cannot be zero. Put $y = \sum_{j=1}^{n} \beta_j x_j$. Since $\{x_1, \ldots, x_n\}$ is a linearly independent set, we have $y \neq 0$. Since $\mu_{y_m}(t) \to 1$, by the assumption, we have $\mu_{y_{n,m}}(t) \to 1$. Hence, we have

$$\begin{aligned}\mu_y(t) &= \mu_{(y-y_{n,m})+y_{n,m}}(t)\\ &\geq T(\mu_{y-y_{n,m}}t/2), \mu_{y_{n,m}}(t/2)) \to 1\end{aligned}$$

and so $y = 0$, which is a contradiction. This completes the proof. □

Definition 2.3.2 Let (X, μ, T) and (X, ν, T') be two RN-spaces. Then two random norms μ and ν are said to be *equivalent* whenever $x_n \xrightarrow{\mu} x$ in (X, μ, T) If and only if $x_n \xrightarrow{\nu} x$ in (X, ν, T').

Theorem 2.3.3 *In a finite dimensional vector space X, every two random norms μ and ν are equivalent.*

Proof Let $\dim X = n$ and $\{v_1, \dots, v_n\}$ be a basis for X. Then every $x \in X$ has a unique representation $x = \sum_{j=1}^{n} \alpha_j v_j$. Let $x_m \xrightarrow{\mu} x$ in (X, μ, T), but, for each $m \geq 1$, suppose that x_m has a unique representation, that is,

$$x_m = \alpha_{1,m} v_1 + \cdots + \alpha_{n,m} v_n.$$

By Theorem 2.3.1, there exist $c \neq 0$ and the random norm μ' such that (2.3.1) holds. thus we have

$$\mu_{x_m - x}(t) \leq \mu'_{c \sum_{j=1}^{n} |\alpha_{j,m} - \alpha_j|}(t) \leq \mu'_{c|\alpha_{j,m} - \alpha_j|}(t).$$

Now, if $m \to \infty$, then we have

$$\mu_{x_m - x}(t) \to 1$$

for all $t > 0$ and hence $|\alpha_{j,m} - \alpha_j| \to 0$ in $\mathbb{R}$.

On the other hand, by the Lemma 2.2.15 (2), for any $\alpha \in (0, 1)$, there exists $\lambda \in (0, 1)$ such that

$$E_{\alpha,\nu}(x_m - x) \leq \sum_{j=1}^{n} |\alpha_{j,m} - \alpha_j| E_{\lambda,\nu}(v_j).$$

Since $|\alpha_{j,m} - \alpha_j| \to 0$, we have $x_m \xrightarrow{\nu} x$ in (X, ν, T'). Therefore, with the same argument, $x_m \to x$ in (X, ν, T') imply $x_m \to x$ in (X, μ, T). This completes the proof. □

Definition 2.3.4 A linear operator $\Lambda : (X, \mu, T) \to (Y, \nu, T')$ is said to be *random bounded* if there exists a constant $h \in \mathbb{R} - \{0\}$ such that, for all $x \in X$ and $t > 0$,

$$\nu_{\Lambda x}(t) \geq \mu_{hx}(t). \tag{2.3.3}$$

Note that, by Lemma 2.2.15 and the last definition, we have

$$\begin{aligned} E_{\lambda,\nu}(\Lambda x) &= \inf\{t>0:\nu_{\Lambda x}(t)>1-\lambda\} \\ &\le \inf\{t>0:\mu_x(t/|h|)>1-\lambda\} \\ &= |h|\inf\{t>0:\mu_x(t)>1-\lambda\} \\ &= |h|E_{\lambda,\mu}(x). \end{aligned}$$

Theorem 2.3.5 *Every linear operator $\Lambda:(X,\mu,T)\to(Y,\nu,T')$ is random bounded if and only if it is continuous.*

Proof By (2.3.3), every random bounded linear operator is continuous.

Now, we prove the converse. Let the linear operator Λ be continuous, but is not random bounded. Then, for each $n\ge 1$, there exists $x_n\in X$ such that $E_{\lambda,\nu}(\Lambda x_n)\ge nE_{\lambda,\mu}(p_n)$.

If we let

$$y_n=\frac{x_n}{nE_{\lambda,\mu}(x_n)},$$

then it is easy to see $y_n\to 0$, but $\{\Lambda y_n\}$ do not tend to 0. This completes the proof. □

Definition 2.3.6 A linear operator $\Lambda:(X,\mu,T)\to(Y,\nu,T')$ is an *random topological isomorphism* if Λ is one-to-one, onto and both Λ, Λ^{-1} are continuous. The RN-spaces (X,μ,T) and (Y,ν,T') for which such a Λ exists are said to be *random topologically isomorphic*.

Lemma 2.3.7 *A linear operator $\Lambda:(X,\mu,T)\to(Y,\nu,T')$ is random topological isomorphism if Λ is onto and there exist constants $a,b\ne 0$ such that*

$$\mu_{ax}(t)\le\nu_{\Lambda x}(t)\le\mu_{bx}(t).$$

Proof By the hypothesis, Λ is random bounded and, by last theorem, is continuous. Since $\Lambda x=0$ implies that

$$1=\nu_{\Lambda x}(t)\le\mu_x\left(\frac{t}{|b|}\right)$$

and so $x=0$, it follows that Λ is one-to-one. Thus Λ^{-1} exists and, since

$$\nu_{\Lambda x}(t)\le\mu_{bx}(t)$$

is equivalent to

$$\nu_y(t)\le\mu_{b\Lambda^{-1}y}(t)=\mu_{\Lambda^{-1}y}\left(\frac{t}{|b|}\right)$$

or

$$\nu_{\frac{1}{b}y}(t) \le \mu_{\Lambda^{-1}y}(t),$$

where $y = \Lambda x$, we see that Λ^{-1} is random bounded and, by last theorem, is continuous. Therefore, Λ is an random topological isomorphism. This completes the proof. □

Corollary 2.3.8 *Ever random topologically isomorphism preserves completeness.*

Theorem 2.3.9 *Every linear operator $\Lambda : (X, \mu, T) \to (Y, \nu, T')$, where $\dim X < \infty$, but other is not necessarily finite dimensional, is continuous.*

Proof If we define

$$\eta_x(t) = T'\big(\mu_x(t), \nu_{\Lambda x}(t)\big), \tag{2.3.4}$$

where $T' \gg T$. Then (X, η, T) is an RN-space since (RN1) and (RN2) are immediate from the definition and, for the triangle inequality (RN3),

$$\begin{aligned} T\big(\eta_x(t), \eta_z(s)\big) &= T\big[T'\big(\mu_x(t), \nu_{\Lambda x}(t)\big), T'\big(\mu_z(s), \nu_{\Lambda z}(s)\big)\big] \\ &\le T'\big[T\big(\mu_x(t), \mu_z(s)\big)T\big(\nu_{\Lambda x}(t), \nu_{\Lambda z}(s)\big)\big] \\ &\le T'\big(\mu_{x+z}(t+s), \nu_{\Lambda(x+z)}(t+s)\big) \\ &= \eta_{x+z}(t+s). \end{aligned}$$

Now, let $x_n \xrightarrow{\mu} x$. Then, by Theorem 2.3.3, $x_n \xrightarrow{\eta} x$, but, by (2.3.3), since

$$\nu_{\Lambda x}(t) \ge \eta_x(t),$$

we have $\Lambda x_n \xrightarrow{\nu} \Lambda x$. Hence, Λ is continuous. This completes the proof. □

Corollary 2.3.10 *Every linear isomorphism between finite dimensional RN-spaces is a topological isomorphism.*

Corollary 2.3.11 *Every finite dimensional RN-space (X, μ, T) is complete.*

Proof By Corollary 2.3.10, (X, μ, T) and $(\mathbb{R}^n, \Phi, T)$ are random topologically isomorph. Since $(\mathbb{R}^n, \Phi, T)$ is complete and every random topological isomorphism preserves completeness, (X, μ, T) is complete. □

Definition 2.3.12 Let (V, μ, T) be an RN-space, W be a linear manifold in V and $Q : V \to V/W$ be the natural mapping with $Qx = x + W$. For any $t > 0$, we define

$$\bar{\mu}(x + W, t) = \sup\big\{\mu_{x+y}(t) : y \in W\big\}.$$

Theorem 2.3.13 *Let W be a closed subspace of an RN-space (V, μ, T). If $x \in V$ and $\epsilon > 0$, then there exists $x' \in V$ such that*

$$x' + W = x + W, \qquad E_{\lambda,\mu}(x') < E_{\lambda,\bar{\mu}}(x + W) + \epsilon.$$

Proof By the properties of sup, there always exists $y \in W$ such that

$$E_{\lambda,\mathcal{P}}(x + y) < E_{\lambda,\bar{\mu}}(x + W) + \epsilon.$$

Now, it is enough to put $x' = x + y$. □

Theorem 2.3.14 *Let W be a closed subspace of an RN-space (V, μ, T) and $\bar{\mu}$ be given in the above definition. Then we have*

(1) *$\bar{\mu}$ is an RN-space on V/W;*
(2) *$\bar{\mu}_{Qx}(t) \geq \mu_x(t)$;*
(3) *If (V, μ, T) is an random Banach space, then so is $(V/W, \bar{\mu}, T)$.*

Proof (1) It is clear that $\bar{\mu}_{x+W}(t) > 0$. Let $\bar{\mu}_{x+W}(t) = 1$. By the definition, there exists a sequence $\{x_n\}$ in W such that $\mu_{x+x_n}(t) \to 1$. Thus, $x + x_n \to 0$ or, equivalently, $x_n \to (-x)$ and since W is closed, $x \in W$ and $x + W = W$, the zero element of V/W. Now, we have

$$\begin{aligned}\bar{\mu}_{(x+W)+(y+W)}(t) &= \bar{\mu}_{(x+y)+W}(t)\\ &\geq \mu_{(x+m)+(y+n)}(t)\\ &\geq T\big(\mu_{x+m}(t_1), \mu_{y+n}(t_2)\big)\end{aligned}$$

for all $m, n \in W$, $x, y \in V$ and $t_1 + t_2 = t$. Now, if we take the sup, then we have

$$\bar{\mu}_{(x+W)+(y+W)}(t) \geq T\big(\bar{\mu}_{x+W}(t_1), \bar{\mu}_{y+W}(t_2)\big).$$

Therefore, $\bar{\mu}$ is random norm on V/W.

(2) By Definition 2.3.12, we have

$$\bar{\mu}_{Qx}(t) = \bar{\mu}_{x+W}(t) = \sup\big\{\mu_{x+y}(t) : y \in W\big\} \geq \mu_x(t).$$

Note that, by Lemma 2.2.15,

$$\begin{aligned}E_{\lambda,\bar{\mu}}(Qx) &= \inf\big\{t > 0 : \bar{\mu}_{Qx}(t) > 1 - \lambda\big\}\\ &\leq \inf\big\{t > 0 : \mu_x(t) > 1 - \lambda\big\}\\ &= E_{\lambda,\mu}(x).\end{aligned} \tag{2.3.5}$$

(3) Let $\{x_n + W\}$ be a Cauchy sequence in V/W. Then there exists $n_0 \in \mathbb{N}$ such that, for each $n \geq n_0$,

$$E_{\lambda,\bar{\mu}}\big((x_n + W) - (x_{n+1} + W)\big) \leq 2^{-n}.$$

Let $y_1 = 0$ and choose $y_2 \in W$ such that

$$E_{\lambda,\mu}\big(x_1 - (x_2 - y_2), t\big) \le E_{\lambda,\bar{\mu}}\big((x_1 - x_2) + W\big) + \frac{1}{2}.$$

However, $E_{\lambda,\bar{\mu}}((x_1 - x_2) + W) \le \frac{1}{2}$ and so $E_{\lambda,\mu}(x_1 - (x_2 - y_2)) \le (\frac{1}{2})^2$.

Now, suppose that y_{n-1} has been chosen. Then choose $y_n \in W$ such that

$$E_{\lambda,\mu}\big((x_{n-1} + y_{n-1}) - (x_n + y_n)\big) \le E_{\lambda,\bar{\mu}}\big((x_{n-1} - x_n) + W\big) + 2^{-n+1}.$$

Hence, we have

$$E_{\lambda,\mu}\big((x_{n-1} + y_{n-1}) - (x_n + y_n)\big) \le 2^{-n+2}.$$

However, by Lemma 2.2.15, for each positive integer $m > n$ and $\lambda \in (0, 1)$, there exists $\gamma \in (0, 1)$ such that

$$\begin{aligned} E_{\lambda,\mu}\big((x_m + y_m) - (x_n + y_n)\big) &\le E_{\gamma,\mu}\big((x_{n+1} + y_{n+1}) - (x_n + y_n)\big) + \cdots \\ &\quad + E_{\gamma,\mu}\big((x_m + y_m) - (x_{m-1} + y_{m-1})\big) \\ &\le \sum_{i=n}^{m} 2^{-i}. \end{aligned}$$

By Lemma 2.2.15, $\{x_n + y_n\}$ is a Cauchy sequence in V. Since V is complete, there exists x_0 in V such that $x_n + y_n \to x_0$ in V.

On the other hand, we have

$$x_n + W = Q(x_n + y_n) \to Q(x_0) = x_0 + W.$$

Therefore, every Cauchy sequence $\{x_n + W\}$ is convergent in V/W and so V/W is complete. Thus $(V/W, \bar{\mu}, T)$ is a random Banach space. This completes the proof. □

Theorem 2.3.15 *Let W be a closed subspace of an RN-space (V, μ, T). If two of the spaces V, W and V/W are complete, then so is the third one.*

Proof If V is a random Banach space, then so are V/W and W. Hence, the fact that needs to be checked is that V is complete whenever both W and V/W are complete. Suppose that W, V/W are random Banach spaces and $\{x_n\}$ is a Cauchy sequence in V. Since

$$E_{\lambda,\bar{\mu}}\big((x_n - x_m) + W\big) \le E_{\lambda,\mu}(x_n - x_m)$$

for each $m, n \ge 1$, the sequence $\{x_n + W\}$ is a Cauchy sequence in V/W and so converges to $y + W$ for some $y \in W$. Thus, there exists $n_0 \ge 1$ such that, for each $n \ge n_0$,

$$E_{\lambda,\bar{\mu}}\big((x_n - y) + W\big) < 2^{-n}.$$

Now, by the last theorem, there exist a sequence $\{y_n\}$ in V such that

$$y_n + W = (x_n - y) + W, \qquad E_{\lambda,\mu}(y_n) < E_{\lambda,\bar{\mu}}\big((x_n - y) + W\big) + 2^{-n}.$$

Thus, we have

$$\lim_{n\to\infty} E_{\lambda,\mu}(y_n) \le 0$$

and so, by Lemma 2.2.15, $\mu_{y_n}(t) \to 1$ for any $t > 0$, that is, $\lim_{n\to\infty} y_n = 0$. Therefore, $\{x_n - y_n - y\}$ is a Cauchy sequence in W and so it is convergent to a point $z \in W$. This implies that $\{x_n\}$ converges to $z + y$ and hence V is complete. This completes the proof. □

Theorem 2.3.16 (Open Mapping Theorem) *If T is a random bounded linear operator from a RN-space (V, μ, T) onto an RN-space (V', ν, T), then T is an open mapping.*

Proof The theorem will be proved by the following steps:

Step 1: Let E be a neighborhood of the 0 in V. We show that $0 \in (\overline{T(E)})^o$. Let W be a balanced neighborhood of 0 such that $W + W \subset E$. Since $T(V) = V'$ and W is absorbing, it follows that $V' = \bigcap_n T(nW)$ and so there exists $n_0 \ge 1$ such that $\overline{T(n_0 W)}$ has a nonempty interior. Therefore, we have

$$0 \in \big(\overline{T(W)}\big)^o - \big(\overline{T(W)}\big)^o.$$

On the other hand, we have

$$\big(\overline{T(W)}\big)^o - \big(\overline{T(W)}\big)^o \subset \overline{T(W)} - \overline{T(W)} = \overline{T(W) + T(W)}$$
$$\subset \overline{T(E)}.$$

Thus, the set $\overline{T(E)}$ includes the neighborhood $(\overline{T(W)})^o - (\overline{T(W)})^o$ of 0.

Step 2: We show $0 \in (T(E))^o$. Since $0 \in E$ and E is an open set, there exist $0 < \alpha < 1$ and $t_0 \in (0, \infty)$ such that $B_0(\alpha, t_0) \subset E$. However, $0 < \alpha < 1$ and so a sequence $\{\epsilon_n\}$ can be found such that

$$T^{m-n}(1 - \epsilon_{n+1}, \ldots, 1 - \epsilon_m) \to 1$$

and

$$1 - \alpha < \lim_n T^{n-1}(1 - \epsilon_1, 1 - \epsilon_n),$$

in which $m > n$.

On the other hand, $0 \in \overline{T(B_0(\epsilon_n, t_n'))}$, where $t_n' = \frac{1}{2^n} t_0$, and so, by Step 1, there exist $0 < \sigma_n < 1$ and $t_n > 0$ such that

$$B_0(\sigma_n, t_n) \subset \overline{T\big(B_0(\epsilon_n, t_n')\big)}.$$

Since the set $\{B_0(r, 1/n)\}$ is a countable local base at zero and $t'_n \to 0$ as $n \to \infty$, t_n and σ_n can be chosen such that $t_n \to 0$ and $\sigma_n \to 0$ as $n \to \infty$.

Now, we show that

$$B_0(\sigma_1, t_1) \subset \big(T(E)\big)^o.$$

Suppose that $y_0 \in B_0(, \sigma_1, t_1)$. Then $y_0 \in \overline{T(B_0(\epsilon_1, t'_1))}$ and so for any $0 < \sigma_2$ and $t_2 > 0$, the ball $B_{y_0}(\sigma_2, t_2)$ intersects $T(B_0(\epsilon_1, t'_1))$. Therefore, there exists $x_1 \in B_0(\epsilon_1, t'_1)$ such that $Tx_1 \in B_{y_0}(\sigma_2, t_2)$, that is,

$$\nu_{y_0 - Tx_1}(t_2) > 1 - \sigma_2$$

or, equivalently,

$$y_0 - Tx_1 \in B_0(\sigma_2, t_2) \subset \overline{T\big(B_0\big(\epsilon_1, t'_1\big)\big)}.$$

By the similar argument, there exist $x_2 \in B_0(\epsilon_2, t'_2)$ such that

$$\nu_{y_0 - (Tx_1 + Tx_2)}(t_3) = \nu_{(y_0 - Tx_1) - Tx_2}(t_3) > 1 - \sigma_3.$$

If this process is continued, it leads to a sequence $\{x_n\}$ such that

$$x_n \in B_0\big(\epsilon_n, t'_n\big), \quad \nu_{y_0 - \sum_{j=1}^{n-1} Tx_j}(t_n) > 1 - \sigma_n.$$

Now, if $n, m \geq 1$ and $m > n$, then we have

$$\begin{aligned}\mu_{\sum_{j=1}^{n} x_j - \sum_{j=n+1}^{m} x_j}(t) &= \mu_{\sum_{j=n+1}^{m} x_j}(t) \\ &\geq T^{m-n}\big(\mu_{x_{n+1}}(t_{n+1}), \mu_{x_m}(t_m)\big),\end{aligned}$$

where $t_{n+1} + t_{n+2} + \cdots + t_m = t$. Put $t'_0 = \min\{t_{n+1}, t_{n+2}, \ldots, t_m\}$. Since $t'_n \to 0$, there exists $n_0 \geq 1$ such that $0 < t'_n \leq t'_0$ for all $n > n_0$. Therefore, for all $m > n$, we have

$$\begin{aligned}T^{m-n}\big(\mu_{x_{n+1}}\big(t'_0\big), \mu_{x_m}\big(t'_0\big)\big) &\geq T^{m-n}\big(\mu_{x_{n+1}}\big(t'_{n+1}\big), \mu_{x_m}\big(t'_m\big)\big) \\ &\geq T^{m-n}(1 - \epsilon_{n+1}, 1 - \epsilon_m)\end{aligned}$$

and so

$$\lim_{n\to\infty} \mu_{\sum_{j=n+1}^{m} x_j}(t) \geq \lim_{n\to\infty} T^{m-n}(1 - \epsilon_{n+1}, 1 - \epsilon_m) = 1,$$

that is,

$$\mu_{\sum_{j=n+1}^{m} x_j}(t) \to 1$$

for all $t > 0$. Thus, the sequence $\{\sum_{j=1}^{n} x_j\}$ is a Cauchy sequence and so the series $\{\sum_{j=1}^{\infty} x_j\}$ converges to a point $x_0 \in V$ since V is a complete space. For any fixed

$t > 0$, there exists $n_0 \geq 1$ such that $t > t_n$ for all $n > n_0$ since $t_n \to 0$. Thus, we have

$$\begin{aligned} \nu_{y_0 - T(\sum_{j=1}^{n-1} x_j)}(t) &\geq \nu_{y_0 - T(\sum_{j=1}^{n-1} x_j)}(t_n) \\ &\geq 1 - \sigma_n \end{aligned}$$

and so

$$\nu_{y_0 - T(\sum_{j=1}^{n-1} x_j)}(t) \to 1.$$

Therefore, we have

$$y_0 = \lim_{n\to\infty} T\left(\sum_{j=1}^{n-1} x_j\right) = T\left(\lim_{n\to\infty} \sum_{j=1}^{n-1} x_j\right) = Tx_0.$$

But, we have

$$\begin{aligned} \mu_{x_0}(t_0) &= \lim_{n\to\infty} \mu_{\sum_{j=1}^{n} x_j}(t_0) \\ &\geq T^n\Big(\lim_{n\to\infty} \big(\mu_{x_1}(t_1'), \mu_{x_n}(t_n')\big)\Big) \\ &\geq \lim_{n\to\infty} T^{n-1}(1-\epsilon_1, \dots, 1-\epsilon_n) \\ &> 1 - \alpha. \end{aligned}$$

Therefore, $x_0 \in B_0(\alpha, t_0)$.

Step 3: Let G be an open subset of V and $x \in G$. Then we have

$$T(G) = Tx + T(-x+G) \supset Tx + \big(T(-x+G)\big)^o.$$

Hence, $T(G)$ is open since it includes a neighborhood of each of its point. This completes the proof. □

Corollary 2.3.17 *Every one-to-one random bounded linear operator from a random Banach space onto a random Banach space has a random bounded converse.*

Theorem 2.3.18 (Closed Graph Theorem) *Let T be a linear operator from a random Banach space (V, μ, T) into a random Banach space (V', ν, T). Suppose that, for every sequence $\{x_n\}$ in V such that $x_n \to x$ and $Tx_n \to y$ for some elements $x \in V$ and $y \in V'$, it follows that $Tx = y$. Then T is random bounded.*

Proof For any $t > 0$, $x \in X$ and $y \in V'$, define

$$\Phi_{(x,y)}(t) = T'\big(\mu_x(t), \nu_y(t)\big),$$

where $T' \gg T$.

First, we show that $(V \times V', \Phi, T)$ is a complete RN-space. The properties of (RN1) and (RN2) are immediate from the definition. For the triangle inequality (RN3), suppose that $x, z \in V$, $y, u \in V'$ and $t, s > 0$. Then we have

$$\begin{aligned} T\big(\Phi_{(x,y)}(t), \Phi_{(z,u)}(s)\big) &= T\big[T'\big(\mu_x(t), \nu_y(t)\big), T'\big(\mu_z(s), \nu_u(s)\big)\big] \\ &\le T'\big[T\big(\mu_x(t), \mu_z(s)\big), T\big(\nu_y(t), \nu_u(s)\big)\big] \\ &\le T'\big(\mu_{x+z}(t+s), \nu_{y+u}(t+s)\big) \\ &= \Phi_{(x+z,y+u)}(t+s). \end{aligned}$$

Now, if $\{(x_n, y_n)\}$ is a Cauchy sequence in $V \times V'$, then, for any $\epsilon > 0$ and $t > 0$, there exists $n_0 \ge 1$ such that

$$\Phi_{(x_n,y_n)-(x_m,y_m)}(t) > 1 - \epsilon$$

for all $m, n > n_0$. Thus, for all $m, n > n_0$, we have

$$\begin{aligned} T'\big(\mu_{x_n - x_m}(t), \nu_{y_n - y_m}(t)\big) &= \Phi_{(x_n - x_m, y_n - y_m)}(t) \\ &= \Phi_{(x_n,y_n)-(x_m,y_m)}(t) \\ &> 1 - \epsilon. \end{aligned}$$

Therefore, $\{x_n\}$ and $\{y_n\}$ are Cauchy sequences in V and V', respectively, and there exist $x \in V$ and $y \in V'$ such that $x_n \to x$ and $y_n \to y$ and so $(x_n, y_n) \to (x, y)$. Hence, $(V \times V', \Phi, T)$ is a complete RN-space. The remainder of the proof is the same as the classical case. This completes the proof. □

2.4 Non-Archimedean Random Normed Spaces

By a *non-Archimedean field* we mean a field $\mathcal{K}$ equipped with a function (valuation) $|\cdot|$ from $\mathcal{K}$ into $[0, \infty)$ such that

(1) $|r| = 0$ if and only if $r = 0$;
(2) $|rs| = |r||s|$;
(3) $|r + s| \le \max\{|r|, |s|\}$ for all $r, s \in \mathcal{K}$.

Clearly, $|1| = |-1| = 1$ and $|n| \le 1$ for all $n \ge 1$. By the *trivial valuation,* we mean the mapping $|\cdot|$ taking everything but 0 into 1 and $|0| = 0$.

Let X be a vector space over a field $\mathcal{K}$ with a non-Archimedean nontrivial valuation $|\cdot|$, that is, there exists $a_0 \in \mathcal{K}$ such that $|a_0|$ is not in $\{0, 1\}$.

The most important examples of non-Archimedean spaces are p-adic numbers. In 1897, Hensel [106] discovered the p-adic numbers as a number theoretical analogue of power series in complex analysis. Fix a prime number p. For any nonzero rational number x, there exists a unique integer $n_x \in \mathbb{Z}$ such that $x = \frac{a}{b} p^{n_x}$, where a and b are integers not divisible by p. Then $|x|_p := p^{-n_x}$ defines a non-Archimedean norm on $\mathbb{Q}$. The completion of $\mathbb{Q}$ with respect to the metric $d(x, y) = |x - y|_p$ is denoted by $\mathbb{Q}_p$, which is called the *p-adic number field*.

A function $\|\cdot\| : X \to [0, \infty)$ is called a *non-Archimedean norm* if it satisfies the following conditions:

(NAN1) $\|x\| = 0$ if and only if $x = 0$;
(NAN2) for any $r \in \mathcal{K}, x \in X$, $\|rx\| = |r|\|x\|$;
(NAN3) the strong triangle inequality (ultrametric), namely,

$$\|x + y\| \leq \max\{\|x\|, \|y\|\}$$

for all $x, y \in X$.

Then $(X, \|\cdot\|)$ is called a *non-Archimedean normed space*.

Due to the fact that

$$\|x_n - x_m\| \leq \max\{\|x_{j+1} - x_j\| : m \leq j \leq n - 1\}$$

for all $n, m \geq 1$ with $n > m$, a sequence $\{x_n\}$ is a Cauchy sequence in X if and only if $\{x_{n+1} - x_n\}$ converges to zero in a non-Archimedean normed space.

By a *complete non-Archimedean normed space*, we mean one in which every Cauchy sequence is convergent.

Definition 2.4.1 A *non-Archimedean random normed space* (briefly, non-Archimedean RN-space) is a triple (X, μ, T), where X is a linear space over a non-Archimedean field $\mathcal{K}$, T is a continuous t-norm, and μ is a mapping from X into D^+ such that the following conditions hold:

(NA-RN1) $\mu_x(t) = \varepsilon_0(t)$ for all $t > 0$ if and only if $x = 0$;
(NA-RN2) $\mu_{\alpha x}(t) = \mu_x(\frac{t}{|\alpha|})$ for all $x \in X$, $t > 0$ and $\alpha \neq 0$;
(NA-RN3) $\mu_{x+y}(\max\{t, s\}) \geq T(\mu_x(t), \mu_y(s))$ for all $x, y, z \in X$ and $t, s \geq 0$.

It is easy to see that, if (NA-RN3) holds, then so is

(RN3) $\mu_{x+y}(t + s) \geq T(\mu_x(t), \mu_y(s))$.

Example 2.4.2 As a classical example, if $(X, \|\cdot\|)$ is a non-Archimedean normed linear space, then the triple (X, μ, T_M), where

$$\mu_x(t) = \begin{cases} 0, & \text{if } t \leq \|x\|, \\ 1, & \text{if } t > \|x\|, \end{cases}$$

is a non-Archimedean RN-space.

Example 2.4.3 Let $(X, \|\cdot\|)$ be a non-Archimedean normed linear space. Define

$$\mu_x(t) = \frac{t}{t + \|x\|}$$

for all $x \in X$ and $t > 0$. Then (X, μ, T_M) is a non-Archimedean RN-space.

Definition 2.4.4 Let (X, μ, T) be a non-Archimedean RN-space. Let $\{x_n\}$ be a sequence in X.

(1) The sequence $\{x_n\}$ is said to be *convergent* if there exists $x \in X$ such that

$$\lim_{n\to\infty} \mu_{x_n - x}(t) = 1$$

for all $t > 0$. In this case, the point x is called the *limit* of the sequence $\{x_n\}$.

(2) The sequence $\{x_n\}$ in X is called a *Cauchy sequence* if, for any $\varepsilon > 0$ and $t > 0$, there exists $n_0 \geq 1$ such that, for all $n \geq n_0$ and $p > 0$,

$$\mu_{x_{n+p} - x_n}(t) > 1 - \varepsilon.$$

(3) If each Cauchy sequence in X is convergent, then the random normed space is said to be *complete* and the non-Archimedean RN-space (X, μ, T) is called a *non-Archimedean random Banach space.*

Remark 2.4.5 [168] Let (X, μ, T_M) be a non-Archimedean RN-space. Then we have

$$\mu_{x_{n+p} - x_n}(t) \geq \min\{\mu_{x_{n+j+1} - x_{n+j}}(t) : j = 0, 1, 2, \ldots, p - 1\}.$$

Thus, the sequence $\{x_n\}$ is a Cauchy sequence in X if, for any $\varepsilon > 0$ and $t > 0$, there exists $n_0 \geq 1$ such that, for all $n \geq n_0$,

$$\mu_{x_{n+1} - x_n}(t) > 1 - \varepsilon.$$

2.5 Fuzzy Normed Spaces

Now, we define the concept of fuzzy normed spaces and give some examples of these spaces. Here the t-norms notation is denoted by $*$.

Definition 2.5.1 The triple $(X, M, *)$ is called a *fuzzy metric space* if X is an arbitrary set, $*$ is a continuous t-norm and M is a fuzzy set on $X^2 \times (0, \infty)$ satisfying the following conditions: for all $x, y, z \in X$ and $t, s > 0$,

(FM1) $M(x, y, 0) > 0$;
(FM2) $M(x, y, t) = 1$ for all $t > 0$ if and only if $x = y$;
(FM3) $M(x, y, t) = M(y, x, t)$;
(FM4) $M(x, y, t) * M(y, z, s) \leq M(x, z, t + s)$ for all $t, s > 0$;
(FM5) $M(x, y, \cdot) : (0, \infty) \to [0, 1]$ is continuous.

Definition 2.5.2 The triple $(X, N, *)$ is called a *fuzzy normed space* if X is a vector space, $*$ is a continuous t-norm and N is a fuzzy set on $X \times (0, \infty)$ satisfying the following conditions: for all $x, y \in X$ and $t, s > 0$,

(FN1) $N(x, t) > 0$;

(FN2) $N(x,t) = 1$ if and only if $x = 0$;
(FN3) $N(\alpha x, t) = N(x, t/|\alpha|)$ for all $\alpha \neq 0$;
(FN4) $N(x,t) * N(y,s) \le N(x+y, t+s)$;
(FN5) $N(x,\cdot) : (0,\infty) \to [0,1]$ is continuous;
(FN6) $\lim_{t\to\infty} N(x,t) = 1$.

Lemma 2.5.3 *Let N be a fuzzy norm. Then we have*

(1) *$N(x,t)$ is nondecreasing with respect to t for all $x \in X$;*
(2) *$N(x-y,t) = N(y-x,t)$.*

Proof Let $t < s$. Then $k = s - t > 0$ and we have

$$N(x,t) = N(x,t) * 1 = N(x,t) * N(0,k) \le N(x,s),$$

which proves (1).

To prove (2), we have

$$N(x-y,t) = N\big((-1)(y-x),t\big) = N\left(y-x, \frac{t}{|-1|}\right) = N(y-x,t).$$

This completes the proof. □

Example 2.5.4 Let $(X, \|\cdot\|)$ be a normed linear space. Define $a * b = ab$ or $a * b = \min(a,b)$ and

$$N(x,t) = \frac{kt^n}{kt^n + m\|x\|}$$

for all $k, m, n \in \mathbb{R}^+$. Then $(X, N, *)$ is a fuzzy normed space. In particular, if $k = n = m = 1$, then we have

$$N(x,t) = \frac{t}{t + \|x\|},$$

which is called the *standard fuzzy norm* induced by the norm $\|\cdot\|$.

Lemma 2.5.5 *Let $(X, N, *)$ be a fuzzy normed space. If we define*

$$M(x,y,t) = N(x-y,t),$$

then M is a fuzzy metric on X, which is called the fuzzy metric induced by the fuzzy norm N.

We can see that both definition and properties on fuzzy normed spaces are very similar to those of random normed spaces. Then X equipped with $\mu_x(t) = N(x,t)$ and $T = *$ can be regarded as a RN-space.

Now, we extend the definition of fuzzy metric space. In fact, we extend the range of fuzzy sets to arbitrary lattice.

Definition 2.5.6 The triple $(X, \mathcal{P}, \mathcal{T})$ is called an *$\mathcal{L}$-fuzzy normed space* (briefly, $\mathcal{L}F$-normed space) if X is a vector space, $\mathcal{T}$ is a continuous t-norm on $\mathcal{L}$ and $\mathcal{P}$ is an $\mathcal{L}$-fuzzy set on $X \times (0, +\infty)$ satisfying the following conditions: for all $x, y \in X$ and $t, s \in (0, +\infty)$,

(LFN1) $\mathcal{P}(x, t) >_L 0_{\mathcal{L}}$;
(LFN2) $\mathcal{P}(x, t) = 1_{\mathcal{L}}$ if and only if $x = 0$;
(LFN3) $\mathcal{P}(\alpha x, t) = \mathcal{P}(x, \frac{t}{|\alpha|})$ for any $\alpha \neq 0$;
(LFN4) $\mathcal{T}(\mathcal{P}(x, t), \mathcal{P}(y, s)) \leq_L \mathcal{P}(x + y, t + s)$;
(LFN5) $\mathcal{P}(x, \cdot) : (0, \infty) \to L$ is continuous;
(LFN6) $\lim_{t\to\infty} \mathcal{P}(x, t) = 1_{\mathcal{L}}$.

In this case, $\mathcal{P}$ is called an *$\mathcal{L}$-fuzzy norm* (briefly, $\mathcal{L}F$-norm). If $\mathcal{P} = \mathcal{P}_{\mu,\nu}$ is an intuitionistic fuzzy set and the t-norm $\mathcal{T}$ is t-representable, then the triple $(X, \mathcal{P}_{\mu,\nu}, \mathcal{T})$ is said to be an *intuitionistic fuzzy normed space* (briefly, *IF*-normed space).

Example 2.5.7 Let $(X, \|\cdot\|)$ be a normed linear space. Denote $\mathcal{T}(a, b) = (a_1 b_1, \min(a_2 + b_2, 1))$ for all $a = (a_1, a_2), b = (b_1, b_2) \in L^*$ and let M, N be the fuzzy sets on $X \times (0, \infty)$ defined as follows:

$$\mathcal{P}_{M,N}(x, t) = \left(\frac{ht^n}{ht^n + m\|x\|}, \frac{m\|x\|}{ht^n + m\|x\|} \right)$$

for all $t, h, m, n \in \mathbb{R}^+$. Then $(X, \mathcal{P}_{M,N}, \mathcal{T})$ is an *IF*-normed space.

Chapter 3
Stability of Functional Equations in RN-Spaces Under Spacial t-Norm

In this chapter, we consider the stability of some functional equations in random normed spaces under the spacial t-norm, that is, T_M via direct method.

3.1 Cauchy Additive Equations

The first results on the stability of Cauchy equations in the setting of fuzzy normed spaces are given in [166] via direct method and using the t-norm T_M. Here, we improve and generalize the result of [166].

Recall that the functional equation satisfying the following property:

$$f(x+y) = f(x) + f(y) \tag{3.1.1}$$

is called a *Cauchy additive functional equation*.

Theorem 3.1.1 *Let X be a linear space, (Y, μ, T_M) be an RN-space and φ be a mapping from X^2 to D^+ ($\varphi(x, y)$ is denoted by $\varphi_{x,y}$) such that, some $0 < \alpha < 2$,*

$$\varphi_{2x,2x}(\alpha t) \geq \varphi_{x,x}(t) \tag{3.1.2}$$

and

$$\lim_{n\to\infty} \varphi_{2^n x, 2^n y}\big(2^n t\big) = 1 \tag{3.1.3}$$

for all $x, y \in X$ and $t > 0$. If $f : X \to Y$ is a mapping such that

$$\mu_{f(x+y)-f(x)-f(y)}(t) \geq \varphi_{x,y}(t) \tag{3.1.4}$$

for all $x, y \in X$ and $t > 0$, then there exists a unique additive mapping $A : X \to Y$ such that

$$\mu_{f(x)-A(x)}(t) \geq \varphi_{x,x}\big((2-\alpha)t\big). \tag{3.1.5}$$

Y.J. Cho et al., *Stability of Functional Equations in Random Normed Spaces*, Springer Optimization and Its Applications 86, DOI 10.1007/978-1-4614-8477-6_3,

Proof Putting $y = x$ in (3.1.4), we get

$$\mu_{\frac{f(2x)}{2} - f(x)}(t) \geq \varphi_{x,x}(2t) \tag{3.1.6}$$

for all $x \in X$ and $t > 0$. Replacing x by $2^n x$ in (3.1.6) and using (3.1.2), we obtain

$$\mu_{\frac{f(2^{n+1}x)}{2^{n+1}} - \frac{f(2^n x)}{2^n}}(t) \geq \varphi_{2^n x, 2^n x}\left(2 \times 2^n t\right) \geq \varphi_{x,x}\left(\frac{2 \times 2^n}{\alpha^n} t\right). \tag{3.1.7}$$

It follows from

$$\frac{f(2^n x)}{2^n} - f(x) = \sum_{k=0}^{n-1} \left(\frac{f(2^{k+1}x)}{2^{k+1}} - \frac{f(2^k x)}{2^k}\right)$$

and (3.1.7) that

$$\mu_{\frac{f(2^n x)}{2^n} - f(x)}\left(t \sum_{k=0}^{n-1} \frac{\alpha^k}{2 \times 2^k}\right) \geq T_{M,k=0}^{n-1}\left(\varphi_{x,x}(t)\right) = \varphi_{x,x}(t),$$

that is,

$$\mu_{\frac{f(2^n x)}{2^n} - f(x)}(t) \geq \varphi_{x,x}\left(\frac{t}{\sum_{k=0}^{n-1} \frac{\alpha^k}{2 \times 2^k}}\right). \tag{3.1.8}$$

By replacing x with $2^m x$ in (3.1.8), we have

$$\mu_{\frac{f(2^{n+m}x)}{2^{n+m}} - \frac{f(2^m x)}{2^m}}(t) \geq \varphi_{x,x}\left(\frac{t}{\sum_{k=m}^{n+m} \frac{\alpha^k}{2 \times 2^k}}\right). \tag{3.1.9}$$

Since

$$\varphi_{x,x}\left(\frac{t}{\sum_{k=m}^{n+m} \frac{\alpha^k}{2 \times 2^k}}\right)$$

tend to 1 as $m, n \to \infty$, it follows that $\{\frac{f(2^n x)}{2^n}\}$ is a Cauchy sequence in (Y, μ, T_M). Since (Y, μ, T_M) is a complete RN-space, the sequence $\{\frac{f(2^n x)}{2^n}\}$ converges to a point $A(x) \in Y$. Fix $x \in X$ and put $m = 0$ in (3.1.9). Then we obtain

$$\mu_{\frac{f(2^n x)}{2^n} - f(x)}(t) \geq \varphi_{x,x}\left(\frac{t}{\sum_{k=0}^{n-1} \frac{\alpha^k}{2 \times 2^k}}\right)$$

and so, for any $\delta > 0$,

$$\begin{aligned} \mu_{A(x) - f(x)}(t + \delta) &\geq T_M\left(\mu_{A(x) - \frac{f(2^n x)}{2^n}}(\delta), \mu_{\frac{f(2^n x)}{2^n} - f(x)}(t)\right) \\ &\geq T_M\left(\mu_{A(x) - \frac{f(2^n x)}{2^n}}(\delta), \varphi_{x,x}\left(\frac{t}{\sum_{k=0}^{n-1} \frac{\alpha^k}{2 \times 2^k}}\right)\right). \end{aligned} \tag{3.1.10}$$

Taking $n \to \infty$ and using (3.1.10), we get

$$\mu_{A(x)-f(x)}(t+\delta) \geq \varphi_{x,x}\big(t(2-\alpha)\big). \tag{3.1.11}$$

Since δ is arbitrary, by taking $\delta \to 0$ in (3.1.11), we get

$$\mu_{A(x)-f(x)}(t) \geq \varphi_{x,x}\big(t(2-\alpha)\big).$$

Replacing x and y by $2^n x$ and $2^n y$ in (3.1.4), respectively, we get

$$\mu_{\frac{f(2^n(x+y))}{2^n}-\frac{f(2^n x)}{2^n}-\frac{f(2^n y)}{2^n}}(t) \geq \varphi_{2^n x,2^n y}\big(2^n t\big)$$

for all $x, y \in X$ and $t > 0$. Since $\lim_{n\to\infty} \varphi_{2^n x,2^n y}(2^n t) = 1$, we conclude that A satisfies (3.1.1).

To prove the uniqueness of the additive mapping A, assume that there exists an additive mapping $A' : X \to Y$ which satisfies (3.1.5). Fix $x \in X$. Clearly, $A(2^n x) = 2^n A(x)$ and $A'(2^n x) = 2^n A'(x)$ for all $n \geq 1$. It follows from (3.1.5) that

$$\mu_{A(x)-A'(x)}(t) = \lim_{n\to\infty} \mu_{\frac{A(2^n x)}{2^n}-\frac{A'(2^n x)}{2^n}}(t),$$

$$\begin{aligned}
\mu_{\frac{A(2^n x)}{2^n}-\frac{A'(2^n x)}{2^n}}(t) &\geq T_M\left(\mu_{\frac{A(2^n x)}{2^n}-\frac{f(2^n x)}{2^n}}\left(\frac{t}{2}\right), \mu_{\frac{A'(2^n x)}{2^n}-\frac{f(2^n x)}{2^n}}\left(\frac{t}{2}\right)\right)\\
&\geq \varphi_{2^n x,2^n x}\big(2^n(2-\alpha)t\big)\\
&\geq \varphi_{x,x}\left(\frac{2^n(2-\alpha)t}{\alpha^n}\right).
\end{aligned}$$

Since $\lim_{n\to\infty} \frac{2^n(2-\alpha)t}{\alpha^n} = \infty$, we get

$$\lim_{n\to\infty} \varphi_{x,x}\left(\frac{2^n(2-\alpha)t}{\alpha^n}\right) = 1.$$

Therefore, it follows that $\mu_{A(x)-A'(x)}(t) = 1$ for all $x \in X$ and $t > 0$ and so $A(x) = A(x)$. This completes the proof. □

Corollary 3.1.2 *Let $\theta \geq 0$ and p be a real number with $0 < p < 1$. Let X be a normed linear space with the norm $\|\cdot\|$. Let $f : X \to Y$ be an odd mapping satisfying*

$$\mu_{f(x+y)-f(x)-f(y)}(t) \geq \frac{t}{t+\theta(\|x\|^p + \|y\|^p)} \tag{3.1.12}$$

for all $x, y \in X$ and $t > 0$. Then

$$A(x) := \lim_{n\to\infty} 2^n f\left(\frac{x}{2^n}\right)$$

exists for all $x \in X$ and defines an additive mapping $A : X \to Y$ such that

$$\mu_{f(x)-A(x)}(t) \geq \frac{(2-2^p)t}{(2-2^p)t + 2\theta\|x\|^p}$$

for all $x \in X$ and $t > 0$.

Proof The proof follows from Theorem 3.1.1 by taking

$$\varphi_{x,y}(t) := \frac{t}{t + \theta(\|x\|^p + \|y\|^p)}$$

for all $x, y \in X$ and $t > 0$. Then the proof follows from Theorem 3.1.1 by $\alpha = 2^p$. □

Corollary 3.1.3 *Let $\varepsilon > 0$ and $x_0 \in X$. Let X be a normed linear space with the norm $\|\cdot\|$. Let $f : X \to Y$ be an odd mapping satisfying*

$$\mu_{f(x+y)-f(x)-f(y)}(t) \geq \frac{t}{t + \varepsilon\|x_0\|}$$

for all $x, y \in X$ and $t > 0$. Then there exists an unique additive mapping $A : X \to Y$ such that

$$\mu_{f(x)-A(x)}(t) \geq \frac{2t}{2t + \varepsilon\|x_0\|}$$

for all $x \in X$ and $t > 0$.

Proof The proof follows from Theorem 3.1.1 by taking

$$\varphi_{x,y}(t) := \frac{t}{t + \varepsilon\|x_0\|}$$

for all $x, y \in X$ and $t > 0$. Then the proof follows from Theorem 3.1.1 by $\alpha = 1$. □

Theorem 3.1.4 *Let X be a linear space, (Y, μ, T_M) be an RN-space and φ be a mapping from X^2 to D^+ ($\varphi(x, y)$ is denoted by $\varphi_{x,y}$) such that, some $\alpha > 2$,*

$$\varphi_{\frac{x}{2},\frac{x}{2}}(t) \geq \varphi_{x,x}(\alpha t) \tag{3.1.13}$$

and

$$\lim_{n\to\infty} \varphi_{\frac{x}{2^n},\frac{y}{2^n}}\left(\frac{t}{2^n}\right) = 1$$

for all $x, y \in X$ and $t > 0$. If $f : X \to Y$ is a mapping such that

$$\mu_{f(x+y)-f(x)-f(y)}(t) \geq \varphi_{x,y}(t) \tag{3.1.14}$$

for all $x, y \in X$ and $t > 0$, then there exists a unique additive mapping $A : X \to Y$ such that

$$\mu_{f(x)-A(x)}(t) \geq \varphi_{\frac{x}{2},\frac{x}{2}}\big((\alpha - 2)t\big). \tag{3.1.15}$$

Corollary 3.1.5 *Let $\theta \geq 0$ and p be a real number with $p > 1$. Let X be a normed linear space with norm $\|\cdot\|$. Let $f : X \to Y$ be an odd mapping satisfying*

$$\mu_{f(x+y)-f(x)-f(y)}(t) \geq \frac{t}{t + \theta(\|x\|^p + \|y\|^p)} \tag{3.1.16}$$

for all $x, y \in X$ and $t > 0$. Then

$$A(x) := \lim_{n\to\infty} 2^n f\left(\frac{x}{2^n}\right)$$

exists for all $x \in X$ and $t > 0$ and defines an additive mapping $A : X \to Y$ such that

$$\mu_{f(x)-A(x)}(t) \geq \frac{(2^p - 2)t}{(2^p - 2)t + 2\theta\|x\|^p}$$

for all $x \in X$ and $t > 0$.

Proof The proof follows from Theorem 3.1.4 by taking

$$\varphi_{x,y}(t) := \frac{t}{t + \theta(\|x\|^p + \|y\|^p)}$$

for all $x, y \in X$ and $t > 0$. Then the proof follows from Theorem 3.1.4 by $\alpha = 2^p$. □

Remark 3.1.6 In Theorem 3.1.1, if we replace (3.1.2) by

$$\varphi_{2x,2y}(\alpha t) \geq \varphi_{x,y}(t) \tag{3.1.17}$$

for all $x, y \in X$ and $t > 0$, then we can remove (3.1.3). This remark holds for Theorem 3.1.4.

3.2 Quadratic Functional Equations

Let X and Y be linear spaces and $f : X \to Y$ be a mapping. Recall that the functional equation satisfying the following:

$$f(x + y) + f(x - y) = 2f(x) + 2f(y) \tag{3.2.1}$$

for all $x, y \in X$ is called the *quadratic functional equation* (see [48] and [176]).

Theorem 3.2.1 *Let X be a linear space, (Y, μ, T_M) be an RN-space and φ be a mapping from X^2 to D^+ ($\varphi(x, y)$ is denoted by $\varphi_{x,y}$) such that, some $0 < \alpha < 4$,*

$$\varphi_{2x,2x}(\alpha t) \geq \varphi_{x,x}(t) \tag{3.2.2}$$

and

$$\lim_{n\to\infty} \varphi_{2^n x,2^n y}(4^n t) = 1$$

for all $x, y \in X$ and $t > 0$. If $f : X \to Y$ is a mapping such that

$$\mu_{f(x+y)+f(x-y)-2f(x)-2f(y)}(t) \geq \varphi_{x,y}(t) \tag{3.2.3}$$

for all $x, y \in X$ and $t > 0$ and $f(0) = 0$, then there exists a unique quadratic mapping $Q : X \to Y$ such that

$$\mu_{f(x)-Q(x)}(t) \geq \varphi_{x,x}\big((4-\alpha)t\big). \tag{3.2.4}$$

Proof Putting $y = x$ in (3.2.3), we get

$$\mu_{\frac{f(2x)}{4}-f(x)}(t) \geq \varphi_{x,x}(4t) \tag{3.2.5}$$

for all $x \in X$ and $t > 0$. Replacing x by $2^n x$ in (3.2.5) and using (3.2.2), we obtain

$$\mu_{\frac{f(2^{n+1}x)}{4^{n+1}}-\frac{f(2^n x)}{4^n}}(t) \geq \varphi_{2^n x,2^n x}(4 \times 4^n t) \geq \varphi_{x,x}\left(\frac{4 \times 4^n}{\alpha^n} t\right). \tag{3.2.6}$$

It follows from

$$\frac{f(2^n x)}{4^n} - f(x) = \sum_{k=0}^{n-1}\left(\frac{f(2^{k+1}x)}{4^{k+1}} - \frac{f(2^k x)}{4^k}\right)$$

and (3.2.6) that

$$\mu_{\frac{f(2^n x)}{4^n}-f(x)}\left(t\sum_{k=0}^{n-1}\frac{\alpha^k}{4 \times 4^k}\right) \geq T_{M,k=0}^{n-1}\big(\varphi_{x,x}(t)\big) = \varphi_{x,x}(t),$$

that is,

$$\mu_{\frac{f(2^n x)}{4^n}-f(x)}(t) \geq \varphi_{x,x}\left(\frac{t}{\sum_{k=0}^{n-1}\frac{\alpha^k}{4\times 4^k}}\right). \tag{3.2.7}$$

By replacing x with $2^m x$ in (3.2.7), we observe that

$$\mu_{\frac{f(2^{n+m}x)}{4^{n+m}}-\frac{f(2^m x)}{4^m}}(t) \geq \varphi_{x,x}\left(\frac{t}{\sum_{k=m}^{n+m}\frac{\alpha^k}{4\times 4^k}}\right). \tag{3.2.8}$$

Since

$$\varphi_{x,x}\left(\frac{t}{\sum_{k=m}^{n+m}\frac{\alpha^k}{4\times 4^k}}\right)\to 1$$

as $m,n\to\infty$, it follows that $\{\frac{f(2^n x)}{4^n}\}$ is a Cauchy sequence in (Y,μ,T_M). Since (Y,μ,T_M) is a complete RN-space, the sequence $\{\frac{f(2^n x)}{4^n}\}$ converges to a point $Q(x)\in Y$. Fix $x\in X$ and put $m=0$ in (3.2.8). Then we obtain

$$\mu_{\frac{f(2^n x)}{4^n}-f(x)}(t)\geq \varphi_{x,x}\left(\frac{t}{\sum_{k=0}^{n-1}\frac{\alpha^k}{4\times 4^k}}\right)$$

and so, for any $\delta>0$,

$$\begin{aligned}\mu_{Q(x)-f(x)}(t+\delta)&\geq T_M\big(\mu_{Q(x)-\frac{f(2^n x)}{4^n}}(\delta),\mu_{\frac{f(2^n x)}{4^n}-f(x)}(t)\big)\\&\geq T_M\left(\mu_{A(x)-\frac{f(2^n x)}{4^n}}(\delta),\varphi_{x,x}\left(\frac{t}{\sum_{k=0}^{n-1}\frac{\alpha^k}{4\times 4^k}}\right)\right).\end{aligned}\tag{3.2.9}$$

Taking $n\to\infty$ and using (3.2.9), we get

$$\mu_{Q(x)-f(x)}(t+\delta)\geq \varphi_{x,x}\big(t(4-\alpha)\big).\tag{3.2.10}$$

Since δ is arbitrary, by taking $\delta\to 0$ in (3.2.10), we get

$$\mu_{Q(x)-f(x)}(t)\geq \varphi_{x,x}\big(t(4-\alpha)\big).$$

Replacing x and y by $2^n x$ and $2^n y$ in (3.2.3), respectively, we get

$$\mu_{\frac{f(2^n(x+y))}{4^n}+\frac{f(2^n(x-y))}{4^n}-\frac{f(2^n x)}{4^n}-\frac{f(2^n y)}{4^n}}(t)\geq \varphi_{2^n x,2^n y}\big(4^n t\big)$$

for all $x,y\in X$ and $t>0$. Since $\lim_{n\to\infty}\varphi_{2^n x,2^n y}(4^n t)=1$, we conclude that Q satisfies (3.2.1).

To prove the uniqueness of the quadratic mapping Q, assume that there exists a quadratic mapping $Q':X\to Y$ which satisfies (3.2.4). Fix $x\in X$. Clearly, $Q(2^n x)=4^n Q(x)$ and $Q'(2^n x)=4^n Q'(x)$ for all $n\in\mathbb{N}$. It follows from (3.2.4) that

$$\mu_{Q(x)-Q'(x)}(t)=\lim_{n\to\infty}\mu_{\frac{Q(2^n x)}{4^n}-\frac{Q'(2^n x)}{4^n}}(t)$$

and

$$\begin{aligned}\mu_{\frac{Q(2^n x)}{4^n}-\frac{Q'(2^n x)}{4^n}}(t)&\geq T_M\left(\mu_{\frac{Q(2^n x)}{4^n}-\frac{f(2^n x)}{4^n}}\left(\frac{t}{2}\right),\mu_{\frac{Q'(2^n x)}{4^n}-\frac{f(2^n x)}{4^n}}\left(\frac{t}{2}\right)\right)\\&\geq \varphi_{2^n x,2^n x}\big(4^n(4-\alpha)t\big)\end{aligned}$$

$$\geq \varphi_{x,x}\left(\frac{4^n(4-\alpha)t}{\alpha^n}\right).$$

Since $\lim_{n\to\infty}\frac{4^n(4-\alpha)t}{\alpha^n}=\infty$, we get

$$\lim_{n\to\infty}\varphi_{x,x}\left(\frac{4^n(4-\alpha)t}{\alpha^n}\right)=1.$$

Therefore, it follows that $\mu_{Q(x)-Q'(x)}(t)=1$ for all $t>0$ and so $Q(x)=Q'(x)$. This completes the proof. □

Corollary 3.2.2 *Let $\theta \geq 0$ and p be a real number with $0<p<1$. Let X be a normed linear space with the norm $\|\cdot\|$. Let $f: X\to Y$ be an odd mapping satisfying*

$$\mu_{f(x+y)+f(x-y)-2f(x)-2f(y)}(t)\geq \frac{t}{t+\theta(\|x\|^p+\|y\|^p)} \tag{3.2.11}$$

for all $x,y\in X$ and $t>0$. Then

$$Q(x):=\lim_{n\to\infty}\frac{f(2^n x)}{4^n}$$

exists for all $x\in X$ and defines a quadratic mapping $Q: X\to Y$ such that

$$\mu_{f(x)-Q(x)}(t)\geq \frac{(4-4^p)t}{(4-4^p)t+2\theta\|x\|^p}$$

for all $x\in X$ and $t>0$.

Proof The proof follows from Theorem 3.2.1 by taking

$$\varphi_{x,y}(t):=\frac{t}{t+\theta(\|x\|^p+\|y\|^p)}$$

for all $x,y\in X$ and $t>0$. Then the proof follows from Theorem 3.2.1 by $\alpha=4^p$. □

Corollary 3.2.3 *Let $\varepsilon>0$ and $x_0\in X$. Let X be a normed linear space with norm $\|\cdot\|$. Let $f: X\to Y$ be an odd mapping satisfying*

$$\mu_{f(x+y)+f(x-y)-2f(x)-2f(y)}(t)\geq \frac{t}{t+\varepsilon\|x_0\|}$$

for all $x,y\in X$ and $t>0$. Then there exists an unique quadratic mapping $Q: X\to Y$ such that

$$\mu_{f(x)-Q(x)}(t)\geq \frac{3t}{3t+\varepsilon\|x_0\|}$$

for all $x\in X$ and $t>0$.

Proof The proof follows from Theorem 3.2.1 by taking

$$\varphi_{x,y}(t) := \frac{t}{t + \varepsilon\|x_0\|}$$

for all $x, y \in X$ and $t > 0$. Then the proof follows from Theorem 3.2.1 by $\alpha = 1$. □

Theorem 3.2.4 *Let X be a linear space, (Y, μ, T_M) be an RN-space and φ be a mapping from X^2 to D^+ ($\varphi(x, y)$ is denoted by $\varphi_{x,y}$) such that, some $\alpha > 4$,*

$$\varphi_{\frac{x}{2},\frac{x}{2}}(t) \geq \varphi_{x,x}(\alpha t)$$

and

$$\lim_{n\to\infty} \varphi_{\frac{x}{2^n},\frac{y}{2^n}}\left(\frac{t}{4^n}\right) = 1$$

for all $x, y \in X$ and $t > 0$. If $f : X \to Y$ is a mapping such that

$$\mu_{f(x+y)+f(x-y)-2f(x)-2f(y)}(t) \geq \varphi_{x,y}(t)$$

for all $x, y \in X$ and $t > 0$, then there exists a unique quadratic mapping $Q : X \to Y$ such that

$$\mu_{f(x)-Q(x)}(t) \geq \varphi_{\frac{x}{2},\frac{x}{2}}\big((4 - \alpha)t\big).$$

Corollary 3.2.5 *Let $\theta \geq 0$ and p be a real number with $p > 1$. Let X be a normed linear space with norm $\|\cdot\|$. Let $f : X \to Y$ be an odd mapping satisfying*

$$\mu_{f(x+y)-f(x)-f(y)}(t) \geq \frac{t}{t + \theta(\|x\|^p + \|y\|^p)} \tag{3.2.12}$$

for all $x, y \in X$ and $t > 0$. Then

$$Q(x) := \lim_{n\to\infty} 4^n f\left(\frac{x}{2^n}\right)$$

exists for all $x \in X$ and defines a quadratic mapping $Q : X \to Y$ such that

$$\mu_{f(x)-Q(x)}(t) \geq \frac{(4^p - 4)t}{(4^p - 4)t + 2\theta\|x\|^p}$$

for all $x \in X$ and $t > 0$.

Proof The proof follows from Theorem 3.2.4 by taking

$$\varphi_{x,y}(t) := \frac{t}{t + \theta(\|x\|^p + \|y\|^p)}$$

for all $x, y \in X$ and $t > 0$. Then the proof follows from Theorem 3.2.4 by $\alpha = 4^p$. □

Remark 3.2.6 As in Remark 3.1.6, we can replace and remove some conditions of Theorems 3.2.1 and 3.2.4.

3.3 Cubic Functional Equations

Let X and Y be linear spaces and $f: X \to Y$ be a mapping. Recall that the functional equation satisfying the following:

$$f(2x+y)+f(2x-y)=2f(x+y)+2f(x-y)+12f(x) \qquad (3.3.1)$$

is called the *cubic functional equation*.

Theorem 3.3.1 *Let X be a linear space, (Y, μ, T_M) be an RN-space and φ a mapping from X^2 to D^+ ($\varphi(x, y)$ is denoted by $\varphi_{x,y}$) such that, some $0 < \alpha < 8$,*

$$\varphi_{2x,0}(\alpha t) \geq \varphi_{x,0}(t) \qquad (3.3.2)$$

and

$$\lim_{n\to\infty} \varphi_{2^n x, 2^n y}\big(8^n t\big) = 1$$

for all $x, y \in X$ and $t > 0$. If $f: X \to Y$ is a mapping such that

$$\mu_{f(2x+y)+f(2x-y)-2f(x+y)-2f(x-y)-12f(x)}(t) \geq \varphi_{x,y}(t) \qquad (3.3.3)$$

for all $x, y \in X$ and $t > 0$ and $f(0) = 0$, then there exists a unique cubic mapping $C: X \to Y$ such that

$$\mu_{f(x)-C(x)}(t) \geq \varphi_{x,x}\big(2(8-\alpha)t\big). \qquad (3.3.4)$$

Proof Putting $y = 0$ in (3.3.3), we get

$$\mu_{\frac{f(2x)}{8}-f(x)}(t) \geq \varphi_{x,0}(16t) \qquad (3.3.5)$$

for all $x \in X$ and $t > 0$. Replacing x by $2^n x$ in (3.3.5) and using (3.3.2), we obtain

$$\mu_{\frac{f(2^{n+1}x)}{8^{n+1}}-\frac{f(2^n x)}{8^n}}(t) \geq \varphi_{2^n x,0}\big(16 \times 8^n t\big) \geq \varphi_{x,0}\left(\frac{16 \times 8^n}{\alpha^n} t\right). \qquad (3.3.6)$$

It follows from

$$\frac{f(2^n x)}{8^n} - f(x) = \sum_{k=0}^{n-1}\left(\frac{f(2^{k+1}x)}{8^{k+1}} - \frac{f(2^k x)}{8^k}\right)$$

and (3.3.6) that

$$\mu_{\frac{f(2^n x)}{8^n}-f(x)}\left(t\sum_{k=0}^{n-1}\frac{\alpha^k}{16 \times 8^k}\right) \geq T_{M,k=0}^{n-1}\big(\varphi_{x,0}(t)\big) = \varphi_{x,0}(t),$$

that is,

$$\mu_{\frac{f(2^n x)}{8^n}-f(x)}(t) \geq \varphi_{x,0}\left(\frac{t}{\sum_{k=0}^{n-1}\frac{\alpha^k}{16\times 8^k}}\right). \tag{3.3.7}$$

By replacing x with $2^m x$ in (3.3.7), we observe that

$$\mu_{\frac{f(2^{n+m} x)}{8^{n+m}}-\frac{f(2^m x)}{8^m}}(t) \geq \varphi_{x,0}\left(\frac{t}{\sum_{k=m}^{n+m}\frac{\alpha^k}{16\times 8^k}}\right). \tag{3.3.8}$$

From

$$\varphi_{x,x}\left(\frac{t}{\sum_{k=m}^{n+m}\frac{\alpha^k}{16\times 8^k}}\right) \to 1$$

as $m, n \to \infty$, it follows that $\{\frac{f(2^n x)}{8^n}\}$ is a Cauchy sequence in (Y, μ, T_M). Since (Y, μ, T_M) is a complete RN-space, the sequence $\{\frac{f(2^n x)}{8^n}\}$ converges to a point $C(x) \in Y$. Fix $x \in X$ and put $m = 0$ in (3.3.8). Then we obtain

$$\mu_{\frac{f(2^n x)}{8^n}-f(x)}(t) \geq \varphi_{x,0}\left(\frac{t}{\sum_{k=0}^{n-1}\frac{\alpha^k}{16\times 8^k}}\right)$$

and so, for all $\delta > 0$,

$$\begin{aligned}\mu_{C(x)-f(x)}(t+\delta) &\geq T\left(\mu_{C(x)-\frac{f(2^n x)}{8^n}}(\delta), \mu_{\frac{f(2^n x)}{8^n}-f(x)}(t)\right)\\ &\geq T_M\left(\mu_{A(x)-\frac{f(2^n x)}{8^n}}(\delta), \varphi_{x,0}\left(\frac{t}{\sum_{k=0}^{n-1}\frac{\alpha^k}{16\times 8^k}}\right)\right).\end{aligned} \tag{3.3.9}$$

Taking $n \to \infty$ and using (3.3.9), we get

$$\mu_{C(x)-f(x)}(t+\delta) \geq \varphi_{x,0}\big(2t(8-\alpha)\big). \tag{3.3.10}$$

Since δ is arbitrary, by taking $\delta \to 0$ in (3.3.10), we get

$$\mu_{C(x)-f(x)}(t) \geq \varphi_{x,0}\big(2t(8-\alpha)\big).$$

Replacing x and y by $2^n x$ and $2^n y$ in (3.3.3), respectively, we get

$$\begin{aligned}&\mu_{\frac{f(2^n(2x+y))}{8^n}+\frac{f(2^n(2x-y))}{8^n}-\frac{2f(2^n(x+y))}{8^n}-\frac{2f(2^n(x-y))}{8^n}-\frac{12f(2^n(x))}{8^n}}(t)\\ &\geq \varphi_{2^n x, 2^n y}(8^n t)\end{aligned}$$

for all $x, y \in X$ and $t > 0$. Since $\lim_{n\to\infty}\varphi_{2^n x,2^n y}(8^n t) = 1$, we conclude that C satisfies (3.3.1).

To prove the uniqueness of the cubic mapping C, assume that there exists a cubic mapping $C' : X \to Y$ which satisfies (3.3.4). Fix $x \in X$. Clearly, $C(2^n x) = 8^n Q(x)$ and $C'(2^n x) = 8^n Q'(x)$ for all $n \geq 1$. It follows from (3.3.4) that

$$\mu_{C(x)-C'(x)}(t) = \lim_{n\to\infty} \mu_{\frac{C(2^n x)}{8^n} - \frac{C'(2^n x)}{8^n}}(t)$$

and

$$\begin{aligned}\mu_{\frac{C(2^n x)}{8^n} - \frac{C'(2^n x)}{8^n}}(t) &\geq \min\left\{\mu_{\frac{C(2^n x)}{8^n} - \frac{f(2^n x)}{8^n}}\left(\frac{t}{2}\right), \mu_{\frac{C'(2^n x)}{8^n} - \frac{f(2^n x)}{8^n}}\left(\frac{t}{2}\right)\right\} \\ &\geq \varphi_{2^n x,0}\big(8^n(8-\alpha)t\big) \\ &\geq \varphi_{x,0}\left(\frac{8^n(8-\alpha)t}{\alpha^n}\right).\end{aligned}$$

Since $\lim_{n\to\infty} \frac{8^n(8-\alpha)t}{\alpha^n} = \infty$, we get

$$\lim_{n\to\infty} \varphi_{x,0}\left(\frac{8^n(8-\alpha)t}{\alpha^n}\right) = 1.$$

Therefore, it follows that $\mu_{C(x)-C'(x)}(t) = 1$ for all $t > 0$ and so $C(x) = C'(x)$. This completes the proof. □

Corollary 3.3.2 *Let $\theta \geq 0$ and p be a real number with $0 < p < 1$. Let X be a normed linear space with norm $\|\cdot\|$. Let $f : X \to Y$ be an odd mapping satisfying*

$$\begin{aligned}&\mu_{f(2x+y)+f(2x-y)-2f(x+y)-2f(x-y)-12f(x)}(t) \\ &\quad \geq \frac{t}{t + \theta(\|x\|^p + \|y\|^p)}\end{aligned} \tag{3.3.11}$$

for all $x, y \in X$ and $t > 0$. Then

$$C(x) := \lim_{n\to\infty} \frac{f(2^n x)}{8^n}$$

exists for all $x \in X$ and defines a cubic mapping $Q : X \to Y$ such that

$$\mu_{f(x)-C(x)}(t) \geq \frac{2(8-8^p)t}{2(8-8^p)t + \theta\|x\|^p}$$

for all $x \in X$ and $t > 0$.

Proof The proof follows from Theorem 3.3.1 by taking

$$\varphi_{x,y}(t) := \frac{t}{t + \theta(\|x\|^p + \|y\|^p)}$$

for all $x, y \in X$. Then the proof follows from Theorem 3.3.1 by $\alpha = 8^p$. □

Corollary 3.3.3 *Let $\varepsilon > 0$ and $x_0 \in X$. Let X be a normed linear space with norm $\|\cdot\|$. Let $f : X \to Y$ be an odd mapping satisfying*

$$\mu_{f(2x+y)+f(2x-y)-2f(x+y)-2f(x-y)-12f(x)}(t) \geq \frac{t}{t+\varepsilon\|x_0\|}$$

for all $x, y \in X$ and $t > 0$. Then there exists an unique additive mapping $A : X \to Y$ such that

$$\mu_{f(x)-C(x)}(t) \geq \frac{14t}{14t+\varepsilon\|x_0\|}$$

for all $x \in X$ and $t > 0$.

Proof The proof follows from Theorem 3.3.1 by taking

$$\varphi_{x,y}(t) := \frac{t}{t+\varepsilon\|x_0\|}$$

for all $t > 0$. Then the proof follows from Theorem 3.3.1 by $\alpha = 1$. □

Theorem 3.3.4 *Let X be a linear space, (Y, μ, T_M) be an RN-space and φ be a mapping from X^2 to D^+ ($\varphi(x, y)$ is denoted by $\varphi_{x,y}$) such that, some $\alpha > 8$,*

$$\varphi_{\frac{x}{2},0}(t) \geq \varphi_{x,0}(\alpha t)$$

and

$$\lim_{n\to\infty} \varphi_{\frac{x}{2^n},\frac{y}{2^n}}\left(\frac{t}{8^n}\right) = 1$$

for all $x, y \in X$ and $t > 0$. If $f : X \to Y$ is a mapping such that

$$\mu_{f(2x+y)+f(2x-y)-2f(x+y)-2f(x-y)-12f(x)}(t) \geq \varphi_{x,y}(t)$$

for all $x, y \in X$ and $t > 0$, then there exists a unique cubic mapping $C : X \to Y$ such that

$$\mu_{f(x)-C(x)}(t) \geq \varphi_{\frac{x}{2},0}\big(2(\alpha-8)t\big).$$

Corollary 3.3.5 *Let $\theta \geq 0$ and p be a real number with $p > 1$. Let X be a normed linear space with norm $\|\cdot\|$. Let $f : X \to Y$ be an odd mapping satisfying*

$$\mu_{f(2x+y)+f(2x-y)-2f(x+y)-2f(x-y)-12f(x)}(t) \geq \frac{t}{t+\theta(\|x\|^p+\|y\|^p)} \tag{3.3.12}$$

for all $x, y \in X$ and $t > 0$. Then

$$C(x) := \lim_{n\to\infty} 8^n f\left(\frac{x}{2^n}\right)$$

exists for all $x \in X$ and defines a cubic mapping $C : X \to Y$ such that

$$\mu_{f(x)-C(x)}(t) \geq \frac{2(8^p - 8)t}{2(8^p - 8)t + \theta \|x\|^p}$$

for all $x \in X$ and $t > 0$.

Proof The proof follows from Theorem 3.3.4 by taking

$$\varphi_{x,y}(t) := \frac{t}{t + \theta(\|x\|^p + \|y\|^p)}$$

for all $x, y \in X$. Then the proof follows from Theorem 3.3.4 by $\alpha = 8^p$. □

Remark 3.3.6 As in Remark 3.1.6, we can replace and remove some conditions of Theorems 3.3.1 and 3.3.4.

3.4 Quartic Functional Equations

Let X and Y be linear spaces and $f : X \to Y$ be a mapping. Recall the functional equation satisfying

$$\begin{aligned} &f(2x + y) + f(2x - y) \\ &\quad = 4f(x + y) + 4f(x - y) + 24f(x) - 6f(y) \end{aligned} \tag{3.4.1}$$

for all $x, y \in X$ is called the *quartic functional equation.*

Theorem 3.4.1 *Let X be a linear space, (Y, μ, T_M) be an RN-space and φ be a mapping from X^2 to D^+ ($\varphi(x, y)$ is denoted by $\varphi_{x,y}$) such that, some $0 < \alpha < 16$,*

$$\varphi_{2x,0}(\alpha t) \geq \varphi_{x,0}(t) \tag{3.4.2}$$

and

$$\lim_{n\to\infty} \varphi_{2^n x, 2^n y}\big(16^n t\big) = 1$$

for all $x, y \in X$ and $t > 0$. If $f : X \to Y$ is a mapping such that

$$\mu_{f(2x+y)+f(2x-y)-4f(x+y)-4f(x-y)-24f(x)+6f(y)}(t) \geq \varphi_{x,y}(t) \tag{3.4.3}$$

for all $x, y \in X$ and $t > 0$ and $f(0) = 0$, then there exists a unique quartic mapping $Q : X \to Y$ such that

$$\mu_{f(x)-Q(x)}(t) \geq \varphi_{x,0}\big(2(16 - \alpha)t\big). \tag{3.4.4}$$

Corollary 3.4.2 *Let $\theta \geq 0$ and p be a real number with $0 < p < 1$. Let X be a normed linear space with the norm $\|\cdot\|$. Let $f : X \to Y$ be an odd mapping satisfying*

$$\mu_{f(2x+y)+f(2x-y)-4f(x+y)-4f(x-y)-24f(x)+6f(y)}(t) \geq \frac{t}{t+\theta(\|x\|^p+\|y\|^p)} \tag{3.4.5}$$

for all $x, y \in X$ and $t > 0$. Then

$$Q(x) := \lim_{n\to\infty} \frac{f(2^n x)}{16^n}$$

exists for all $x \in X$ and defines a quartic mapping $Q : X \to Y$ such that

$$\mu_{f(x)-Q(x)}(t) \geq \frac{2(16-16^p)t}{2(16-16^p)t+\theta\|x\|^p}$$

for all $x \in X$ and $t > 0$.

Proof The proof follows by taking

$$\varphi_{x,y}(t) := \frac{t}{t+\theta(\|x\|^p+\|y\|^p)}$$

for all $x, y \in X$, $t > 0$ and $\alpha = 16^p$ in Theorem 3.4.1. □

Corollary 3.4.3 *Let $\varepsilon > 0$ and $x_0 \in X$. Let X be a normed linear space with the norm $\|\cdot\|$. Let $f : X \to Y$ be an odd mapping satisfying*

$$\mu_{f(2x+y)+f(2x-y)-4f(x+y)-4f(x-y)-24f(x)+6f(y)}(t) \geq \frac{t}{t+\varepsilon\|x_0\|}$$

for all $x, y \in X$ and $t > 0$. Then there exists an unique additive mapping $A : X \to Y$ such that

$$\mu_{f(x)-Q(x)}(t) \geq \frac{30t}{30t+\varepsilon\|x_0\|}$$

for all $x \in X$ and $t > 0$.

Proof The proof follows by taking

$$\varphi_{x,y}(t) := \frac{t}{t+\varepsilon\|x_0\|}$$

for all $x, y \in X$, $t > 0$ and $\alpha = 1$ in Theorem 3.4.1. □

Chapter 4
Stability of Functional Equations in RN-Spaces Under Arbitrary t-Norms

In this chapter, we prove the stability of some functional equations in random, $\mathcal{L}$-random, intuitionistic random and fuzzy normed spaces under the arbitrary continuous t-norms.

4.1 Cauchy Additive Equations

In this section, we prove the stability of the Cauchy functional equations in random normed spaces.

Theorem 4.1.1 *Let X be a real linear space, (Y, μ, T) be a complete RN-space and $f : X \to Y$ be a mapping with $f(0) = 0$ for which there is $\xi : X^2 \to D^+$ ($\xi(x, y)$ is denoted by $\xi_{x,y}$) with the property*:

$$\mu_{f(x+y)-f(x)-f(y)}(t) \geq \xi_{x,y}(t) \tag{4.1.1}$$

for all $x, y \in X$ and $t > 0$. If

$$\lim_{n\to\infty} \mathrm{T}_{i=1}^{\infty}\left(\xi_{2^{n+i-1}x,2^{n+i-1}x}(t)\right) = 1 \tag{4.1.2}$$

and

$$\lim_{n\to\infty} \xi_{2^n x,2^n y}\left(2^n t\right) = 1 \tag{4.1.3}$$

for all $x, y \in X$ and $t > 0$, then there exists a unique additive mapping $A : X \to Y$ such that

$$\mu_{f(x)-A(x)}(t) \geq \mathrm{T}_{i=1}^{\infty}\left(\xi_{2^{i-1}x,2^{i-1}x}(t)\right) \tag{4.1.4}$$

for all $x \in X$ and $t > 0$.

Proof Putting $y = x$ in (4.1.1), we have

$$\mu_{\frac{f(2x)}{2}-f(x)}(t) \geq \xi_{x,x}(2t) \tag{4.1.5}$$

Y.J. Cho et al., *Stability of Functional Equations in Random Normed Spaces*, Springer Optimization and Its Applications 86, DOI 10.1007/978-1-4614-8477-6_4,

for all $x \in X$ and $t > 0$. Therefore, we have

$$\mu_{\frac{f(2^{k+1}x)}{2^{(k+1)}} - \frac{f(2^k x)}{2^k}}\left(\frac{t}{2^k}\right) \geq \xi_{2^k x, 2^k x}(2t)$$

for all $x \in X$, $k \in \mathbf{N}$ and $t > 0$. which implies that

$$\mu_{\frac{f(2^{k+1}x)}{2^{(k+1)}} - \frac{f(2^k x)}{2^k}}(t) \geq \xi_{2^k x, 2^k x}(2^{k+1}t)$$

for all $x \in X$, $k \in \mathbf{N}$ and $t > 0$, that is,

$$\mu_{\frac{f(2^{k+1}x)}{2^{(k+1)}} - \frac{f(2^k x)}{2^k}}\left(\frac{t}{2^{k+1}}\right) \geq \xi_{2^k x, 2^k x}(t)$$

for all $k \geq 1$, $k \in \mathbf{N}$ and $t > 0$. Since $1 > \frac{1}{2} + \cdots + \frac{1}{2^n}$, by the triangle inequality, it follows that

$$\begin{aligned}
\mu_{\frac{f(2^n x)}{2^n} - f(x)}(t) &\geq T_{k=0}^{n-1}\left(\mu_{\frac{f(2^{k+1}x)}{2^{(k+1)}} - \frac{f(2^k x)}{2^k}}\left(\sum_{k=0}^{n-1} \frac{1}{2^{k+1}} t\right)\right) \\
&\geq T_{k=0}^{n-1}\left(\xi_{2^k x, 2^k x}(t)\right) \\
&= T_{i=1}^{n}\left(\xi_{2^{i-1}x, 2^{i-1}x}(t)\right)
\end{aligned} \tag{4.1.6}$$

for all $x \in X$, $k \in \mathbf{N}$ and $t > 0$.

Now, in order to prove the convergence of the sequence $\{\frac{f(2^n x)}{2^n}\}$, we replace x with $2^m x$ in (4.1.6) to find that, for all $m, n \in \mathbf{N}$,

$$\mu_{\frac{f(2^{n+m}x)}{2^{(n+m)}} - \frac{f(2^m x)}{2^m}}(t) \geq T_{i=1}^{n}\left(\xi_{2^{i+m-1}x, 2^{i+m-1}x}(t)\right) \tag{4.1.7}$$

for all $x \in X$ and $t > 0$. Since the right-hand side of the inequality tends to 1 as $m, n \to \infty$, the sequence $\{\frac{f(2^n x)}{2^n}\}$ is a Cauchy sequence. Therefore, we can define $A(x) = \lim_{n\to\infty} \frac{f(2^n x)}{2^n}$ for all $x \in X$. Replacing x, y with $2^n x$ and $2^n y$, respectively, in (4.1.1), it follows that

$$\begin{aligned}
&\mu_{\frac{f(2^{n+1}x + 2^n y)}{2^n} + \frac{f(2^{n+1}x - 2^{n+1}y)}{2^n} - \frac{f(2^n x)}{2^n} - \frac{f(2^n y)}{2^n}}(t) \\
&\geq \xi_{2^n x, 2^n y}(t)
\end{aligned} \tag{4.1.8}$$

for all $x, y \in X$, $n \in \mathbf{N}$ and $t > 0$. Taking $n \to \infty$, it follows that A is an additive mapping. For (4.1.4), taking $n \to \infty$ in (4.1.6), we can get (4.1.4).

Finally, to prove the uniqueness of the Cauchy function A subject to (4.1.4), let us assume that there exists a Cauchy function A' which satisfies (4.1.4). Since $A(2^n x) = 2^n A(x)$ and $A'(2^n x) = 2^n A'(x)$ for all $x \in X$ and $n \geq 1$, from (4.1.4), it follows that

$$
\begin{aligned}
&\mu_{A(x)-A'(x)}(t)\\
&= \mu_{A(2^n x)-A'(2^n x)}(2^n t)\\
&\geq T\big(\mu_{A(2^n x)-f(2^n x)}(2^{n-1}t), \mu_{f(2^n x)-A'(2^n x)}(2^{n-1}t)\big)\\
&\geq T\big(T_{i=1}^{\infty}(\xi_{2^{n+i-1}x,2^{n+i-1}x}(t)), T_{i=1}^{\infty}(\xi_{2^{n+i-1}x,2^{n+i-1}x}(t))\big)
\end{aligned}
$$

for all $x \in X$, $n \in \mathbf{N}$ and $t > 0$. By letting $n \to \infty$ in (4.1.4), we have $A = A'$. This completes the proof. □

4.2 Cubic Functional Equations

Let X and Y be linear spaces and $f : X \to Y$ be a mapping. The functional equation $f : X \to Y$ defined by

$$3f(x+3y) + f(3x-y) = 15f(x+y) + 15f(x-y) + 80f(y) \tag{4.2.1}$$

for all $x, y \in X$ is called the *cubic functional equation.*

Theorem 4.2.1 *Let X be a linear space and $(Y, \mathcal{P}_{\mu,\nu}, \mathcal{T})$ be a complete IRN-space. Let $f : X \to Y$ be a mapping with $f(0) = 0$ for which there exist mappings $\xi, \zeta : X^2 \to D^+$, where $\xi(x, y)$ is denoted by $\xi_{x,y}$, $\zeta(x, y)$ is denoted by $\zeta_{x,y}$ and, further, $(\xi_{x,y}(t), \zeta_{x,y}(t))$ is denoted by $Q_{\xi,\zeta}(x, y, t)$, such that*

$$
\begin{aligned}
&\mathcal{P}_{\mu,\nu}\big(3f(x+3y) + f(3x-y) - 15f(x+y) - 15f(x-y) - 80f(y), t\big)\\
&\quad \geq_{L^*} Q_{\xi,\zeta}(x, y, t)
\end{aligned}
\tag{4.2.2}
$$

for all $x, y \in X$ and $t > 0$. If

$$3T_{i=1}^{\infty}\big(Q_{\xi,\zeta}(3^{n+i-1}x, 0, 3^{3n+2i+1}t)\big) = 1_{L^*} \tag{4.2.3}$$

and

$$\lim_{n\to\infty} Q_{\xi,\zeta}(3^n x, 3^n y, 3^{3n}t) = 1_{L^*} \tag{4.2.4}$$

for all $x, y \in X$ and $t > 0$, then there exists a unique cubic mapping $C : X \to Y$ such that

$$\mathcal{P}_{\mu,\nu}\big(f(x) - C(x), t\big) \geq_{L^*} T_{i=1}^{\infty}\big(Q_{\xi,\zeta}(3^{i-1}x, 0, 3^{2i+2}t)\big). \tag{4.2.5}$$

Proof Putting $y = 0$ in (4.3.10), we have

$$\mathcal{P}_{\mu,\nu}\left(\frac{f(3x)}{27} - f(x), t\right) \geq_{L^*} Q_{\xi,\zeta}(x, 0, 3^3 t) \tag{4.2.6}$$

for all $x \in X$ and $t > 0$. Therefore, it follows that

$$\mathcal{P}_{\mu,\nu}\left(\frac{f(3^{k+1}x)}{3^{3(k+1)}} - \frac{f(3^k x)}{3^{3k}}, \frac{t}{3^{3k}}\right) \geq_{L^*} \mathcal{Q}_{\xi,\zeta}(3^k x, 0, 3^3 t) \tag{4.2.7}$$

for all $x \in X$, $k \in \mathbf{N}$ and $t > 0$, which implies that

$$\mathcal{P}_{\mu,\nu}\left(\frac{f(3^{k+1}x)}{3^{3(k+1)}} - \frac{f(3^k x)}{3^{3k}}, t\right) \geq_{L^*} \mathcal{Q}_{\xi,\zeta}(3^k x, 0, 3^{3(k+1)} t) \tag{4.2.8}$$

for all $x \in X$, $k \in \mathbf{N}$ and $t > 0$, that is,

$$\mathcal{P}_{\mu,\nu}\left(\frac{f(3^{k+1}x)}{3^{3(k+1)}} - \frac{f(3^k x)}{3^{3k}}, \frac{t}{3^{k+1}}\right) \geq_{L^*} \mathcal{Q}_{\xi,\zeta}(3^k x, 0, 3^{2(k+1)} t) \tag{4.2.9}$$

for all $x \in X$, $k \in \mathbf{N}$ and $t > 0$. Since $1 > \frac{1}{3} + \cdots + \frac{1}{3^n}$, by the triangle inequality, it follows that

$$\begin{aligned}
\mathcal{P}_{\mu,\nu}\left(\frac{f(3^n x)}{27^n} - f(x), t\right) &\geq_{L^*} \mathcal{P}_{\mu,\nu}\left(\frac{f(3^n x)}{27^n} - f(x), \sum_{k=0}^{n-1} \frac{1}{3^{k+1}} t\right) \\
&\geq_{L^*} \mathcal{T}_{k=0}^{n-1}\left(\mathcal{P}_{\mu,\nu}\left(\frac{f(3^{k+1}x)}{3^{3(k+1)}} - \frac{f(3^k x)}{3^{3k}}, \frac{t}{3^{k+1}}\right)\right) \\
&\geq_{L^*} \mathcal{T}_{i=1}^{n}\left(\mathcal{Q}_{\xi,\zeta}(3^{i-1}x, 0, 3^{2i+2} t)\right)
\end{aligned} \tag{4.2.10}$$

for all $x \in X$, $k \in \mathbf{N}$ and $t > 0$.

Now, in order to prove convergence of the sequence $\{\frac{f(3^n x)}{27^n}\}$, we replace x with $3^m x$ in (4.2.22) to show that, for any $m, n > 0$,

$$\begin{aligned}
&\mathcal{P}_{\mu,\nu}\left(\frac{f(3^{n+m}x)}{27^{(n+m)}} - \frac{f(3^m x)}{27^m}, t\right) \\
&\quad \geq_{L^*} \mathcal{T}_{i=1}^{n}\left(\mathcal{Q}_{\xi,\zeta}(3^{i+m-1}x, 0, 3^{2i+3m+2} t)\right)
\end{aligned} \tag{4.2.11}$$

for all $x \in X$ and $t > 0$. Since the right-hand side of the inequality tends 1_{L^*} as $m \to \infty$, the sequence $\{\frac{f(3^n x)}{3^{3n}}\}$ is a Cauchy sequence. Therefore, we can define $C(x) = \lim_{n\to\infty} \frac{f(3^n x)}{3^{3n}}$ for all $x \in X$.

Now, we show that C is a cubic mapping. Replacing x, y with $3^n x$ and $3^n y$, respectively, in (4.3.10), it follows that

$$\begin{aligned}
&\mathcal{P}_{\mu,\nu}\left(\frac{f(3^n(x+3y))}{27^n} + \frac{f(3^n(3x-y))}{27^n} - \frac{15 f(3^n(x+y))}{27^n}\right. \\
&\qquad \left. - \frac{15 f(3^n(x-y))}{27^n} - \frac{80 f(3^n(y))}{27^n}, t\right) \\
&\geq_{L^*} \mathcal{Q}_{\xi,\zeta}(3^n x, 3^n y, 3^{3n} t)
\end{aligned} \tag{4.2.12}$$

for all $x, y \in X$, $n \in \mathbf{N}$ and $t > 0$. Taking $n \to \infty$, it follows that C satisfies (4.2.1) for all $x, y \in X$. To prove (4.3.13), taking $n \to \infty$ in (4.2.22), we get (4.3.13).

To prove the uniqueness of the cubic mapping C subject to (4.3.13), assume that there exists a cubic mapping C' which satisfies (4.3.13). Obviously, we have $C(3^n x) = 3^{3n} C(x)$ and $C'(3^n x) = 3^{3n} C'(x)$ for all $x \in X$ and $n \geq 1$. Hence, it follows from (4.3.13) and (4.3.11) that

$$\begin{aligned}
&\mathcal{P}_{\mu,\nu}\big(C(x) - C'(x), t\big) \\
&\quad \geq_{L^*} \mathcal{P}_{\mu,\nu}\big(C(3^n x) - C'(3^n x), 3^{3n} t\big) \\
&\quad \geq_{L^*} \mathcal{T}\big(\mathcal{P}_{\mu,\nu}\big(C(3^n x) - f(3^n x), 3^{3n-1} t\big), \mathcal{P}_{\mu,\nu}\big(f(3^n x) - C'(3^n x), 3^{3n-1} t\big)\big) \\
&\quad \geq_{L^*} \mathcal{T}\big(\mathcal{T}_{i=1}^{\infty}\big(\mathcal{Q}_{\xi,\varsigma}\big(3^{n+i-1} x, 0, 3^{3n+2i+2} t\big)\big), \mathcal{T}_{i=1}^{\infty}\big(\mathcal{Q}_{\xi,\varsigma}\big(3^{n+i-1} x, 0, 3^{3n+2i+2} t\big)\big) \\
&\quad = \mathcal{T}(1_{L^*}, 1_{L^*}) = 1_{L^*}
\end{aligned}$$

for all $x \in X$, $n \in \mathbf{N}$ and $t > 0$. This proves the uniqueness of C. This completes the proof. □

Corollary 4.2.2 *Let $(X, \mathcal{P}'_{\mu',\nu'}, \mathcal{T})$ be IRN-space and $(Y, \mathcal{P}_{\mu,\nu}, \mathcal{T})$ be a complete IRN-space. Let $f : X \to Y$ be a mapping such that*

$$\begin{aligned}
&\mathcal{P}_{\mu,\nu}\big(3f(x+3y) + f(3x-y) - 15f(x+y) - 15f(x-y) - 80f(y), t\big) \\
&\quad \geq_{L^*} \mathcal{P}'_{\mu',\nu'}(x+y, t)
\end{aligned}$$

for all $x, y \in X$ and $t > 0$ in which

$$\lim_{n\to\infty} \mathcal{T}_{i=1}^{\infty}\big(\mathcal{P}'_{\mu',\nu'}\big(3^{n+i-1} x, 3^{3n+2i+1} t\big)\big) = 1_{L^*}$$

for all $x, y \in X$ and $t > 0$. Then there exists a unique cubic mapping $C : X \to Y$ such that

$$\mathcal{P}_{\mu,\nu}\big(f(x) - C(x), t\big) \geq_{L^*} \mathcal{T}_{i=1}^{\infty}\big(\mathcal{P}'_{\mu',\nu'}\big(3^{i-1} x, 3^{2i+2} t\big)\big)$$

for all $x \in X$ and $t > 0$.

Now, we give one example to illustrate Theorem 4.2.1 as follows:

Example 4.2.3 Let $(X, \|\cdot\|)$ be a Banach algebra space, $(X, \mathcal{P}_{\mu,\nu}, \mathbf{M})$ be IRN-space in which

$$\mathcal{P}_{\mu,\nu}(x, t) = \left(\frac{t}{t + \|x\|}, \frac{\|x\|}{t + \|x\|}\right)$$

and $(Y, \mathcal{P}_{\mu,\nu}, \mathbf{M})$ be a complete IRN-space for all $x \in X$ and $t > 0$. Define a mapping $f : X \to Y$ by $f(x) = x^3 + x_0$, where x_0 is a unit vector in X. A straightforward computation shows that

$$\mathcal{P}_{\mu,\nu}\big(3f(x+3y)+f(3x-y)-15f(x+y)-15f(x-y)-80f(y),t\big) \\ \geq_{L^*} \mathcal{P}_{\mu,\nu}(x+y,t)$$

for all $x, y \in X$ and $t > 0$. Also, we have

$$\begin{aligned} &\lim_{n\to\infty} \mathbf{M}_{i=1}^{\infty}\big(\mathcal{P}_{\mu,\nu}\big(3^{n+i-1}x, 3^{3n+2i+1}t\big)\big) \\ &= \lim_{n\to\infty}\lim_{m\to\infty} \mathbf{M}_{i=1}^{m}\big(\mathcal{P}_{\mu,\nu}\big(x, 3^{2n+i+2}t\big)\big) \\ &= \lim_{n\to\infty}\lim_{m\to\infty} \mathcal{P}_{\mu,\nu}\big(x, 3^{2n+3}t\big) \\ &= \lim_{n\to\infty} \mathcal{P}_{\mu,\nu}\big(x, 3^{2n+3}t\big) \\ &= 1_{L^*} \end{aligned}$$

for all $x \in X$ and $t > 0$. Therefore, all the conditions of Theorem 4.2.1 hold and so there exists a unique cubic mapping $C : X \to Y$ such that

$$\mathcal{P}_{\mu,\nu}\big(f(x)-C(x),t\big) \geq_{L^*} \mathcal{P}_{\mu,\nu}\big(x, 3^4 t\big)$$

for all $x \in X$ and $t > 0$.

Now, we consider another version of cubic functional equation. The functional equation satisfying the following:

$$f(2x+y)+f(2x-y)=2f(x+y)+2f(x-y)+12f(x) \tag{4.2.13}$$

is called the *cubic functional equation*.

Theorem 4.2.4 *Let X be a real linear space and $(Y, \mathcal{P}_{\mu,\nu}, \mathcal{T})$ be a complete IRN-space. Let $f : X \to Y$ be a mapping with $f(0) = 0$ for which there exist mappings $\xi, \zeta : X^2 \to D^+$ ($\xi(x, y)$ is denoted by $\xi_{x,y}$, $\zeta(x, y)$ is denoted by $\zeta_{x,y}$ and, further, $(\xi_{x,y}(t), \zeta_{x,y}(t))$ is denoted by $Q_{\xi,\zeta}(x, y, t)$) such that*

$$\mathcal{P}_{\mu,\nu}\big(f(2x+y)+f(2x-y)-2f(x+y)-2f(x-y)-12f(x),t\big) \\ \geq_{L^*} Q_{\xi,\zeta}(x,y,t). \tag{4.2.14}$$

If

$$\mathcal{T}_{i=1}^{\infty}\big(Q_{\xi,\zeta}\big(2^{n+i-1}x, 0, 2^{3n+2i+1}t\big)\big) = 1_{L^*} \tag{4.2.15}$$

and

$$\lim_{n\to\infty} Q_{\xi,\zeta}\big(2^n x, 2^n y, 2^{3n}t\big) = 1_{L^*} \tag{4.2.16}$$

for all $x, y \in X$ and $t > 0$, then there exists a unique cubic mapping $C : X \to Y$ such that

$$\mathcal{P}_{\mu,\nu}(f(x) - C(x), t) \geq_{L^*} \mathcal{T}_{i=1}^{\infty}(\mathcal{Q}_{\xi,\zeta}(2^{i-1}x, 0, 2^{2i+1}t)). \tag{4.2.17}$$

Proof Putting $y = 0$ in (4.3.10), we have

$$\mathcal{P}_{\mu,\nu}\left(\frac{f(2x)}{8} - f(x), t\right) \geq_{L^*} \mathcal{Q}_{\xi,\zeta}(x, 0, 2^4 t) \geq_{L^*} \mathcal{Q}_{\xi,\zeta}(x, 0, 2^3 t) \tag{4.2.18}$$

for all $x \in X$ and $t > 0$. Therefore, it follows that

$$\mathcal{P}_{\mu,\nu}\left(\frac{f(2^{k+1}x)}{2^{3(k+1)}} - \frac{f(2^k x)}{2^{3k}}, \frac{t}{2^{3k}}\right) \geq_{L^*} \mathcal{Q}_{\xi,\zeta}(2^k x, 0, 2^3 t) \tag{4.2.19}$$

for all $x \in X$, $k \in \mathbf{N}$ and $t > 0$, which implies that

$$\mathcal{P}_{\mu,\nu}\left(\frac{f(2^{k+1}x)}{2^{3(k+1)}} - \frac{f(2^k x)}{2^{3k}}, t\right) \geq_{L^*} \mathcal{Q}_{\xi,\zeta}(2^k x, 0, 2^{3(k+1)} t) \tag{4.2.20}$$

for all $x \in X$, $k \in \mathbf{N}$ and $t > 0$, that is,

$$\mathcal{P}_{\mu,\nu}\left(\frac{f(2^{k+1}x)}{2^{3(k+1)}} - \frac{f(2^k x)}{2^{3k}}, \frac{t}{2^{k+1}}\right) \geq_{L^*} \mathcal{Q}_{\xi,\zeta}(2^k x, 0, 2^{2(k+1)} t) \tag{4.2.21}$$

for all $x \in X$, $k \in \mathbf{N}$ and $t > 0$. Since $1 > \frac{1}{2} + \cdots + \frac{1}{2^n}$, by the triangle inequality, it follows

$$\begin{aligned}
&\mathcal{P}_{\mu,\nu}\left(\frac{f(2^n x)}{8^n} - f(x), t\right) \\
&\quad \geq_{L^*} \mathcal{T}_{k=0}^{n-1}\left(\mathcal{P}_{\mu,\nu}\left(\frac{f(2^{k+1}x)}{2^{3(k+1)}} - \frac{f(2^k x)}{2^{3k}}, \sum_{k=0}^{n-1} \frac{1}{2^{k+1}} t\right)\right) \\
&\quad \geq_{L^*} \mathcal{T}_{i=1}^{n}(\mathcal{Q}_{\xi,\zeta}(2^{i-1}x, 0, 2^{2i+1}t))
\end{aligned} \tag{4.2.22}$$

for all $x \in X$, $k \in \mathbf{N}$ and $t > 0$,

In order to prove convergence of the sequence $\{\frac{f(2^n x)}{8^n}\}$, we replace x with $2^m x$ in (4.2.22) to show that, for any $m, n > 0$,

$$\begin{aligned}
&\mathcal{P}_{\mu,\nu}\left(\frac{f(2^{n+m}x)}{8^{(n+m)}} - \frac{f(2^m x)}{8^m}, t\right) \\
&\quad \geq_{L^*} \mathcal{T}_{i=1}^{n}(\mathcal{Q}_{\xi,\zeta}(2^{i+m-1}x, 2^{i+m-1}x, 2^{2i+m+1}t))
\end{aligned} \tag{4.2.23}$$

for all $x \in X$ and $t > 0$. Since the right hand side of the inequality tend to 1_{L^*} as $m \to \infty$, the sequence $\{\frac{f(2^n x)}{2^{3n}}\}$ is a Cauchy sequence. Therefore, we can define $C(x) = \lim_{n\to\infty} \frac{f(2^n x)}{2^{3n}}$ for all $x \in X$.

Now, we show that C is a cubic mapping. Replacing x, y with $2^n x$ and $2^n y$, respectively, in (4.3.10), it follows that

$$\begin{aligned}
\mathcal{P}_{\mu,\nu}\bigg(&\frac{f(2^{n+1}x+2^n y)}{2^{3n}}+\frac{f(2^{n+1}x-2^n y)}{2^{3n}}-2\frac{f(2^n x+2^n y)}{2^{3n}}\\
&-2\frac{f(2^n x-2^n y)}{2^{3n}}-12\frac{f(2^n x)}{2^{3n}},t\bigg)\\
&\geq_{L^*} \mathcal{Q}_{\xi,\varsigma}\big(2^n x, 2^n y, 2^{3n}t\big) \qquad (4.2.24)
\end{aligned}$$

for all $x, y \in X$, $n \in \mathbf{N}$ and $t > 0$. Taking $n \to \infty$, we find that C satisfies (4.2.13) for all $x, y \in X$. To prove (4.3.13), taking $n \to \infty$ in (4.2.22), we get (4.3.13).

To prove the uniqueness of the cubic mapping C subject to (4.3.13), let us assume that there exists a cubic mapping C' which satisfies (4.3.13). Obviously, we have $C(2^n x) = 2^{3n}C(x)$ and $C'(2^n x) = 2^{3n}C'(x)$ for all $x \in X$ and $n \geq 1$. Hence, it follows from (4.3.13) that

$$\begin{aligned}
&\mathcal{P}_{\mu,\nu}\big(C(x)-C'(x),t\big)\\
&\quad\geq_{L^*} \mathcal{P}_{\mu,\nu}\big(C\big(2^n x\big)-C'\big(2^n x\big),2^{3n}t\big)\\
&\quad\geq_{L^*} \mathcal{T}\big(\mathcal{P}_{\mu,\nu}\big(C\big(2^n x\big)-f\big(2^n x\big),2^{3n-1}t\big),\mathcal{P}_{\mu,\nu}\big(f\big(2^n x\big)-Q'\big(2^n x\big),2^{3n-1}t\big)\big)\\
&\quad\geq_{L^*} \mathcal{T}\big(\mathcal{T}_{i=1}^{\infty}\big(\mathcal{Q}_{\xi,\varsigma}\big(2^{n+i-1}x,0,2^{3n+2i+1}t\big)\big),\mathcal{T}_{i=1}^{\infty}\big(\mathcal{Q}_{\xi,\varsigma}\big(2^{n+i-1}x,0,2^{3n+2i+1}t\big)\big)
\end{aligned}$$

for all $x \in X$, $n \in \mathbf{N}$ and $t > 0$. By letting $n \to \infty$ in (4.3.13), we show the uniqueness of C. This completes the proof. □

Corollary 4.2.5 *Let $(X, \mathcal{P}'_{\mu',\nu'}, \mathcal{T})$ be an IRN-space and $(Y, \mathcal{P}_{\mu,\nu}, \mathcal{T})$ be a complete IRN-space. Let $f : X \to Y$ be a mapping such that*

$$\begin{aligned}
&\mathcal{P}_{\mu,\nu}\big(f(2x+y)+f(2x-y)-2f(x+y)-2f(x-y)-12f(x),t\big)\\
&\quad\geq_{L^*} \mathcal{P}'_{\mu',\nu'}(x+y,t)
\end{aligned}$$

for all $x, y \in X$ and $t > 0$ in which

$$\lim_{n\to\infty} \mathcal{T}_{i=1}^{\infty}\big(\mathcal{P}'_{\mu',\nu'}\big(x,2^{2n+i+2}t\big)\big)=1_{L^*}$$

for all $x, y \in X$ and $t > 0$. Then there exists a unique cubic mapping $C : X \to Y$ such that

$$\mathcal{P}_{\mu,\nu}\big(f(x)-C(x),t\big)\geq_{L^*} \mathcal{T}_{i=1}^{\infty}\big(\mathcal{P}'_{\mu',\nu'}\big(2^{i+1}t\big)\big).$$

Now, we give one example to validate Theorem 4.2.4 as follows:

Example 4.2.6 Let $(X, \|\cdot\|)$ be a real Banach algebra and $(X, \mathcal{P}_{\mu,\nu}, \mathbf{M})$ be a complete IRN-space in which

$$\mathcal{P}_{\mu,\nu}(x,t) = \left(\frac{t}{t+\|x\|}, \frac{\|x\|}{t+\|x\|}\right)$$

for all $x \in X$ and $t > 0$. Define a mapping $f : X \to X$ by $f(x) = x^3 + x_0$, where x_0 is a unit vector in X. A straightforward computation shows that

$$\begin{aligned}&\mathcal{P}_{\mu,\nu}\big(f(2x+y)+f(2x-y)-2f(x+y)-2f(x-y)-12f(x), t\big)\\&\quad\geq_{L^*} \mathcal{P}_{\mu,\nu}(x+y,t)\end{aligned}$$

for all $x, y \in X$ and $t > 0$. Also, we have

$$\begin{aligned}\lim_{n\to\infty} \mathbf{M}_{i=1}^{\infty}\big(\mathcal{P}_{\mu,\nu}\big(x, 2^{2n+i+1}t\big)\big) &= \lim_{n\to\infty}\lim_{m\to\infty} \mathbf{M}_{i=1}^{m}\big(\mathcal{P}_{\mu,\nu}\big(x, 2^{2n+i+1}t\big)\big)\\&= \lim_{n\to\infty}\lim_{m\to\infty} \mathcal{P}_{\mu,\nu}\big(x, 2^{2n+2}t\big)\\&= \lim_{n\to\infty} \mathcal{P}_{\mu,\nu}\big(x, 2^{2n+2}t\big)\\&= 1_{L^*}\end{aligned}$$

for all $x \in X$ and $t > 0$. Therefore, all the conditions of Theorem 4.2.4 hold and so there exists a unique cubic mapping $C : X \to X$ such that

$$\mathcal{P}_{\mu,\nu}\big(f(x)-C(x), t\big) \geq_{L^*} \mathcal{P}_{\mu,\nu}\big(x, 2^2t\big)$$

for all $x \in X$ and $t > 0$.

4.3 Quartic Functional Equations

Let X and Y be linear spaces and $f : X \to Y$ be a mapping. The functional equation satisfying the following:

$$f(2x+y)+f(2x-y) = 4f(x+y)+4f(x-y)+24f(x)-6f(y) \qquad (4.3.1)$$

for all $x, y \in X$ is called the *quartic functional equation* and every solution of the quartic functional equation is called the *quartic function.*

Theorem 4.3.1 *Let X be a real linear space and (Y, μ, T) be a complete RN-space and $f : X \to Y$ be a mapping with $f(0) = 0$ for which there is $\xi : X^2 \to D^+$ ($\xi(x, y)$ is denoted by $\xi_{x,y}$) such that*

$$\mu_{f(2x+y)+f(2x-y)-4f(x+y)-4f(x-y)-24f(x)+6f(y)}(t) \geq \xi_{x,y}(t) \qquad (4.3.2)$$

for all $x, y \in X$ and $t > 0$. If

$$\lim_{n \to \infty} T_{i=1}^{\infty}\left(\xi_{2^{n+i-1}x,0}\left(2^{4n+3i}t\right)\right) = 1 \tag{4.3.3}$$

and

$$\lim_{n \to \infty} \xi_{2^n x, 2^n y}\left(2^{4n}t\right) = 1 \tag{4.3.4}$$

for all $x, y \in X$ and $t > 0$, then there exists a unique quartic mapping $Q : X \to Y$ such that

$$\mu_{f(x)-Q(x)}(t) \geq T_{i=1}^{\infty}\left(\xi_{2^{i-1}x,0}\left(2^{3i+1}t\right)\right) \tag{4.3.5}$$

for all $x \in X$ and $t > 0$.

Proof Putting $y = 0$ in (4.3.2), we have

$$\mu_{\frac{f(2x)}{2^4}-f(x)}(t) \geq \xi_{x,0}\left(2^5 t\right) \geq \xi_{x,0}\left(2^4 t\right) \tag{4.3.6}$$

for all $x \in X$ and $t > 0$. Therefore, it follows that

$$\mu_{\frac{f(2^{k+1}x)}{2^{4(k+1)}}-\frac{f(2^k x)}{2^{4k}}}\left(\frac{t}{2^{4k}}\right) \geq \xi_{2^k x,0}\left(2^4 t\right)$$

for all $x \in X$, $k \in \mathbf{N}$ and $t > 0$, which implies that

$$\mu_{\frac{f(2^{k+1}x)}{2^{4(k+1)}}-\frac{f(2^k x)}{2^{4k}}}(t) \geq \xi_{2^k x,0}\left(2^{4(k+1)}t\right)$$

for all $x \in X$, $k \in \mathbf{N}$ and $t > 0$, that is,

$$\mu_{\frac{f(2^{k+1}x)}{2^{4(k+1)}}-\frac{f(2^k x)}{2^{4k}}}\left(\frac{t}{2^{k+1}}\right) \geq \xi_{2^k x,0}\left(2^{3(k+1)}t\right)$$

for all $x \in X$, $k \in \mathbf{N}$ and $t > 0$. Since $1 > \frac{1}{2} + \cdots + \frac{1}{2^n}$, by the triangle inequality, it follows that

$$\mu_{\frac{f(2^n x)}{2^{4n}}-f(x)}(t) \geq T_{k=0}^{n-1}\left(\mu_{\frac{f(2^{k+1}x)}{2^{4(k+1)}}-\frac{f(2^k x)}{2^{4k}}}\left(\sum_{k=0}^{n-1}\frac{1}{2^{k+1}}t\right)\right)$$

$$\geq T_{k=0}^{n-1}\left(\xi_{2^k x,0}\left(2^{3(k+1)}t\right)\right)$$

$$= T_{i=1}^{n}\left(\xi_{2^{i-1}x,0}\left(2^{3i}t\right)\right) \tag{4.3.7}$$

for all $x \in X$ and $t > 0$.

In order to prove the convergence of the sequence $\{\frac{f(2^n x)}{2^{4n}}\}$, we replace x with $2^m x$ in (4.1.6) to show that, for all $m, n \in \mathbf{N}$,

$$\mu_{\frac{f(2^{n+m}x)}{2^{4(n+m)}}-\frac{f(2^m x)}{2^{4m}}}(t) \geq T_{i=1}^{n}\left(\xi_{2^{i+m-1}x,0}\left(2^{3i+4m}t\right)\right) \tag{4.3.8}$$

for all $x \in X$ and $t > 0$. Since the right-hand side of the inequality tends to 1 as $m, n \to \infty$, the sequence $\{\frac{f(2^n x)}{2^{4n}}\}$ is a Cauchy sequence. Therefore, we can define $Q(x) = \lim_{n\to\infty} \frac{f(2^n x)}{2^{4n}}$ for all $x \in X$. Replacing x, y with $2^n x$ and $2^n y$, respectively, in (4.3.2), it follows that

$$\begin{aligned} &\mu_{\frac{f(2^{n+1}x+2^n y)}{2^{4n}}+\frac{f(2^{n+1}x-2^n y)}{2^{4n}}-4\frac{f(2^n x+2^n y)}{2^{4n}}-4\frac{f(2^n x-2^n y)}{2^{4n}}-24\frac{f(2^n x)}{2^{4n}}+6\frac{f(2^n y)}{2^{4n}}}(t) \\ &\quad \geq \xi_{2^n x, 2^n y}(2^{4n} t) \end{aligned} \tag{4.3.9}$$

for all $x, y \in X$, $n \in \mathbf{N}$ and $t > 0$. Taking $n \to \infty$, we show that Q satisfies (4.3.1) for all $x, y \in X$, that is, Q is a quartic mapping. To prove (4.3.5), taking $n \to \infty$ in (4.3.7), we get (4.3.5).

Finally, to prove the uniqueness of the quartic mapping Q subject to (4.3.5), let us assume that there exists a quartic mapping Q' which satisfies (4.3.5). Since $Q(2^n x) = 2^{4n} Q(x)$ and $Q'(2^n x) = 2^{4n} Q'(x)$ for all $x \in X$ and $n \geq 1$, from (4.3.5) it follows that

$$\begin{aligned} &\mu_{Q(x)-Q'(x)}(t) \\ &\quad = \mu_{Q(2^n x)-Q'(2^n x)}(2^{4n} t) \\ &\quad \geq T\big(\mu_{Q(2^n x)-f(2^n x)}(2^{4n-1} t), \mu_{f(2^n x)-Q'(2^n x)}(2^{4n-1} t)\big) \\ &\quad \geq T\big(T_{i=1}^{\infty}\big(\xi_{2^{n+i-1}x,0}(2^{4n+3i} t)\big), T_{i=1}^{\infty}\big(\xi_{2^{n+i-1}x,0}(2^{4n+3i} t)\big)\big) \end{aligned}$$

for all $x \in X$, $n \in \mathbf{N}$ and $t > 0$. By letting $n \to \infty$ in (4.3.5), we show that $Q = Q'$. This completes the proof. □

Example 4.3.2 Let $(X, \|\cdot\|)$ be a Banach algebra and

$$\mu_x(t) = \begin{cases} \max\{1 - \frac{\|x\|}{t}, 0\}, & \text{if } t > 0, \\ 0, & \text{if } t \leq 0. \end{cases}$$

For all $x, y \in X$ and $t > 0$, let

$$\xi_{x,y}(t) = \max\left\{1 - \frac{128\|x\| + 128\|y\|}{t}, 0\right\}$$

and $\xi_{x,y}(t) = 0$ if $t \leq 0$. We note that $\xi_{x,y}$ is a distribution function and

$$\lim_{n\to\infty} \xi_{2^n x, 2^n y}(2^{4n} t) = 1$$

for all $x, y \in X$ and $t > 0$. It is easy to show that (X, μ, T_L) is a RN-space. Indeed, we have

$$\mu_x(t) = 1 \quad \Longrightarrow \quad \frac{\|x\|}{t} = 0 \quad \Longrightarrow \quad x = 0$$

for all $x \in X$ and $t > 0$ and, obviously, $\mu_{\lambda x}(t) = \mu_x(\frac{t}{\lambda})$ for all $x \in X$ and $t > 0$.

Next, for all $x, y \in X$ and $t, s > 0$, we have

$$\begin{aligned}
\mu_{x+y}(t+s) &= \max\left\{1 - \frac{\|x+y\|}{t+s}, 0\right\} \\
&= \max\left\{1 - \left\|\frac{x+y}{t+s}\right\|, 0\right\} \\
&= \max\left\{1 - \left\|\frac{x}{t+s} + \frac{y}{t+s}\right\|, 0\right\} \\
&\geq \max\left\{1 - \left\|\frac{x}{t}\right\| - \left\|\frac{y}{s}\right\|, 0\right\} \\
&= T_L\big(\mu_x(t), \mu_y(s)\big).
\end{aligned}$$

It is also easy to see that (X, μ, T_L) is complete,

$$\mu_{x-y}(t) \geq 1 - \frac{\|x-y\|}{t}$$

for all $x, y \in X$ and $t > 0$ and $(X, \|\cdot\|)$ is complete.

Define a mapping $f : X \to X$ by $f(x) = x^4 + \|x\|x_0$ for all $x \in X$, where x_0 is a unit vector in X. A simple computation shows that

$$\begin{aligned}
&\big\|f(2x+y) + f(2x-y) - 4f(x+y) - 4f(x-y) - 24f(x) + 6f(y)\big\| \\
&= \big|\|2x+y\| + \|2x-y\| - 4\|x+y\| - 4\|x-y\| - 24\|x\| + 6\|y\|\big|\|x_0\| \\
&\leq 128\|x\| + 128\|y\|
\end{aligned}$$

for all $x, y \in X$ and hence

$$\mu_{f(2x+y)+f(2x-y)-4f(x+y)-4f(x-y)-24f(x)+6f(y)}(t) \geq \xi_{x,y}(t)$$

for all $x, y \in X$ and $t > 0$. Fix $x \in X$ and $t > 0$. Then it follows that

$$\begin{aligned}
(T_L)_{i=1}^{\infty}\big(\xi_{2^{n+i-1}x,0}(2^{4n+3i}t)\big) &= \max\left\{\sum_{i=1}^{\infty}\big(\xi_{2^{n+i-1}x,0}(2^{4n+3i}t) - 1\big) + 1, 0\right\} \\
&= \max\left\{1 - \frac{64\|x\|}{3 \cdot 2^{3n}t}, 0\right\}
\end{aligned}$$

for all $x \in X$, $n \in \mathbf{N}$ and $t > 0$ and hence

$$\lim_{n\to\infty}(T_L)_{i=1}^{\infty}\big(\xi_{2^{n+i-1}x,0}(2^{4n+3i}t)\big) = 1$$

for all $x \in X$ and $t > 0$. Thus, all the conditions of Theorem 4.3.1 hold. Since

$$(T_L)_{i=1}^{\infty}\big(\xi_{2^{i-1}x,0}(2^{3i+1}t)\big) = \max\left\{\sum_{i=1}^{\infty}\big(\xi_{2^{i-1}x,0}(2^{3i+1}t) - 1\big) + 1, 0\right\}$$
$$= \max\left\{1 - \frac{32\|x\|}{3t}, 0\right\}$$

for all $x \in X$ and $t > 0$, we can deduce that $Q(x) = x^4$ is the unique quartic mapping $Q : X \to X$ such that

$$\mu_{f(x)-Q(x)}(t) \geq \max\left\{1 - \frac{32\|x\|}{3t}, 0\right\}$$

for all $x \in X$ and $t > 0$.

Theorem 4.3.3 *Let X be a linear space and $(Y, \mathcal{P}_{\mu,\nu}, \mathcal{T})$ be a complete IRN-space. If $f : X \to Y$ be a mapping with $f(0) = 0$ for which there exist mappings $\xi, \zeta : X^2 \to D^+$ ($\xi(x, y)$ is denoted by $\xi_{x,y}$, $\zeta(x, y)$ is denoted by $\zeta_{x,y}$ and, further, $(\xi_{x,y}(t), \zeta_{x,y}(t))$ is denoted by $\mathcal{Q}_{\xi,\zeta}(x, y, t)$) such that*

$$\begin{aligned}&\mathcal{P}_{\mu,\nu}\big(f(2x + y) + f(2x - y) - 4f(x + y) - 4f(x - y)\\&\quad - 24f(x) + 6f(y), t\big)\\&\geq_{L^*} \mathcal{Q}_{\xi,\zeta}(x, y, t)\end{aligned} \tag{4.3.10}$$

for all $x, y \in X$ and $t > 0$. If

$$\mathcal{T}_{i=1}^{\infty}\big(\mathcal{Q}_{\xi,\zeta}(2^{n+i-1}x, 0, 2^{4n+3i+1}t)\big) = 1_{L^*} \tag{4.3.11}$$

and

$$\lim_{n\to\infty} \mathcal{Q}_{\xi,\zeta}\big(2^n x, 2^n y, 2^{4n}t\big) = 1_{L^*} \tag{4.3.12}$$

for all $x, y \in X$ and $t > 0$, then there exists a unique quartic mapping $Q : X \to Y$ such that

$$\mathcal{P}_{\mu,\nu}\big(f(x) - Q(x), t\big) \geq_{L^*} \mathcal{T}_{i=1}^{\infty}\big(\mathcal{Q}_{\xi,\zeta}(2^{i-1}x, 0, 2^{3i+1}t)\big). \tag{4.3.13}$$

Corollary 4.3.4 *Let $(X, \mathcal{P}'_{\mu',\nu'}, \mathcal{T})$ be an IRN-space and $(Y, \mathcal{P}_{\mu,\nu}, \mathcal{T})$ be a complete IRN-space. If $f : X \to Y$ be a mapping such that*

$$\begin{aligned}&\mathcal{P}_{\mu,\nu}\big(f(2x + y) + f(2x - y) - 4f(x + y) - 4f(x - y)\\&\quad - 24f(x) + 6f(y), t\big)\\&\geq_{L^*} \mathcal{P}'_{\mu',\nu'}(x + y, t)\end{aligned}$$

all $x, y \in X$ and $t > 0$ in which

$$\lim_{n\to\infty} \mathcal{T}_{i=1}^{\infty}\big(\mathcal{P}'_{\mu',\nu'}\big(x, 2^{3n+2i+2}t\big)\big) = 1_{L^*}$$

for all $x, y \in X$ and $t > 0$. Then there exists a unique quartic mapping $Q : X \to Y$ such that

$$\mathcal{P}_{\mu,\nu}\big(f(x) - Q(x), t\big) \geq_{L^*} \mathcal{T}_{i=1}^{\infty}\big(\mathcal{P}'_{\mu',\nu'}\big(x, 2^{2i+2}t\big)\big).$$

Proof Putting $\mathcal{Q}_{\xi,\zeta}(x, y, t) = \mathcal{P}'_{\mu',\nu'}(x + y, t)$ for all $x, y \in X$ and $t > 0$, the corollary immediate from Theorem 4.3.3. □

Example 4.3.5 Let $(X, \|\cdot\|)$ be a Banach algebra space and $(X, \mathcal{P}_{\mu,\nu}, \mathbf{M})$ be an IRN-space in which

$$\mathcal{P}_{\mu,\nu}(x, t) = \left(\frac{t}{t + \|x\|}, \frac{\|x\|}{t + \|x\|}\right)$$

for all $x \in X$ and $t > 0$ and $(Y, \mathcal{P}_{\mu,\nu}, \mathbf{M})$ be a complete IRN-space. Define a mapping $f : X \to Y$ by $f(x) = x^4 + x_0$ for all $x \in X$, where x_0 is a unit vector in X. A straightforward computation shows that

$$\begin{aligned}&\mathcal{P}_{\mu,\nu}\big(f(2x + y) + f(2x - y) - 4f(x + y) - 4f(x - y)\\&\quad - 24f(x) + 6f(y), t\big)\\&\geq_{L^*} \mathcal{P}_{\mu,\nu}(x + y, t)\end{aligned}$$

for all $x, y \in X$ and $t > 0$. Also, we have

$$\begin{aligned}\lim_{n\to\infty} \mathbf{M}_{i=1}^{\infty}\big(\mathcal{P}_{\mu,\nu}\big(x, 2^{3n+2i+2}t\big)\big) &= \lim_{n\to\infty}\lim_{m\to\infty} \mathbf{M}_{i=1}^{m}\big(\mathcal{P}_{\mu,\nu}\big(x, 2^{3n+2i+2}t\big)\big)\\&= \lim_{n\to\infty}\lim_{m\to\infty} \mathcal{P}_{\mu,\nu}\big(x, 2^{2n+4}t\big)\\&= \lim_{n\to\infty} \mathcal{P}_{\mu,\nu}\big(x, 2^{2n+4}t\big)\\&= 1_{L^*}\end{aligned}$$

for all $x \in X$ and $t > 0$. Therefore, all the conditions of Theorem 4.3.3 hold and so there exists a unique quartic mapping $Q : X \to Y$ such that

$$\mathcal{P}_{\mu,\nu}\big(f(x) - Q(x), t\big) \geq_{L^*} \mathcal{P}_{\mu,\nu}\big(x, 2^4 t\big)$$

for all $x \in X$ and $t > 0$.

Next, we consider another quartic functional equation. In fact, we study the stability of the following functional equation:

$$16f(x + 4y) + f(4x - y)$$

$$= 306\left[9f\left(x+\frac{y}{3}\right)+f(x+2y)\right]$$
$$+136f(x-y)-1394f(x+y)+425f(y)-1530f(x). \quad (4.3.14)$$

Theorem 4.3.6 *Let X be a linear space and $(X, \mathcal{P}_{\mu,\nu}, \mathcal{T})$ be a complete IRN-space. Let $f : X \to Y$ be a mapping with $f(0) = 0$ for which there exist mappings $\xi, \zeta : X^2 \to D^+$, where $\xi(x, y)$ is denoted by $\xi_{x,y}$, $\zeta(x, y)$ is denoted by $\zeta_{x,y}$ and, further, $(\xi_{x,y}(t), \zeta_{x,y}(t))$ is denoted by $Q_{\xi,\zeta}(x, y, t)$, such that*

$$\mathcal{P}_{\mu,\nu}\bigg(16f(x+4y)+f(4x-y)-306\left[9f\left(x+\frac{y}{3}\right)+f(x+2y)\right]$$
$$-136f(x-y)+1394f(x+y)-425f(y)+1530f(x),t\bigg)$$
$$\geq_{L^*} Q_{\xi,\zeta}(x,y,t) \quad (4.3.15)$$

for all $x, y \in X$ and $t > 0$. If

$$\mathcal{T}_{i=1}^{\infty}\big(Q_{\xi,\zeta}\big(4^{n+i-1}x,0,4^{4n+3i+3}t\big)\big)=1_{L^*} \quad (4.3.16)$$

and

$$\lim_{n\to\infty} Q_{\xi,\zeta}\big(4^n x,4^n y,4^{4n}t\big)=1_{L^*} \quad (4.3.17)$$

for all $x, y \in X$ and $t > 0$, then there exists a unique quartic mapping $Q : X \to Y$ such that

$$\mathcal{P}_{\mu,\nu}\big(f(x)-Q(x),t\big)\geq_{L^*} \mathcal{T}_{i=1}^{\infty}\big(Q_{\xi,\zeta}\big(4^{i-1}x,0,4^{3i+3}t\big)\big) \quad (4.3.18)$$

for all $x \in X$ and $t > 0$.

Proof Putting $y = 0$ in (4.3.15), we have

$$\mathcal{P}_{\mu,\nu}\left(\frac{f(4x)}{256}-f(x),t\right)\geq_{L^*} Q_{\xi,\zeta}\big(x,0,4^4 t\big) \quad (4.3.19)$$

for all $x \in X$ and $t > 0$. Therefore, it follows that

$$\mathcal{P}_{\mu,\nu}\left(\frac{f(4^{k+1}x)}{4^{4(k+1)}}-\frac{f(4^k x)}{4^{4k}},\frac{t}{4^{4k}}\right)\geq_{L^*} Q_{\xi,\zeta}\big(4^k x,0,4^4 t\big) \quad (4.3.20)$$

for all $x \in X$, $k \in \mathbf{N}$ and $t > 0$, which implies that

$$\mathcal{P}_{\mu,\nu}\left(\frac{f(4^{k+1}x)}{4^{4(k+1)}}-\frac{f(4^k x)}{4^{4k}},t\right)\geq_{L^*} Q_{\xi,\zeta}\big(4^k x,0,4^{4(k+1)}t\big) \quad (4.3.21)$$

for all $x \in X$, $k \in \mathbf{N}$ and $t > 0$, that is,

$$\mathcal{P}_{\mu,\nu}\left(\frac{f(4^{k+1}x)}{4^{4(k+1)}} - \frac{f(4^k x)}{4^{4k}}, \frac{t}{4^{k+1}}\right) \geq_{L^*} \mathcal{Q}_{\xi,\zeta}\left(4^k x, 0, 4^{4(k+1)}t\right) \tag{4.3.22}$$

for all $x \in X$, $k \in \mathbf{N}$ and $t > 0$. Since $1 > \frac{1}{4} + \frac{1}{4^n}$, from the triangle inequality, it follows that

$$\begin{aligned}
&\mathcal{P}_{\mu,\nu}\left(\frac{f(4^n x)}{256^n} - f(x), t\right) \\
&\quad \geq_{L^*} \mathcal{T}_{k=0}^{n-1}\left(\mathcal{P}_{\mu,\nu}\left(\frac{f(4^{k+1}x)}{4^{4(k+1)}} - \frac{f(4^k x)}{4^{4k}}, \sum_{k=0}^{n-1}\frac{1}{4^{k+1}}t\right)\right) \\
&\quad \geq_{L^*} \mathcal{T}_{i=1}^{n}\left(\mathcal{Q}_{\xi,\zeta}\left(4^{i-1}x, 0, 4^{3i+3}t\right)\right)
\end{aligned} \tag{4.3.23}$$

for all $x \in X$, $k \in \mathbf{N}$ and $t > 0$.

In order to prove convergence of the sequence $\{\frac{f(4^n x)}{256^n}\}$, replacing x with $4^m x$ in (4.3.23), we show that, for any $m, n > 0$,

$$\begin{aligned}
&\mathcal{P}_{\mu,\nu}\left(\frac{f(4^{n+m}x)}{256^{(n+m)}} - \frac{f(4^m x)}{256^m}, t\right) \\
&\quad \geq_{L^*} \mathcal{T}_{i=1}^{n}\left(\mathcal{Q}_{\xi,\zeta}\left(4^{i+m-1}x, 0, 4^{3i+4m+3}t\right)\right)
\end{aligned} \tag{4.3.24}$$

for all $t > 0$. Since the right-hand side of the inequality tends 1_{L^*} as $m \to \infty$, the sequence $\{\frac{f(4^n x)}{4^{4n}}\}$ is a Cauchy sequence. So we can define $Q(x) = \lim_{n\to\infty} \frac{f(4^n x)}{4^{4n}}$ for all $x \in X$.

Now, we show that Q is a quartic mapping. Replacing x, y with $4^n x$ and $4^n y$, respectively, in (4.3.15), we obtain

$$\begin{aligned}
&\mathcal{P}_{\mu,\nu}\left(\frac{f(4^n(x+4y))}{256^n} + \frac{f(4^n(4x-y))}{256^n}\right. \\
&\quad - \frac{306[9f(4^n(x+\frac{y}{3})) + f(4^n(x+2y))]}{256^n} \\
&\quad - \frac{136f(4^n(x-y))}{256^n} + \frac{1394f(4^n(x+y))}{256^n} \\
&\quad \left. - \frac{425f(4^n(y))}{256^n} + \frac{1530f(4^n(x))}{256^n}, t\right) \\
&\geq_{L^*} \mathcal{Q}_{\xi,\zeta}\left(4^n x, 4^n y, 4^{4n}t\right)
\end{aligned} \tag{4.3.25}$$

for all $x, y \in X$, $n \in \mathbf{N}$ and $t > 0$. Taking $n \to \infty$, we show that Q satisfies (4.3.14) for all $x, y \in X$. Taking $n \to \infty$ in (4.3.23), we obtain (4.3.18).

To prove the uniqueness of the quartic mapping Q subject to (4.3.18), let us assume that there exists another quartic mapping Q' which satisfies (4.3.18). Obviously, we have $Q(4^n x) = 4^{4n} Q(x)$ and $Q'(4^n x) = 4^{4n} Q'(x)$ for all $x \in X$ and $n \geq 1$. Hence, it follows from (4.3.18) that

$$\begin{aligned} &\mathcal{P}_{\mu,\nu}\big(Q(x) - Q'(x), t\big) \\ &\geq_{L^*} \mathcal{P}_{\mu,\nu}\big(Q(4^n x) - Q'(4^n x), 4^{4n} t\big) \\ &\geq_{L^*} \mathcal{T}\big(\mathcal{P}_{\mu,\nu}\big(Q(4^n x) - f(4^n x), 4^{4n-1} t\big), \mathcal{P}_{\mu,\nu}\big(f(4^n x) - Q'(4^n x), 4^{4n-1} t\big)\big) \\ &\geq_{L^*} \mathcal{T}\big(\mathcal{T}_{i=1}^{\infty}\big(Q_{\xi,\varsigma}\big(4^{n+i-1} x, 0, 4^{4n+3i+3} t\big)\big), \mathcal{T}_{i=1}^{\infty}\big(Q_{\xi,\varsigma}\big(4^{n+i-1} x, 0, 4^{4n+3i+3} t\big)\big)\big) \end{aligned}$$

for all $x \in X$ and $t > 0$. By letting $n \to \infty$ in (4.3.18), we prove the uniqueness of Q. This completes the proof. □

Corollary 4.3.7 *Let $(X, \mathcal{P}'_{\mu',\nu'}, \mathcal{T})$ be an IRN-space and $(Y, \mathcal{P}_{\mu,\nu}, \mathcal{T})$ be a complete IRN-space. Let $f : X \to Y$ be a mapping such that*

$$\begin{aligned} &\mathcal{P}_{\mu,\nu}\bigg(16 f(x+4y) + f(4x-y) - 306\bigg[9 f\bigg(x + \frac{y}{3}\bigg) + f(x+2y)\bigg] \\ &\qquad - 136 f(x-y) + 1394 f(x+y) - 425 f(y) + 1530 f(x), t\bigg) \\ &\quad \geq_{L^*} \mathcal{P}'_{\mu',\nu'}(x+y, t) \end{aligned}$$

for all $x, y \in X$ and $t > 0$ in which

$$\lim_{n\to\infty} \mathcal{T}_{i=1}^{\infty}\big(\mathcal{P}'_{\mu',\nu'}\big(x, 4^{4n+3i+3} t\big)\big) = 1_{L^*}$$

for all $x, y \in X$ and $t > 0$. Then there exists a unique quartic mapping $Q : X \to Y$ such that

$$\mathcal{P}_{\mu,\nu}\big(f(x) - Q(x), t\big) \geq_{L^*} \mathcal{T}_{i=1}^{\infty}\big(\mathcal{P}'_{\mu',\nu'}\big(x, 4^{3i+3} t\big)\big)$$

for all $x \in X$ and $t > 0$.

Now, we give an example to illustrate the main result of Theorem 4.3.6 as follows.

Example 4.3.8 Let $(X, \|\cdot\|)$ be a Banach algebra, $(X, \mathcal{P}_{\mu,\nu}, \mathbf{M})$ an IRN-space in which

$$\mathcal{P}_{\mu,\nu}(x, t) = \left(\frac{t}{t + \|x\|}, \frac{\|x\|}{t + \|x\|}\right)$$

for all $x \in X$ and $t > 0$ and $(Y, \mathcal{P}_{\mu,\nu}, \mathbf{M})$ be a complete IRN-space. Define a mapping $f : X \to X$ by $f(x) = x^4 + x_0$, where x_0 is a unit vector in X. A straightforward computation shows that

$$\mathcal{P}_{\mu,\nu}\bigg(16f(x+4y)+f(4x-y)-306\bigg[9f\bigg(x+\frac{y}{3}\bigg)+f(x+2y)\bigg]$$

$$-136f(x-y)+1394f(x+y)-425f(y)+1530f(x),t\bigg)$$

$$\geq_{L^*}\mathcal{P}_{\mu,\nu}(x+y,t)$$

for all $x, y \in X$ and $t > 0$. Also, we have

$$\begin{aligned}\lim_{n\to\infty}\mathbf{M}_{i=1}^{\infty}\big(\mathcal{P}_{\mu,\nu}\big(4^{n+i-1}x,4^{4n+3i+3}t\big)\big) &= \lim_{n\to\infty}\lim_{m\to\infty}\mathbf{M}_{i=1}^{m}\big(\mathcal{P}_{\mu,\nu}\big(x,4^{3n+2i+4}t\big)\big)\\ &= \lim_{n\to\infty}\lim_{m\to\infty}\mathcal{P}_{\mu,\nu}\big(x,4^{3n+6}t\big)\\ &= \lim_{n\to\infty}\mathcal{P}_{\mu,\nu}\big(x,4^{3n+6}t\big)\\ &= 1_{L^*}\end{aligned}$$

for all $x \in X$ and $t > 0$. Therefore, all the conditions of Theorem 4.3.6 hold and so there exists a unique quartic mapping $Q : X \to Y$ such that

$$\mathcal{P}_{\mu,\nu}\big(f(x)-Q(x),t\big)\geq_{L^*}\mathcal{P}_{\mu,\nu}\big(x,4^6t\big)$$

for all $x \in X$ and $t > 0$.

Chapter 5
Stability of Functional Equations in RN-Spaces via Fixed Point Methods

In this chapter, we consider some functional equations and prove their stability via *fixed point methods* in various random normed spaces.

In [156], Luxemburg introduced the notion of a generalized metric space by allowing the value $+\infty$ for the distance mapping as follows:

Let X be a set. A mapping $d : X \times X \to [0, \infty]$ is called a *generalized metric* on X if and only if d satisfies the following conditions:

(1) $d(x, y) = 0$ if and only if $x = y$;
(2) $d(x, y) = d(y, x)$ for all $x, y \in X$;
(3) $d(x, z) \leq d(x, y) + d(y, z)$ for all $x, y, z \in X$.

In the proof of our main theorem, we use the following fixed point theorem in generalized metric spaces [130] (also see [158, 162]).

Luxemburg–Jung's Theorem [130, 158] *Let (X, d) be a complete generalized metric space and $A : X \to X$ be a strict contraction with the* Lipschitz constant $L \in (0, 1)$ *such that $d(x_0, A(x_0)) < +\infty$ for some $x_0 \in X$. Then we have the following*:

(a) *A has a unique fixed point in the set $Y := \{y \in X : d(x_0, y) < \infty\}$*;
(b) *for all $x \in Y$, the sequence $\{A^n(x)\}$ converges to the fixed point x^**;
(c) *$d(x_0, A(x_0)) \leq \delta$ implies $d(x^*, x_0) \leq \frac{\delta}{1-L}$.*

Let X be a linear space, (Y, ν, T_M) be a complete Menger probabilistic φ-normed space and G be a mapping from $X \times \mathbb{R}$ into $[0, 1]$ such that $G(x, \cdot) \in D_+$ for all $x \in X$. Consider the set $E := \{g : X \to Y : g(0) = 0\}$ and the mapping d_G defined on $E \times E$ by

$$d_G(g, h) = \inf\{a \in \mathbb{R}^+ : \nu_{g(x)-h(x)}(at) \geq G(x, t),\ \forall x \in X,\ t > 0\},$$

where, as usual, $\inf \emptyset = +\infty$.

In [161, 163], Miheţ proved that d_G is a complete generalized metric on E.

Y.J. Cho et al., *Stability of Functional Equations in Random Normed Spaces*, Springer Optimization and Its Applications 86, DOI 10.1007/978-1-4614-8477-6_5,

5.1 m-Variable Additive Functional Equations

In this section, we apply the fixed point method to investigate the Hyers–Ulam–Rassias stability for the following functional equation:

$$\sum_{i=1}^{m} f\left(mx_i + \sum_{j=1, j\neq i}^{m} x_j\right) + f\left(\sum_{i=1}^{m} x_i\right) = 2f\left(\sum_{i=1}^{m} mx_i\right) \tag{5.1.1}$$

for all $m \geq 1$ with $m \geq 2$, where a mapping f from a real linear space to a class of random normed spaces is unknown.

As a particular case, we obtain the Hyers–Ulam–Rassias stability result for this equation when X is a quasi-normed space and Y is a p-Banach space, which is similar to that in [83]. We note that a mapping $f : X \to Y$ satisfies (5.1.1) if and only if it is additive [83].

For convenience, we use the following abbreviation for a given mapping $f : X \to Y$:

$$Df(x_1, \ldots, x_m) := \sum_{i=1}^{m} f\left(mx_i + \sum_{j=1, j\neq i}^{m} x_j\right) + f\left(\sum_{i=1}^{m} x_i\right) - 2f\left(\sum_{i=1}^{m} mx_i\right)$$

for all $x_j \in X$ $(1 \leq j \leq m)$.

Theorem 5.1.1 *Let X be a real linear space, (Y, ν, T_M) be a complete random φ-normed space and $f : X \to Y$ be a Φ-approximate solution of* (5.1.1) *in the sense that*

$$\nu_{Df(x_1,\ldots,x_m)}(t) \geq \Phi(x_1, \ldots, x_m)(t) \tag{5.1.2}$$

for all $x_1, \ldots, x_m \in X$, where Φ is a mapping from X^m to D^+. If there exists $\alpha \in (0, \varphi(m))$ such that

$$\Phi(mx_1, \ldots, mx_m)(\alpha t) \geq \Phi(x_1, \ldots, x_m)(t) \tag{5.1.3}$$

for all $x_1, \ldots, x_m \in X$ and $t > 0$ and

$$\lim_{n\to\infty} \alpha^n \varphi\left(\frac{1}{m^n}\right) = 0, \tag{5.1.4}$$

then there exists a unique additive mapping $g : X \to Y$ such that

$$\nu_{g(x)-f(x)}(t) \geq \Phi(x, 0, \ldots, 0)\big((\varphi(m) - \alpha)t\big) \tag{5.1.5}$$

for all $x \in X$ and $t > 0$. Moreover,

$$g(x) = \lim_{n\to\infty} \frac{f(m^n x)}{m^n}$$

for all $x \in X$.

Proof By setting $x_1 = x$ and $x_j = 0$ $(2 \le j \le m)$ in (5.1.2), we obtain

$$\nu_{mf(x)-f(mx)}(t) \ge \Phi(x, 0, \ldots, 0)(t)$$

for all $x \in X$ and hence

$$\nu_{m^{-1}f(mx)-f(x)}(t) \ge \Phi(x, 0, \ldots, 0)\big(\varphi(m)t\big)$$

for all $x \in X$ and $t > 0$. Let $G(x,t) := \Phi(x, 0, \ldots, 0)(\varphi(m)t)$. Consider the set $E := \{g : X \to Y : g(0) = 0\}$ and the mapping d_G defined on $E \times E$. (E, d_G) is a complete generalized metric space (see the proof of [161, Lemma 2.1]).

Now, let us consider the linear mapping $J : E \to E$ defined by $Jg(x) := \frac{1}{m}g(mx)$. It is easy to check that J is a strictly contractive self-mapping of E with the Lipschitz constant $L = \frac{\alpha}{\varphi(m)}$. Indeed, let $g, h \in E$ be such that $d_G(g,h) < \epsilon$. Then we have

$$\nu_{g(x)-h(x)}(\epsilon t) \ge G(x,t)$$

for all $x \in X$ and $t > 0$ and hence

$$\nu_{Jg(x)-Jh(x)}\left(\frac{\alpha}{\varphi(m)}\epsilon t\right) = \nu_{g(mx)-h(mx)}(\alpha\epsilon t) \ge G(mx, \alpha t)$$

for all $x \in X$ and $t > 0$. Since $G(mx, \alpha t) \ge G(x,t)$ for all $x \in X$ and $t > 0$, we have

$$\nu_{Jg(x)-Jh(x)}\left(\frac{\alpha}{\varphi(m)}\epsilon t\right) \ge G(x,t),$$

that is,

$$d_G(g,h) < \epsilon \quad \Longrightarrow \quad d_G(Jg, Jh) \le \frac{\alpha}{\varphi(m)}\epsilon.$$

This means that

$$d_G(Jg, Jh) \le \frac{\alpha}{\varphi(m)} d_G(g,h)$$

for all g, h in E.

Next, from $\nu_{f(x)-m^{-1}f(mx)}(t) \ge G(x,t)$, it follows that $d_G(f, Jf) \le 1$. Thus, from Luxemburg–Jung's theorem, we can show the existence of a fixed point of J, that is, the existence of a mapping $g : X \to Y$ such that $g(mx) = mg(x)$ for all $x \in X$. Also, it follows that

$$d_G(f,g) \le \frac{1}{1-L} d(f, Jf) \quad \Longrightarrow \quad d_G(f,g) \le \frac{1}{1 - \frac{\alpha}{\varphi(m)}},$$

which immediately implies that

$$\nu_{g(x)-f(x)}\left(\frac{\varphi(m)}{\varphi(m)-\alpha}t\right) \ge G(x,t)$$

for all $x \in X$ and $t > 0$. This means that

$$\nu_{g(x)-f(x)}(t) \geq G\left(x, \frac{\varphi(m) - \alpha}{\varphi(m)} t\right)$$

for all $x \in X$ and $t > 0$ and hence

$$\nu_{g(x)-f(x)}(t) \geq \Phi(x, 0)\big((\varphi(m) - \alpha)t\big)$$

for all $x \in X$ and $t > 0$. Since

$$d_G(u, v) < \epsilon \quad \Longrightarrow \quad \nu_{u(x)-v(x)}(t) \geq G\left(x, \frac{t}{\epsilon}\right)$$

for any $x \in X$ and $t > 0$, it follows from $d_G(J^n f, g) \to 0$ that

$$\lim_{n\to\infty} \frac{f(m^n x)}{m^n} = g(x)$$

for all $x \in X$.

The additivity of g can be proven in the standard way (see [162] and [167]). In fact, since T_M is a continuous t-norm, $z \to \nu_z$ is continuous and thus (see [241, Chap. 12])

$$\begin{aligned}
\nu_{Dg(x_1,\dots,x_m)}(t) &= \lim_{n\to\infty} \nu_{\frac{Df(m^n x_1,\dots,m^n x_m)}{m^n}}(t) \\
&= \lim_{n\to\infty} \nu_{Df(m^n x_1,\dots,m^n x_m)}\left(\frac{t}{\varphi(\frac{1}{m^n})}\right) \\
&\geq \lim_{n\to\infty} \Phi(x_1, \dots, x_m)\left(\frac{t}{\alpha^n \varphi(\frac{1}{m^n})}\right) \\
&= 1.
\end{aligned}$$

Thus it follows that

$$\nu_{Dg(x_1,\dots,x_m)}(t) = 1$$

for all $t > 0$, which implies

$$Dg(x_1, \dots, x_m) = 0.$$

The uniqueness of g follows from the fact that g is the unique fixed point of J in the set $\{h \in E : g_G(f, h) < \infty\}$, that is, the only one with the property: there exists $C \in (0, \infty)$ such that

$$\nu_{g(x)-f(x)}(Ct) \geq G(x, t)$$

for all $x \in X$ and $t > 0$. This completes the proof. □

The following result provides the Hyers–Ulam–Rassias stability result for (5.1.1), which is similar to Theorem 2.2 in [83].

Corollary 5.1.2 *Let X be a real linear space, f be a mapping from X into a Banach p-normed space $(Y, \|\cdot\|_p)$ $(p \in (0,1])$ and $\Psi : X^m \to \mathbb{R}^+$ be a mapping such that there exists $\alpha \in (0, m^p)$ such that*

$$\Psi(mx_1, \dots, mx_m) \le \alpha \Psi(x_1, \dots, x_m)$$

for all $x_1, \dots, x_m \in X$. If

$$\big\|Df(x_1, \dots, x_m)\big\|_p \le \Psi(x_1, \dots, x_m)$$

for all $x_1, \dots, x_m \in X$, then there exists a unique additive mapping $g : X \to Y$ such that

$$\big\|g(x) - f(x)\big\|_p \le \frac{1}{|m|^p - \alpha} \Psi(x, 0, \dots, 0)$$

for all $x \in X$. Moreover, we have

$$g(x) = \lim_{n \to \infty} \frac{f(m^n x)}{m^n}.$$

Proof Recall that a p-normed space $(0 < p \le 1)$ is a pair $(Y, \|\cdot\|_p)$, where Y is real linear space and $\|\cdot\|_p$ is a real valued function on Y (which is called a *p-norm*) satisfying the following conditions:

(pn1) $\|x\|_p \ge 0$ for all $x \in Y$ and $\|x\|_p = 0$ if and only if $x = 0$;
(pn2) $\|\lambda x\|_p = |\lambda|^p \|x\|_p$ for all $x \in Y$ and $\lambda \in \mathbb{R}$;
(pn3) $\|x + y\|_p \le \|x\|_p + \|y\|_p$ for all $x, y \in Y$.

A p-normed space $(Y, \|\cdot\|_p)$ induces a random φ-normed space (Y, ν, T_M) from

$$\nu_x(t) = \frac{t}{t + \|x\|_p}, \qquad \varphi(t) = |t|^p.$$

Indeed, (RN1) is obviously verified and (RN2) follows from

$$\begin{aligned}
\nu_{\alpha x}(t) &= \frac{t}{t + \|\alpha x\|_p} \\
&= \frac{t}{t + |\alpha|^p \|x\|_p} \\
&= \frac{t/|\alpha|^p}{t/|\alpha|^p + \|x\|_p} \\
&= \nu_x\left(\frac{t}{|\alpha|^p}\right) \\
&= \nu_x\left(\frac{t}{\varphi(\alpha)}\right).
\end{aligned}$$

Finally, if $\frac{t}{t+\|x\|_p} \le \frac{s}{s+\|y\|_p}$, then the inequality

$$\frac{t+s}{t+s+\|x+y\|_p} \ge \frac{t}{t+\|x\|_p}$$

follows from $t\|x\|_p + s\|x\|_p \ge t\|x\|_p + t\|y\|_p \ge t\|x+y\|_p$.

If we consider the induced random φ-normed space (Y, ν, T_M) and the mapping Φ on X^m defined by $\Phi(x_1, \ldots, x_m)(t) = \frac{t}{t+\Psi(x_1,\ldots,x_m)}$, then the condition

$$\Phi(mx_1, \ldots, mx_m)(\alpha t) \ge \Phi(x_1, \ldots, x_m)(t)$$

for all $t > 0$ is equivalent to

$$\Psi(mx_1, \ldots, mx_m) \le \alpha \Psi(x_1, \ldots, x_m)$$

and the condition

$$\lim_{n\to\infty} \alpha^n \varphi\left(\frac{1}{m^n}\right) = 0 \tag{5.1.6}$$

reduces to

$$\lim_{n\to\infty} \left(\frac{\alpha}{m^p}\right)^n = 0. \tag{5.1.7}$$

This completes the proof. □

We end this section by stating a slightly improved version of Theorem 5.1.1.

Theorem 5.1.3 *Let X be a real linear space, f be a mapping from X into a complete random φ-normed space (Y, ν, T_M) and $\Phi : X^m \to D^+$ be a mapping with the properties: there exists $\alpha \in (0, \varphi(m))$ such that*

$$\Phi(mx, 0, \ldots, 0)(\alpha t) \ge \Phi(x, 0, \ldots, 0)(t) \tag{5.1.8}$$

for all $x \in X$ and $t > 0$ and

$$\lim_{n\to\infty} \Phi(m^n x_1, \ldots, m^n x_n)\left(\frac{t}{\varphi(\frac{1}{m^n})}\right) = 1 \tag{5.1.9}$$

for all $x_1, \ldots, x_m \in X$ and $t > 0$. If

$$\nu_{Df(x_1,\ldots,x_m)}(t) \ge \Phi(x_1, \ldots, x_m)(t) \tag{5.1.10}$$

for all $x_1, \ldots, x_m \in X$ and $t > 0$, then there exists a unique additive mapping $g : X \to Y$ such that

$$\nu_{g(x)-f(x)}(t) \ge \Phi(x, 0, \ldots, 0)\big((\varphi(m) - \alpha)t\big) \tag{5.1.11}$$

for all $x \in X$ and $t > 0$.

Proof The condition (5.1.8) implies $G(mx, \alpha t) \geq G(x, t)$ for all $x \in X$ and $t > 0$ (see the proof of the preceding theorem) and so it follows that J is a strictly contractive self-mapping of E with the Lipschitz constant $L = \alpha/\varphi(m)$, while the condition (5.1.9) ensures the additivity of g.

Since the proof can now easily be reproduced from that of Theorem 5.1.1, we omit further details. □

5.2 Quartic Functional Equations

In this section, we apply the fixed point method to investigate the Hyers–Ulam–Rassias stability for quartic functional equations in random normed spaces.

Theorem 5.2.1 *Let X be a real linear space, f be a mapping from X into a complete RN-space (Y, μ, T_M) with $f(0) = 0$ and $\Phi : X^2 \to D^+$ be a mapping with the property: there exists $\alpha \in (0, 256)$ such that*

$$\Phi_{4x,4y}(\alpha t) \geq \Phi_{x,y}(t) \tag{5.2.1}$$

for all $x, y \in X$ and $t > 0$. If

$$\mu_{Df(x,y)}(t) \geq \Phi_{x,y}(t) \tag{5.2.2}$$

for all $x, y \in X$ and $t > 0$, where

$$\begin{aligned} Df(x, y) = 16f(x+4y) + f(4x-y) - 306\left[9f\left(x+\frac{y}{3}\right) + f(x+2y)\right] \\ - 136f(x-y) + 1394f(x+y) - 425f(y) + 1530f(x), \end{aligned}$$

then there exists a unique quartic mapping $g : X \to Y$ such that

$$\mu_{g(x)-f(x)}(t) \geq \Phi_{x,0}(Mt) \tag{5.2.3}$$

for all $x \in X$ and $t > 0$, where $M = (256 - \alpha)$. Moreover, we have

$$g(x) = \lim_{n\to\infty} \frac{f(4^n x)}{4^{4n}}.$$

Proof By setting $y = 0$ in (5.2.2), we obtain

$$\mu_{f(4x)-256f(x)}(t) \geq \Phi_{x,0}(t)$$

for all $x \in X$ and $t > 0$ and hence

$$\begin{aligned} \mu_{\frac{1}{256}f(4x)-f(x)}(t) &= \mu_{\frac{1}{256}(f(4x)-256f(x))}(t) \\ &= \mu_{f(4x)-256f(x)}(256t) \end{aligned}$$

$$\geq \Phi_{x,0}(256t)$$

for all $x \in X$ and $t > 0$. Let $G(x,t) := \Phi_{x,0}(256t)$. Consider the set

$$E := \{g : X \to Y : g(0) = 0\}$$

and the mapping d_G defined on $E \times E$ by

$$d_G(g,h) = \inf\{u \in R^+ : \mu_{g(x)-h(x)}(ut) \geq G(x,t),\ \forall x \in X,\ t > 0\}.$$

Then (E, d_G) is a complete generalized metric space. Now, let us consider the linear mapping $J : E \to E$ defined by

$$Jg(x) := \frac{1}{256} g(4x).$$

Now, we show that J is a strictly contractive self-mapping of E with the Lipschitz constant $k = \frac{\alpha}{256}$. Indeed, let $g, h \in E$ be the mappings such that $d_G(g,h) < \epsilon$. Then we have

$$\mu_{g(x)-h(x)}(\epsilon t) \geq G(x,t)$$

for all $x \in X$ and $t > 0$ and hence

$$\begin{aligned}\mu_{Jg(x)-Jh(x)}\left(\frac{\alpha}{256}\epsilon t\right) &= \mu_{\frac{1}{256}(g(4x)-h(4x))}\left(\frac{\alpha}{256}\epsilon t\right)\\ &= \mu_{g(4x)-h(4x)}(\alpha \epsilon t)\\ &\geq G(4x, \alpha t)\end{aligned}$$

for all $x \in X$ and $t > 0$. Since $G(4x, \alpha t) \geq G(x,t)$, we have

$$\mu_{Jg(x)-Jh(x)}\left(\frac{\alpha}{256}\epsilon t\right) \geq G(x,t),$$

that is,

$$d_G(g,h) < \epsilon \quad \Rightarrow \quad d_G(Jg, Jh) \leq \frac{\alpha}{256}\epsilon.$$

This means that

$$d_G(Jg, Jh) \leq \frac{\alpha}{256} d_G(g,h)$$

for all g, h in E.

Next, from $\mu_{f(x)-\frac{1}{256}f(4x)}(t) \geq G(x,t)$, it follows that $d_G(f, Jf) \leq 1$. Using the Luxemburg–Jung's theorem, we show the existence of a fixed point of J, that is, the existence of a mapping $g : X \to Y$ such that $g(4x) = 256g(x)$ for all $x \in X$. Since, for all $x \in X$ and $t > 0$,

$$d_G(u,v) < \epsilon \quad \Rightarrow \quad \mu_{u(x)-v(x)}(t) \geq G\left(x, \frac{t}{\epsilon}\right),$$

it follows from $d_G(J^n f, g) \to 0$ that $\lim_{n\to\infty} \frac{f(4^n x)}{4^{4n}} = g(x)$ for all $x \in X$.

Also, from

$$d_G(f,g) \leq \frac{1}{1-L} d(f, Jf) \quad \Longrightarrow \quad d_G(f,g) \leq \frac{1}{1-\frac{\alpha}{256}},$$

it immediately follows that

$$\nu_{g(x)-f(x)}\left(\frac{256}{256-\alpha} t\right) \geq G(x,t)$$

for all $t > 0$ and $x \in X$. This means that

$$\mu_{g(x)-f(x)}(t) \geq G\left(x, \frac{256-\alpha}{256} t\right)$$

for all $x \in X$ and $t > 0$ and so it follows that

$$\mu_{g(x)-f(x)}(t) \geq \Phi_{x,0}\big((256-\alpha)t\big)$$

for all $x \in X$ and $t > 0$.

Finally, the uniqueness of g follows from the fact that g is the unique fixed point of J such that there exists $C \in (0, \infty)$ such that

$$\mu_{g(x)-f(x)}(Ct) \geq G(x,t)$$

for all $x \in X$ and $t > 0$. This completes the proof. □

5.3 ACQ Functional Equations

In this section, we prove the generalized Hyers–Ulam stability of the following additive–cubic–quartic functional equation

$$\begin{aligned} &11f(x+2y) + 11f(x-2y) \\ &= 44f(x+y) + 44f(x-y) + 12f(3y) - 48f(2y) + 60f(y) - 66f(x) \end{aligned} \tag{5.3.1}$$

in complete latticetic random normed spaces.

5.3.1 The Generalized Hyers–Ulam Stability of the Functional Equation (5.3.1): An Odd Case

We can easily show that an even mapping $f : X \to Y$ satisfies (5.3.1) if and only if the even mapping $f : X \to Y$ is a quartic mapping, that is,

$$f(2x+y) + f(2x-y) = 4f(x+y) + 4f(x-y) + 24f(x) - 6f(y)$$

for all $x, y \in X$ and an odd mapping $f : X \to Y$ satisfies (5.3.1) if and only if the odd mapping $f : X \to Y$ is an additive-cubic mapping, that is,

$$f(x+2y) + f(x-2y) = 4f(x+y) + 4f(x-y) - 6f(x).$$

for all $x, y \in X$.

It was shown in Lemma 2.2 of [66] that $g(x) := f(2x) - 2f(x)$ and $h(x) := f(2x) - 8f(x)$ are cubic and additive, respectively, and $f(x) = \frac{1}{6}g(x) - \frac{1}{6}h(x)$.

For any mapping $f : X \to Y$, we define

$$\begin{aligned} Df(x, y) :&= 11f(x+2y) + 11f(x-2y) - 44f(x+y) - 44f(x-y) \\ &\quad - 12f(3y) + 48f(2y) - 60f(y) + 66f(x) \end{aligned}$$

for all $x, y \in X$. Using the fixed point method, we prove the generalized Hyers–Ulam stability of the functional equation $Df(x, y) = 0$ in complete LRN-spaces: an odd case.

Theorem 5.3.1 *Let X be a linear space, $(Y, \mu, \mathcal{T}_\wedge)$ be a complete* LRN-*space and Φ be a mapping from X^2 to D_L^+ ($\Phi(x, y)$ is denoted by $\Phi_{x,y}$) such that, for some $0 < \alpha < \frac{1}{8}$,*

$$\Phi_{2x,2y}(t) \leq_L \Phi_{x,y}(\alpha t) \tag{5.3.2}$$

for all $x, y \in X$ and $t > 0$. Let $f : X \to Y$ be an odd mapping satisfying

$$\mu_{Df(x,y)}(t) \geq_L \Phi_{x,y}(t) \tag{5.3.3}$$

for all $x, y \in X$ and $t > 0$. Then

$$C(x) := \lim_{n\to\infty} 8^n \left(f\left(\frac{x}{2^{n-1}}\right) - 2f\left(\frac{x}{2^n}\right)\right)$$

exists for all $x \in X$ and defines a cubic mapping $C : X \to Y$ such that

$$\begin{aligned} &\mu_{f(2x)-2f(x)-C(x)}(t) \\ &\quad \geq \mathcal{T}_\wedge\left(\Phi_{0,x}\left(\frac{(33-264\alpha)}{17\alpha}t\right), \Phi_{2x,x}\left(\frac{(33-264\alpha)}{17\alpha}t\right)\right) \end{aligned} \tag{5.3.4}$$

for all $x \in X$ and $t > 0$.

Proof Letting $x = 0$ in (5.3.3), we get

$$\mu_{12f(3y)-48f(2y)+60f(y)}(t) \geq_L \Phi_{0,y}(t) \tag{5.3.5}$$

for all $y \in X$ and $t > 0$. Replacing x by $2y$ in (5.3.3), we get

$$\mu_{11f(4y)-56f(3y)+114f(2y)-104f(y)}(t) \geq_L \Phi_{2y,y}(t) \tag{5.3.6}$$

for all $y \in X$ and $t > 0$. By (5.3.5) and (5.3.6), we have

$$\begin{aligned}
&\mu_{f(4y)-10f(2y)+16f(y)}\left(\frac{14}{33}t + \frac{1}{11}t\right)\\
&\geq_L \mathcal{T}_\wedge\left(\mu_{\frac{14}{33}(12f(3y)-48f(2y)+60f(y))}\left(\frac{14}{33}t\right),\right.\\
&\qquad \left.\mu_{\frac{1}{11}(11f(4y)-56f(3y)+114f(2y)-104f(y))}\left(\frac{1}{11}t\right)\right)\\
&\geq_L \mathcal{T}_\wedge\big(\Phi_{0,y}(t), \Phi_{2y,y}(t)\big) \qquad (5.3.7)
\end{aligned}$$

for all $y \in X$ and $t > 0$. Letting $y := \frac{x}{2}$ and $g(x) := f(2x) - 2f(x)$ for all $x \in X$, we get

$$\mu_{g(x)-8g(\frac{x}{2})}\left(\frac{17}{33}t\right) \geq_L \mathcal{T}_\wedge\big(\Phi_{0,\frac{x}{2}}(t), \Phi_{x,\frac{x}{2}}(t)\big) \qquad (5.3.8)$$

for all $x \in X$ and $t > 0$.

Consider the set $S := \{g : X \to Y\}$ and introduce the generalized metric on S:

$$\begin{aligned}
&d(g,h)\\
&\quad = \inf\big\{u \in \mathbb{R}^+ : \mu_{g(x)-h(x)}(ut) \geq_L \mathcal{T}_\wedge\big(\Phi_{0,x}(t), \Phi_{2x,x}(t)\big),\ \forall x \in X,\ t > 0\big\},
\end{aligned}$$

where, as usual, $\inf \emptyset = +\infty$. It is easy to show that (S, d) is complete.

Now, we consider the linear mapping $J : S \to S$ such that

$$Jg(x) := 8g\left(\frac{x}{2}\right)$$

for all $x \in X$. Let $g, h \in S$ be such that $d(g,h) = \varepsilon$. Then we have

$$\mu_{g(x)-h(x)}(\varepsilon t) \geq_L \mathcal{T}_\wedge\big(\Phi_{0,x}(t), \Phi_{2x,x}(t)\big)$$

for all $x \in X$ and $t > 0$. Hence, we have

$$\begin{aligned}
\mu_{Jg(x)-Jh(x)}(8\alpha\varepsilon t) &= \mu_{8g(\frac{x}{2})-8h(\frac{x}{2})}(8\alpha\varepsilon t)\\
&= \mu_{g(\frac{x}{2})-h(\frac{x}{2})}(\alpha\varepsilon t)\\
&\geq_L \mathcal{T}_\wedge\big(\Phi_{0,\frac{x}{2}}(\alpha t), \Phi_{x,\frac{x}{2}}(\alpha t)\big)\\
&\geq_L \mathcal{T}_\wedge\big(\Phi_{0,x}(t), \Phi_{2x,x}(t)\big)
\end{aligned}$$

for all $x \in X$ and $t > 0$. So $d(g,h) = \varepsilon$ implies that

$$d(Jg, Jh) \leq 8\alpha\varepsilon,$$

which means that

$$d(Jg, Jh) \leq 8\alpha d(g, h)$$

for all $g, h \in S$. It follows from (5.3.8) that

$$\mu_{g(x)-8g(\frac{x}{2})}\left(\frac{17}{33}\alpha t\right) \geq_L \mathcal{T}_\wedge\big(\Phi_{0,x}(t), \Phi_{2x,x}(t)\big)$$

for all $x \in X$ and all $t > 0$ and so $d(g, Jg) \leq \frac{17}{33}\alpha$.

Now, there exists a mapping $C : X \to Y$ satisfying the following:

(1) C is a fixed point of J, i.e.,

$$C\left(\frac{x}{2}\right) = \frac{1}{8}C(x) \tag{5.3.9}$$

for all $x \in X$. Since $g : X \to Y$ is odd, $C : X \to Y$ is an odd mapping. The mapping C is a unique fixed point of J in the set

$$M = \{g \in S : d(f, g) < \infty\}.$$

This implies that C is a unique mapping satisfying (5.3.9) such that there exists $u \in (0, \infty)$ satisfying

$$\mu_{g(x)-C(x)}(ut) \geq_L \mathcal{T}_\wedge\big(\Phi_{0,x}(t), \Phi_{2x,x}(t)\big)$$

for all $x \in X$ and $t > 0$;

(2) $d(J^n g, C) \to 0$ as $n \to \infty$. This implies the equality

$$\lim_{n\to\infty} 8^n g\left(\frac{x}{2^n}\right) = C(x)$$

for all $x \in X$;

(3) $d(g, C) \leq \frac{1}{1-8\alpha} d(g, Jg)$, which implies the inequality

$$d(g, C) \leq \frac{17\alpha}{33 - 264\alpha},$$

which implies that the inequality (5.3.4) holds.

From $Dg(x, y) = Df(2x, 2y) - 2Df(x, y)$ and (5.3.3), it follows that

$$\mu_{Df(2x,2y)}(t) \geq_L \Phi_{2x,2y}(t),$$

$$\mu_{-2Df(x,y)}(t) = \mu_{Df(x,y)}\left(\frac{t}{2}\right) \geq_L \Phi_{x,y}\left(\frac{t}{2}\right)$$

and so, by (LRN3) and (5.3.2),

$$\begin{aligned}\mu_{Dg(x,y)}(3t) &\geq_L \mathcal{T}_\wedge\big(\mu_{Df(2x,2y)}(t), \mu_{-2Df(x,y)}(2t)\big)\\ &\geq_L \mathcal{T}_\wedge\big(\Phi_{2x,2y}(t), \Phi_{x,y}(t)\big)\\ &\geq_L \Phi_{2x,2y}(t)\end{aligned}$$

for all $x, y \in X$ and $t > 0$. It follows that

$$\begin{aligned}\mu_{8^n Dg(\frac{x}{2^n}, \frac{y}{2^n})}(3t) &= \mu_{Dg(\frac{x}{2^n}, \frac{y}{2^n})}\left(3\frac{t}{8^n}\right)\\ &\geq_L \Phi_{\frac{x}{2^{n-1}}, \frac{y}{2^{n-1}}}\left(\frac{t}{8^n}\right)\\ &\geq_L \cdots\\ &\geq_L \Phi_{x,y}\left(\frac{1}{8}\frac{t}{(8\alpha)^{n-1}}\right)\end{aligned}$$

for all $x, y \in X$, $t > 0$ and $n \geq 1$. Since $0 < 8\alpha < 1$, we have

$$\lim_{n\to\infty} \Phi_{x,y}\left(\frac{t}{(8\alpha)^n}\right) = 1_{\mathcal{L}}$$

for all $x, y \in X$ and $t > 0$. Then it follows that

$$\mu_{DC(x,y)}(t) = 1_{\mathcal{L}}$$

for all $x, y \in X$ and $t > 0$. Thus, the mapping $C : X \to Y$ is cubic. This completes the proof. □

Corollary 5.3.2 *Let $\theta \geq 0$ and p be a real number with $p > 3$. Let X be a normed linear space with norm $\|\cdot\|$ and $(X, \mu, \mathcal{T}_\wedge)$ be an LRN-space in which $L = [0, 1]$ and $\mathcal{T}_\wedge = \min$. Let $f : X \to Y$ be an odd mapping satisfying*

$$\mu_{Df(x,y)}(t) \geq \frac{t}{t + \theta(\|x\|^p + \|y\|^p)} \tag{5.3.10}$$

for all $x, y \in X$ and $t > 0$. Then

$$C(x) := \lim_{n\to\infty} 8^n\left(f\left(\frac{x}{2^{n-1}}\right) - 2f\left(\frac{x}{2^n}\right)\right)$$

exists for all $x \in X$ and defines a cubic mapping $C : X \to Y$ such that

$$\mu_{f(2x)-2f(x)-C(x)}(t) \geq \frac{33(2^p - 8)t}{33(2^p - 8)t + 17(1 + 2^p)\theta\|x\|^p}$$

for all $x \in X$ and $t > 0$.

Proof The proof follows from Theorem 5.3.1 by taking

$$\Phi_{x,y}(t) := \frac{t}{t+\theta(\|x\|^p + \|y\|^p)}$$

for all $x, y \in X$ and $t > 0$. Then we can choose $\alpha = 2^{-p}$ and so we get the desired result. □

Theorem 5.3.3 *Let X be a linear space, $(Y, \mu, \mathcal{T}_\wedge)$ be a complete LRN-space and Φ be a mapping from X^2 to D_L^+ ($\Phi(x, y)$ is denoted by $\Phi_{x,y}$) such that, for some $0 < \alpha < 8$,*

$$\Phi_{x,y}(\alpha t) \geq_L \Phi_{\frac{x}{2},\frac{y}{2}}(t)$$

for all $x, y \in X$ and $t > 0$. Let $f : X \to Y$ be an odd mapping satisfying (5.3.1). *Then*

$$C(x) := \lim_{n\to\infty} \frac{1}{8^n}\big(f\big(2^{n+1}x\big) - 2f\big(2^n x\big)\big)$$

exists for all $x \in X$ and defines a cubic mapping $C : X \to Y$ such that

$$\mu_{f(2x)-2f(x)-C(x)}(t) \geq \mathcal{T}_\wedge\left(\Phi_{0,x}\left(\frac{(264-33\alpha)}{17}t\right), \Phi_{2x,x}\left(\frac{(264-33\alpha)}{17}t\right)\right) \qquad (5.3.11)$$

for all $x \in X$ and $t > 0$.

Proof Let (S, d) be the generalized metric space defined in the proof of Theorem 5.3.1. Consider the linear mapping $J : S \to S$ such that

$$Jg(x) := \frac{1}{8}g(2x)$$

for all $x \in X$. Let $g, h \in S$ be given such that $d(g, h) = \varepsilon$. Then we have

$$\mu_{g(x)-h(x)}(\varepsilon t) \geq_L \mathcal{T}_\wedge\big(\Phi_{0,x}(t), \Phi_{2x,x}(t)\big)$$

for all $x \in X$ and $t > 0$. Hence, we have

$$\begin{aligned}
\mu_{Jg(x)-Jh(x)}\left(\frac{\alpha}{8}\varepsilon t\right) &= \mu_{\frac{1}{8}g(2x)-\frac{1}{8}h(2x)}\left(\frac{\alpha}{8}\varepsilon t\right) \\
&= \mu_{g(2x)-h(2x)}(\alpha\varepsilon t) \\
&\geq_L \mathcal{T}_\wedge\big(\Phi_{0,2x}(\alpha t), \Phi_{4x,2x}(\alpha t)\big) \\
&\geq_L \mathcal{T}_\wedge\big(\Phi_{0,x}(t), \Phi_{2x,x}(t)\big)
\end{aligned}$$

for all $x \in X$ and $t > 0$. So $d(g, h) = \varepsilon$ implies that

$$d(Jg, Jh) \leq \frac{\alpha}{8}\varepsilon.$$

This means that $d(Jg, Jh) \le \frac{\alpha}{8} d(g, h)$ for all $g, h \in S$. It follows from (5.3.8) that

$$\mu_{g(x)-\frac{1}{8}g(2x)}\left(\frac{17}{264}t\right) \ge_L \mathcal{T}_\wedge\big(\Phi_{0,x}(t), \Phi_{2x,x}(t)\big)$$

for all $x \in X$ and $t > 0$. So $d(g, Jg) \le \frac{17}{264}$.

Now, there exists a mapping $C : X \to Y$ satisfying the following:

(1) C is a fixed point of J, that is,

$$C(2x) = 8C(x) \tag{5.3.12}$$

for all $x \in X$. Since $g : X \to Y$ is odd, $C : X \to Y$ is an odd mapping. The mapping C is a unique fixed point of J in the set

$$M = \{g \in S : d(f, g) < \infty\}.$$

This implies that C is a unique mapping satisfying (5.3.12) such that there exists $u \in (0, \infty)$ satisfying

$$\mu_{g(x)-C(x)}(ut) \ge_L \mathcal{T}_\wedge\big(\Phi_{0,x}(t), \Phi_{2x,x}(t)\big)$$

for all $x \in X$ and $t > 0$;

(2) $d(J^n g, C) \to 0$ as $n \to \infty$. This implies the equality

$$\lim_{n\to\infty} \frac{1}{8^n} g(2^n x) = C(x)$$

for all $x \in X$;

(3) $d(g, C) \le \frac{1}{1-\frac{\alpha}{8}} d(g, Jg)$, which implies the inequality

$$d(g, C) \le \frac{17}{264 - 33\alpha}.$$

This implies that the inequality (5.3.11) holds. The rest of the proof is similar to the proof of Theorem 5.3.1. This completes the proof. □

Corollary 5.3.4 *Let $\theta \ge 0$ and p be a real number with $0 < p < 3$. Let X be a normed linear space with norm $\|\cdot\|$ and $(X, \mu, \mathcal{T}_\wedge)$ be an LRN-space in which $L = [0, 1]$ and $\mathcal{T}_\wedge = \min$. Let $f : X \to Y$ be an odd mapping satisfying* (5.3.10). *Then*

$$C(x) := \lim_{n\to\infty} \frac{1}{8^n}\big(f(2^{n+1}x) - 2f(2^n x)\big)$$

exists for all $x \in X$ and defines a cubic mapping $C : X \to Y$ such that

$$\mu_{f(2x)-2f(x)-C(x)}(t) \ge \frac{33(8-2^p)t}{33(8-2^p)t + 17(1+2^p)\theta\|x\|^p}$$

for all $x \in X$ and $t > 0$.

Proof The proof follows from Theorem 5.3.3 by taking

$$\Phi_{x,y}(t) := \frac{t}{t + \theta(\|x\|^p + \|y\|^p)}$$

for all $x, y \in X$ and $t > 0$. Then we can choose $\alpha = 2^p$ and so we get the desired result. □

Theorem 5.3.5 *Let X be a linear space, $(X, \mu, \mathcal{T}_\wedge)$ be an* LRN*-space and Φ be a mapping from X^2 to D_L^+ ($\Phi(x, y)$ is denoted by $\Phi_{x,y}$) such that, for some $0 < \alpha < \frac{1}{2}$,*

$$\Phi_{x,y}(\alpha t) \geq_L \Phi_{2x,2y}(t)$$

for all $x, y \in X$ and $t > 0$. Let $f : X \to Y$ be an odd mapping satisfying (5.3.3). *Then*

$$A(x) := \lim_{n\to\infty} 2^n \left(f\left(\frac{x}{2^{n-1}}\right) - 8f\left(\frac{x}{2^n}\right)\right)$$

exists for all $x \in X$ and defines an additive mapping $A : X \to Y$ such that

$$\mu_{f(2x)-8f(x)-A(x)}(t) \geq_L \mathcal{T}_\wedge\left(\Phi_{0,x}\left(\frac{(33-66\alpha)}{17\alpha}t\right), \Phi_{2x,x}\left(\frac{(33-66\alpha)}{17\alpha}t\right)\right) \tag{5.3.13}$$

for all $x \in X$ and $t > 0$.

Proof Let (S, d) be the generalized metric space defined in the proof of Theorem 5.3.1. Letting $y := \frac{x}{2}$ and $h(x) := f(2x) - 8f(x)$ for all $x \in X$ in (5.3.7), we get

$$\mu_{h(x)-2h(\frac{x}{2})}\left(\frac{17}{33}t\right) \geq_L \mathcal{T}_\wedge\left(\Phi_{0,\frac{x}{2}}(t), \Phi_{x,\frac{x}{2}}(t)\right) \tag{5.3.14}$$

for all $x \in X$ and $t > 0$. Now, we consider the linear mapping $J : S \to S$ such that

$$Jh(x) := 2h\left(\frac{x}{2}\right)$$

for all $x \in X$. Let $g, h \in S$ be such that $d(g, h) = \varepsilon$. Then we have

$$\mu_{g(x)-h(x)}(\varepsilon t) \geq_L \mathcal{T}_\wedge\left(\Phi_{0,x}(t), \Phi_{2x,x}(t)\right)$$

for all $x \in X$ and $t > 0$. Hence, it follows that

$$\begin{aligned}\mu_{Jg(x)-Jh(x)}(2\alpha\varepsilon t) &= \mu_{2g(\frac{x}{2})-2h(\frac{x}{2})}(2\alpha\varepsilon t)\\ &= \mu_{g(\frac{x}{2})-h(\frac{x}{2})}(\alpha\varepsilon t)\end{aligned}$$

$$\geq_L \mathcal{T}_\wedge\big(\Phi_{0,\frac{x}{2}}(\alpha t), \Phi_{x,\frac{x}{2}}(\alpha t)\big)$$
$$\geq_L \mathcal{T}_\wedge\big(\Phi_{0,x}(t), \Phi_{2x,x}(t)\big)$$

for all $x \in X$ and $t > 0$. So $d(g,h) = \varepsilon$ implies that $d(Jg, Jh) \leq 2\alpha\varepsilon$. This means that

$$d(Jg, Jh) \leq 2\alpha d(g,h)$$

for all $g, h \in S$. It follows from (5.3.14) that

$$\mu_{h(x)-2h(\frac{x}{2})}\left(\frac{17}{33}\alpha t\right) \geq_L \mathcal{T}_\wedge\big(\Phi_{0,x}(t), \Phi_{2x,x}(t)\big)$$

for all $x \in X$ and $t > 0$ and so $d(h, Jh) \leq \frac{17\alpha}{33}$.

Now, there exists a mapping $A : X \to Y$ satisfying the following:

(1) A is a fixed point of J, that is,

$$A\left(\frac{x}{2}\right) = \frac{1}{2}A(x) \tag{5.3.15}$$

for all $x \in X$. Since $h : X \to Y$ is odd, $A : X \to Y$ is an odd mapping. The mapping A is a unique fixed point of J in the set

$$M = \{g \in S : d(f, g) < \infty\}.$$

This implies that A is a unique mapping satisfying (5.3.15) such that there exists $u \in (0, \infty)$ satisfying

$$\mu_{h(x)-A(x)}(ut) \geq_L \mathcal{T}_\wedge\big(\Phi_{0,x}(t), \Phi_{2x,x}(t)\big)$$

for all $x \in X$ and $t > 0$;

(2) $d(J^n h, A) \to 0$ as $n \to \infty$. This implies the equality

$$\lim_{n\to\infty} 2^n h\left(\frac{x}{2^n}\right) = A(x)$$

for all $x \in X$;

(3) $d(h, A) \leq \frac{1}{1-2\alpha} d(h, Jh)$, which implies the inequality

$$d(h, A) \leq \frac{17\alpha}{33 - 66\alpha}.$$

This implies that the inequality (5.3.13) holds.

The rest of the proof is similar to the proof of Theorem 5.3.1. This completes the proof. □

Corollary 5.3.6 *Let $\theta \geq 0$ and p be a real number with $p > 1$. Let X be a normed linear space with norm $\|\cdot\|$ and $(X, \mu, \mathcal{T}_\wedge)$ be an LRN-space in which $L = [0, 1]$ and $\mathcal{T}_\wedge = \min$. Let $f : X \to Y$ be an odd mapping satisfying* (5.3.10). *Then*

$$A(x) := \lim_{n\to\infty} 2^n \left(f\left(\frac{x}{2^{n-1}}\right) - 8f\left(\frac{x}{2^n}\right)\right)$$

exists for all $x \in X$ and defines an additive mapping $A : X \to Y$ such that

$$\mu_{f(2x)-8f(x)-A(x)}(t) \geq \frac{33(2^p - 2)t}{33(2^p - 2)t + 17(1 + 2^p)\theta\|x\|^p}$$

for all $x \in X$ and $t > 0$.

Proof The proof follows from Theorem 5.3.5 by taking

$$\Phi_{x,y}(t) := \frac{t}{t + \theta(\|x\|^p + \|y\|^p)}$$

for all $x, y \in X$ and $t > 0$. Then we can choose $\alpha = 2^{-p}$ and so we get the desired result. □

Theorem 5.3.7 *Let X be a linear space, $(X, \mu, \mathcal{T}_\wedge)$ be an LRN-space and Φ be a mapping from X^2 to D_L^+ ($\Phi(x, y)$ is denoted by $\Phi_{x,y}$) such that, for some $0 < \alpha < 2$,*

$$\Phi_{x,y}(\alpha t) \geq_L \Phi_{\frac{x}{2},\frac{y}{2}}(t)$$

for all $x, y \in X$ and $t > 0$. Let $f : X \to Y$ be an odd mapping satisfying (5.3.3). *Then*

$$A(x) := \lim_{n\to\infty} \frac{1}{2^n}\left(f(2^{n+1}x) - 8f(2^n x)\right)$$

exists for all $x \in X$ and defines an additive mapping $A : X \to Y$ such that

$$\mu_{f(2x)-8f(x)-A(x)}(t) \geq_L 7\mathcal{T}_\wedge\left(\Phi_{0,x}\left(\frac{(66 - 33\alpha)}{17}t\right), \Phi_{2x,x}\left(\frac{(66 - 33\alpha)}{17}t\right)\right) \tag{5.3.16}$$

for all $x \in X$ and $t > 0$.

Proof Let (S, d) be the generalized metric space defined in the proof of Theorem 5.3.1. Consider the linear mapping $J : S \to S$ such that

$$Jh(x) := \frac{1}{2}h(2x)$$

for all $x \in X$. Let $g, h \in S$ be given such that $d(g, h) = \varepsilon$. Then we have

$$\mu_{g(x)-h(x)}(\varepsilon t) \geq_L \mathcal{T}_\wedge(\Phi_{0,x}(t), \Phi_{2x,x}(t))$$

for all $x \in X$ and $t > 0$. Hence, it follows that

$$\begin{aligned}\mu_{Jg(x)-Jh(x)}(L\varepsilon t) &= \mu_{\frac{1}{2}g(2x)-\frac{1}{2}h(2x)}\left(\frac{\alpha}{2}\varepsilon t\right)\\ &= \mu_{g(2x)-h(2x)}(\alpha\varepsilon t)\\ &\geq_L \mathcal{T}_\wedge\big(\Phi_{0,2x}(\alpha t), \Phi_{4x,2x}(\alpha t)\big)\\ &\geq_L \mathcal{T}_\wedge\big(\Phi_{0,x}(t), \Phi_{2x,x}(t)\big)\end{aligned}$$

for all $x \in X$ and $t > 0$. So $d(g,h) = \varepsilon$ implies that $d(Jg, Jh) \leq \frac{\alpha}{2}\varepsilon$. This means that

$$d(Jg, Jh) \leq \frac{\alpha}{2} d(g,h)$$

for all $g, h \in S$. It follows from (5.3.14) that

$$\mu_{h(x)-\frac{1}{2}h(2x)}\left(\frac{17}{66}t\right) \geq_L \mathcal{T}_\wedge\big(\Phi_{0,x}(t), \Phi_{2x,x}(t)\big)$$

for all $x \in X$ and $t > 0$. So $d(h, Jh) \leq \frac{17}{66}$.

Now, there exists a mapping $A : X \to Y$ satisfying the following:

(1) A is a fixed point of J, that is,

$$A(2x) = 2A(x) \tag{5.3.17}$$

for all $x \in X$. Since $h : X \to Y$ is odd, $A : X \to Y$ is an odd mapping. The mapping A is a unique fixed point of J in the set

$$M = \{g \in S : d(f, g) < \infty\}.$$

This implies that A is a unique mapping satisfying (5.3.17) such that there exists a $u \in (0, \infty)$ satisfying

$$\mu_{h(x)-A(x)}(ut) \geq_L \mathcal{T}_\wedge\big(\Phi_{0,x}(t), \Phi_{2x,x}(t)\big)$$

for all $x \in X$ and $t > 0$;

(2) $d(J^n h, A) \to 0$ as $n \to \infty$. This implies the equality

$$\lim_{n\to\infty} \frac{1}{2^n} h(2^n x) = A(x)$$

for all $x \in X$;

(3) $d(h, A) \leq \frac{1}{1-\frac{\alpha}{2}} d(h, Jh)$, which implies the inequality

$$d(h, A) \leq \frac{17}{66 - 33\alpha}.$$

This implies that the inequality (5.3.16) holds.

The rest of the proof is similar to the proof of Theorem 5.3.1. This completes the proof. □

Corollary 5.3.8 *Let $\theta \geq 0$ and p be a real number with $0 < p < 1$. Let X be a normed linear space with norm $\|\cdot\|$ and $(X, \mu, \mathcal{T}_\wedge)$ be an LRN-space in which $L = [0, 1]$ and $\mathcal{T}_\wedge = \min$. Let $f : X \to Y$ be an odd mapping satisfying* (5.3.10). *Then*

$$A(x) := \lim_{n\to\infty} \frac{1}{2^n}\left(f\left(2^{n+1}x\right) - 8f\left(2^n x\right)\right)$$

exists for all $x \in X$ and defines an additive mapping $A : X \to Y$ such that

$$\mu_{f(2x)-8f(x)-A(x)}(t) \geq \frac{33(2-2^p)t}{33(2-2^p)t + 17(1+2^p)\theta\|x\|^p}$$

for all $x \in X$ and $t > 0$.

Proof The proof follows from Theorem 5.3.7 by taking

$$\Phi_{x,y}(t) := \frac{t}{t + \theta(\|x\|^p + \|y\|^p)}$$

for all $x, y \in X$ and $t > 0$. Then we can choose $\alpha = 2^p$ and so we get the desired result. □

5.3.2 The Generalized Hyers–Ulam Stability of the Functional Equation (5.3.1): An Even Case

Using the fixed point method, we prove the generalized Hyers–Ulam stability of the functional equation $Df(x, y) = 0$ in complete RN-spaces: an even case.

Theorem 5.3.9 *Let X be a linear space, $(X, \mu, \mathcal{T}_\wedge)$ be an LRN-space and Φ be a mapping from X^2 to D_L^+ ($\Phi(x, y)$ is denoted by $\Phi_{x,y}$) such that, for some $0 < \alpha < 1/16$,*

$$\Phi_{x,y}(\alpha t) \geq_L \Phi_{2x,2y}(t)$$

for all $x, y \in X$ and $t > 0$. Let $f : X \to Y$ be an even mapping satisfying $f(0) = 0$ and (5.3.3). *Then*

$$Q(x) := \lim_{n\to\infty} 16^n f\left(\frac{x}{2^n}\right)$$

exists for all $x \in X$ and defines a quartic mapping $Q : X \to Y$ such that

$$\mu_{f(x)-Q(x),}(t) \geq_L \mathcal{T}_\wedge\left(\Phi_{0,x}\left(\frac{(22-352\alpha)}{13\alpha}t\right), \Phi_{x,x}\left(\frac{(22-352\alpha)}{13\alpha}t\right)\right) \quad (5.3.18)$$

for all $x \in X$ and $t > 0$.

Proof Letting $x = 0$ in (5.3.3), we get

$$\mu_{12f(3y)-70f(2y)+148f(y)}(t) \geq_L \Phi_{0,y}(t) \tag{5.3.19}$$

for all $y \in X$ and $t > 0$. Letting $x = y$ in (5.3.3), we get

$$\mu_{f(3y)-4f(2y)-17f(y)}(t) \geq_L \Phi_{y,y}(t) \tag{5.3.20}$$

for all $y \in X$ and $t > 0$. By (5.4.23) and (5.3.20), it follows that

$$\begin{aligned}
&\mu_{f(2y)-16f(y)}\left(\frac{1}{22}t + \frac{12}{22}t\right) \\
&\quad \geq_L \mathcal{T}_\wedge\left(\mu_{\frac{1}{22}(12f(3y)-70f(2y)+148f(y))}\left(\frac{1}{22}t\right),\right. \\
&\qquad\qquad \left.\mu_{\frac{12}{22}(f(3y)-4f(2y)-17f(y))}\left(\frac{12}{22}t\right)\right) \\
&\quad \geq_L \mathcal{T}_\wedge\big(\Phi_{0,y}(t), \Phi_{y,y}(t)\big)
\end{aligned} \tag{5.3.21}$$

for all $y \in X$ and $t > 0$. Consider the set

$$S := \{g : X \to Y\}$$

and introduce the generalized metric on S:

$$\begin{aligned}
&d(g,h) \\
&\quad = \inf\big\{u \in \mathbb{R}_+ : N\big(g(x) - h(x), ut\big) \geq_L \mathcal{T}_\wedge\big(\Phi_{0,x}(t), \Phi_{x,x}(t)\big), \ \forall x \in X,\ t > 0\big\},
\end{aligned}$$

where, as usual, $\inf \emptyset = +\infty$. It is easy to show that (S, d) is complete (see the proof of Lemma 2.1 of [162]).

Now, we consider the linear mapping $J : S \to S$ such that

$$Jg(x) := 16g\left(\frac{x}{2}\right)$$

for all $x \in X$. Let $g, h \in S$ be given such that $d(g,h) = \varepsilon$. Then we have

$$\mu_{g(x)-h(x)}(\varepsilon t) \geq_L \mathcal{T}_\wedge\big(\Phi_{0,x}(t), \Phi_{x,x}(t)\big)$$

for all $x \in X$ and $t > 0$ and hence

$$\begin{aligned}
\mu_{Jg(x)-Jh(x)}(16\alpha\varepsilon t) &= \mu_{16g(\frac{x}{2})-16h(\frac{x}{2})}(16\alpha\varepsilon t) \\
&= \mu_{g(\frac{x}{2})-h(\frac{x}{2})}(\alpha\varepsilon t) \\
&\geq_L \mathcal{T}_\wedge\big(\Phi_{0,\frac{x}{2}}(\alpha t), \Phi_{\frac{x}{2},\frac{x}{2}}(\alpha t)\big) \\
&\geq_L \mathcal{T}_\wedge\big(\Phi_{0,x}(t), \Phi_{x,x}(t)\big)
\end{aligned}$$

for all $x \in X$ and $t > 0$. So $d(g, h) = \varepsilon$ implies that $d(Jg, Jh) \le 16\alpha\varepsilon$. This means that

$$d(Jg, Jh) \le 16\alpha d(g, h)$$

for all $g, h \in S$. It follows from (5.3.21) that

$$\mu_{f(x)-16f(\frac{x}{2})}\left(\frac{13}{22}\alpha t\right) \ge_L \mathcal{T}_\wedge\big(\Phi_{0,x}(t), \Phi_{x,x}(t)\big)$$

for all $x \in X$ and $t > 0$ and so $d(f, Jf) \le \frac{13\alpha}{22}$.

Now, there exists a mapping $Q : X \to Y$ satisfying the following:

(1) Q is a fixed point of J, that is,

$$Q\left(\frac{x}{2}\right) = \frac{1}{16}Q(x) \tag{5.3.22}$$

for all $x \in X$. Since $f : X \to Y$ is even, $Q : X \to Y$ is an even mapping. The mapping Q is a unique fixed point of J in the set

$$M = \big\{g \in S : d(f, g) < \infty\big\}.$$

This implies that Q is a unique mapping satisfying (5.3.22) such that there exists $u \in (0, \infty)$ satisfying

$$\mu_{f(x)-Q(x)}(ut) \ge_L \mathcal{T}_\wedge\big(\Phi_{0,x}(t), \Phi_{x,x}(t)\big)$$

for all $x \in X$ and $t > 0$;

(2) $d(J^n f, Q) \to 0$ as $n \to \infty$. This implies the equality

$$\lim_{n\to\infty} 16^n f\left(\frac{x}{2^n}\right) = Q(x)$$

for all $x \in X$;

(3) $d(f, Q) \le \frac{1}{1-16\alpha} d(f, Jf)$, which implies the inequality

$$d(f, Q) \le \frac{13\alpha}{22 - 352\alpha}.$$

This implies that the inequality (5.3.18) holds.

The rest of the proof is similar to the proof of Theorem 5.3.1. This completes the proof. □

Corollary 5.3.10 *Let $\theta \ge 0$ and p be a real number with $p > 4$. Let X be a normed linear space with norm $\|\cdot\|$ and $(X, \mu, \mathcal{T}_\wedge)$ be an LRN-space in which $L = [0, 1]$ and $\mathcal{T}_\wedge = \min$. Let $f : X \to Y$ be an even mapping satisfying $f(0) = 0$ and (5.3.10). Then*

$$Q(x) := \lim_{n\to\infty} 16^n f\left(\frac{x}{2^n}\right)$$

exists for all $x \in X$ and defines a quartic mapping $Q : X \to Y$ such that

$$\mu_{f(x)-Q(x)}(t) \geq \frac{11(2^p - 16)t}{11(2^p - 16)t + 13\theta \|x\|^p}$$

for all $x \in X$ and $t > 0$.

Proof The proof follows from Theorem 5.3.9 by taking

$$\Phi_{x,y}(t) := \frac{t}{t + \theta(\|x\|^p + \|y\|^p)}$$

for all $x, y \in X$ and $t > 0$. Then we can choose $\alpha = 2^{-p}$ and we get the desired result. □

Similarly, we can obtain the following and so we omit the proof.

Theorem 5.3.11 *Let X be a linear space, $(X, \mu, \mathcal{T}_\wedge)$ be an* LRN*-space and Φ be a mapping from X^2 to D_L^+ ($\Phi(x, y)$ is denoted by $\Phi_{x,y}$) such that, for some* $0 < \alpha < 16$,

$$\Phi_{x,y}(\alpha t) \geq_L \Phi_{\frac{x}{2}, \frac{y}{2}}(t)$$

for all $x, y \in X$ and $t > 0$. Let $f : X \to Y$ be an even mapping satisfying $f(0) = 0$ *and* (5.3.3). *Then*

$$Q(x) := \lim_{n\to\infty} \frac{1}{16^n} f(2^n x)$$

exists for all $x \in X$ and defines a quartic mapping $Q : X \to Y$ such that

$$\mu_{f(x)-Q(x)}(t) \geq_L \mathcal{T}_\wedge\left(\Phi_{0,x}\left(\frac{(352 - 22\alpha)}{13}t\right), \Phi_{x,x}\left(\frac{(352 - 22\alpha)}{13}t\right)\right)$$

for all $x \in X$ and $t > 0$.

Corollary 5.3.12 *Let $\theta \geq 0$ and p be a real number with* $0 < p < 4$. *Let X be a normed linear space with norm $\| \cdot \|$ and $(X, \mu, \mathcal{T}_\wedge)$ be an LRN-space in which $L = [0, 1]$ and $\mathcal{T}_\wedge = \min$. Let $f : X \to Y$ be an even mapping satisfying* $f(0) = 0$ *and* (5.3.10). *Then*

$$Q(x) := \lim_{n\to\infty} \frac{1}{16^n} f(2^n x)$$

exists for each $x \in X$ and defines a quartic mapping $Q : X \to Y$ such that

$$\mu_{f(x)-Q(x)}(t) \geq \frac{11(16 - 2^p)t}{11(16 - 2^p)t + 13\theta \|x\|^p}$$

for all $x \in X$ and $t > 0$.

Proof The proof follows from Theorem 5.3.11 by taking

$$\Phi_{x,y}(t) := \frac{t}{t + \theta(\|x\|^p + \|y\|^p)}$$

for all $x, y \in X$ and $t > 0$. Then we can choose $\alpha = 2^p$ and so we can get the desired result. □

5.4 AQCQ Functional Equations

In this section, we prove the generalized Hyers–Ulam stability of the following additive-quadratic-cubic-quartic functional equation:

$$\begin{aligned} f(x+2y) + f(x-2y) = 4f(x+y) + 4f(x-y) - 6f(x) + f(2y) \\ + f(-2y) - 4f(y) - 4f(-y) \end{aligned} \tag{5.4.1}$$

in complete random normed spaces.

One can easily show that an even mapping $f : X \to Y$ satisfies (5.4.1) if and only if it is a quadratic-quartic mapping, that is,

$$f(x+2y) + f(x-2y) = 4f(x+y) + 4f(x-y) - 6f(x) + 2f(2y) - 8f(y)$$

for all $x, y \in X$.

It was shown in Lemma 2.1 of [79] that $g(x) := f(2x) - 4f(x)$ and $h(x) := f(2x) - 16f(x)$ for all $x \in X$ are quartic and quadratic, respectively, and $f(x) = \frac{1}{12}g(x) - \frac{1}{12}h(x)$ for all $x \in X$.

For any mapping $f : \mathcal{X} \to \mathcal{Y}$, we define

$$\begin{aligned} Df(x,y) := f(x+2y) + f(x-2y) - 4f(x+y) - 4f(x-y) + 6f(x) \\ - f(2y) - f(-2y) + 4f(y) + 4f(-y) \end{aligned}$$

for all $x, y \in X$.

5.4.1 *The Generalized Hyers–Ulam Stability of the Functional Equation (5.4.1): An Odd Case via Fixed Point Method*

Using the fixed point method, we prove the generalized Hyers–Ulam stability of the functional equation $Df(x, y) = 0$ in random Banach spaces: an odd case.

Theorem 5.4.1 *Let X be a linear space, $(Y, \mu, T_\wedge)$ be a complete LRN-space and Φ be a mapping from X^2 to D_L^+ ($\Phi(x, y)$ is denoted by $\Phi_{x,y}$) such that, for some $0 < \alpha < 1/8$,*

$$\Phi_{2x,2y}(t) \leq_L \Phi_{x,y}(\alpha t) \tag{5.4.2}$$

for all $x, y \in X$ *and* $t > 0$. *Let* $f : X \to Y$ *be an odd mapping satisfying*

$$\mu_{Df(x,y)}(t) \geq_L \Phi_{x,y}(t) \tag{5.4.3}$$

for all $x, y \in X$ *and* $t > 0$. *Then*

$$C(x) := \lim_{n\to\infty} 8^n\left(f\left(\frac{x}{2^{n-1}}\right) - 2f\left(\frac{x}{2^n}\right)\right)$$

exists for all $x \in X$ *and defines a cubic mapping* $C : X \to Y$ *such that*

$$\mu_{f(2x)-2f(x)-C(x)}(t) \geq_L \mathcal{T}_\wedge\left(\Phi_{x,x}\left(\frac{1-8\alpha}{5\alpha}t\right), \Phi_{2x,x}\left(\frac{1-8\alpha}{5\alpha}t\right)\right) \tag{5.4.4}$$

for all $x \in X$ *and* $t > 0$.

Proof Letting $x = y$ in (5.4.3), we get

$$\mu_{f(3y)-4f(2y)+5f(y)}(t) \geq_L \Phi_{y,y}(t) \tag{5.4.5}$$

for all $y \in X$ and $t > 0$. Replacing x by $2y$ in (5.4.3), we get

$$\mu_{f(4y)-4f(3y)+6f(2y)-4f(y)}(t) \geq_L \Phi_{2y,y}(t) \tag{5.4.6}$$

for all $y \in X$ and $t > 0$. By (5.4.5) and (5.4.6), we have

$$\begin{aligned}
&\mu_{f(4y)-10f(2y)+16f(y)}(5t)\\
&\quad\geq_L \mathcal{T}_\wedge\big(\mu_{4(f(3y)-4f(2y)+5f(y))}(4t), \mu_{f(4y)-4f(3y)+6f(2y)-4f(y)}(t)\big)\\
&\quad= \mathcal{T}_\wedge\big(\mu_{f(3y)-4f(2y)+5f(y)}(t), \mu_{f(4y)-4f(3y)+6f(2y)-4f(y)}(t)\big)\\
&\quad\geq_L \mathcal{T}_\wedge\big(\Phi_{y,y}(t), \Phi_{2y,y}(t)\big)
\end{aligned} \tag{5.4.7}$$

for all $y \in X$ and $t > 0$. Letting $y := \frac{x}{2}$ and $g(x) := f(2x) - 2f(x)$ for all $x \in X$, we get

$$\mu_{g(x)-8g(\frac{x}{2})}(5t) \geq_L \mathcal{T}_\wedge\big(\Phi_{\frac{x}{2},\frac{x}{2}}(t), \Phi_{x,\frac{x}{2}}(t)\big) \tag{5.4.8}$$

for all $x \in X$ and $t > 0$. Consider the set

$$S := \{h : X \to Y : h(0) = 0\}$$

and introduce the generalized metric on S:

$$\begin{aligned}
&d(h,k)\\
&\quad= \inf\{u \in \mathbb{R}^+ : \mu_{h(x)-k(x)}(ut) \geq_L \mathcal{T}_\wedge(\Phi_{x,x}(t), \Phi_{2x,x}(t)), \forall x \in X, t > 0\},
\end{aligned}$$

where, as usual, $\inf \emptyset = +\infty$. It is easy to show that (S, d) is complete (see the proof of Lemma 2.1 of [161]).

Now, we consider the linear mapping $J : S \to S$ such that

$$Jh(x) := 8h\left(\frac{x}{2}\right)$$

for all $x \in X$ and we prove that J is a strictly contractive mapping with the Lipschitz constant 8α. Let $h, k \in S$ be given such that $d(h, k) < \varepsilon$. Then we have

$$\mu_{h(x)-k(x)}(\varepsilon t) \geq_L \mathcal{T}_\wedge\big(\Phi_{x,x}(t), \Phi_{2x,x}(t)\big)$$

for all $x \in X$ and $t > 0$ and hence

$$\begin{aligned}\mu_{Jh(x)-Jk(x)}(8\alpha\varepsilon t) &= \mu_{8h(\frac{x}{2})-8k(\frac{x}{2})}(8\alpha\varepsilon t)\\ &= \mu_{h(\frac{x}{2})-k(\frac{x}{2})}(\alpha\varepsilon t)\\ &\geq \mathcal{T}_\wedge\big(\Phi_{\frac{x}{2},\frac{x}{2}}(\alpha t), \Phi_{x,\frac{x}{2}}(\alpha t)\big)\\ &\geq_L \mathcal{T}_\wedge\big(\Phi_{x,x}(t), \Phi_{2x,x}(t)\big)\end{aligned}$$

for all $x \in X$ and $t > 0$. So $d(h, k) < \varepsilon$ implies that $d(Jh, Jk) \leq 8\alpha\varepsilon$. This means that

$$d(Jh, Jk) \leq 8\alpha d(h, k)$$

for all $h, k \in S$. It follows from (5.4.8) that

$$\mu_{g(x)-8g(\frac{x}{2})}(5\alpha t) \geq_L \mathcal{T}_\wedge\big(\Phi_{x,x}(t), \Phi_{2x,x}(t)\big)$$

for all $x \in X$ and $t > 0$ and so $d(g, Jg) \leq 5\alpha \leq \frac{5}{8}$.

Now, there exists a mapping $C : X \to Y$ satisfying the following:

(1) C is a fixed point of J, that is,

$$C\left(\frac{x}{2}\right) = \frac{1}{8}C(x) \tag{5.4.9}$$

for all $x \in X$. Since $g : X \to Y$ is odd, $C : X \to Y$ is an odd mapping. The mapping C is a unique fixed point of J in the set $M = \{h \in S : d(h, g) < \infty\}$. This implies that C is a unique mapping satisfying (5.4.9) such that there exists $u \in (0, \infty)$ satisfying

$$\mu_{g(x)-C(x)}(ut) \geq_L \mathcal{T}_\wedge\big(\Phi_{x,x}(t), \Phi_{2x,x}(t)\big)$$

for all $x \in X$ and $t > 0$;

(2) $d(J^n g, C) \to 0$ as $n \to \infty$. This implies the equality

$$\lim_{n\to\infty} 8^n g\left(\frac{x}{2^n}\right) = C(x)$$

for all $x \in X$;

(3) $d(h, C) \leq \frac{1}{1-8\alpha} d(h, Jh)$ with $h \in M$, which implies the inequality

$$d(g, C) \leq \frac{5\alpha}{1-8\alpha}$$

and so it follows that

$$\mu_{g(x)-C(x)}\left(\frac{5\alpha}{1-8\alpha}t\right) \geq_L \mathcal{T}_\wedge\big(\Phi_{x,x}(t), \Phi_{2x,x}(t)\big).$$

This implies that the inequality (5.4.4) holds.

From $Dg(x, y) = Df(2x, 2y) - 2Df(x, y)$ and (5.4.3), it follows that

$$\mu_{Df(2x,2y)}(t) \geq_L \Phi_{2x,2y}(t),$$
$$\mu_{-2Df(x,y)}(t) = \mu_{Df(x,y)}\left(\frac{t}{2}\right) \geq_L \Phi_{x,y}\left(\frac{t}{2}\right)$$

for all $x, y \in X$ and $t > 0$ and so, by (LRN3) and (5.4.2),

$$\begin{aligned}
\mu_{Dg(x,y)}(3t) &\geq_L \mathcal{T}_\wedge\big(\mu_{Df(2x,2y)}(t), \mu_{-2Df(x,y)}(2t)\big) \\
&\geq_L \mathcal{T}_\wedge\big(\Phi_{2x,2y}(t), \Phi_{x,y}(t)\big) \\
&\geq_L \Phi_{2x,2y}(t).
\end{aligned}$$

Also, it follows that

$$\begin{aligned}
\mu_{8^n Dg(\frac{x}{2^n}, \frac{y}{2^n})}(3t) &= \mu_{Dg(\frac{x}{2^n}, \frac{y}{2^n})}\left(3\frac{t}{8^n}\right) \\
&\geq \Phi_{\frac{x}{2^{n-1}}, \frac{y}{2^{n-1}}}\left(\frac{t}{8^n}\right) \\
&\geq_L \cdots \\
&\geq_L \Phi_{x,y}\left(\frac{1}{8}\frac{t}{(8\alpha)^{n-1}}\right)
\end{aligned}$$

for all $x, y \in X$, $t > 0$ and $n \geq 1$. Since $\lim_{n\to\infty} \Phi_{x,y}(\frac{3}{8}\frac{t}{(8\alpha)^{n-1}}) = 1$ for all $x, y \in X$ and $t > 0$, it follows that

$$\mu_{DC(x,y)}(3t) = 1_{\mathcal{L}}$$

for all $x, y \in X$ and $t > 0$. Thus, the mapping $C : X \to Y$ satisfies (5.4.1).

Now, we have

$$C(2x) - 8C(x) = \lim_{n\to\infty}\left[8^n g\left(\frac{x}{2^{n-1}}\right) - 8^{n+1} g\left(\frac{x}{2^n}\right)\right]$$

$$= 8 \lim_{n\to\infty}\left[8^{n-1} g\left(\frac{x}{2^{n-1}}\right) - 8^n g\left(\frac{x}{2^n}\right)\right]$$
$$= 0$$

for all $x \in X$. Since the mapping $x \to C(2x) - 2C(x)$ is cubic (see Lemma 2.2 of [66]), from the equality $C(2x) = 8C(x)$, we deduce that the mapping $C : X \to Y$ is cubic. This completes the proof. □

Corollary 5.4.2 *Let $\theta \geq 0$ and p be a real number with $p > 3$. Let X be a normed linear space with norm $\|\cdot\|$. Let $f : X \to Y$ be an odd mapping satisfying*

$$\mu_{Df(x,y)}(t) \geq \frac{t}{t + \theta(\|x\|^p + \|y\|^p)} \tag{5.4.10}$$

for all $x, y \in X$ and $t > 0$, where (X, μ, T_M) is a complete LRN-space, in which $L = [0, 1]$. Then

$$C(x) := \lim_{n\to\infty} 8^n\left(f\left(\frac{x}{2^{n-1}}\right) - 2f\left(\frac{x}{2^n}\right)\right)$$

exists for all $x \in X$ and defines a cubic mapping $C : X \to Y$ such that

$$\mu_{f(2x)-2f(x)-C(x)}(t) \geq \frac{(2^p - 8)t}{(2^p - 8)t + 5(1 + 2^p)\theta\|x\|^p}$$

for all $x \in X$ and $t > 0$.

Proof The proof follows from Theorem 5.4.1 by taking

$$\Phi_{x,y}(t) := \frac{t}{t + \theta(\|x\|^p + \|y\|^p)}$$

for all $x, y \in X$ and $t > 0$. Then we can choose $\alpha = 2^{-p}$ and, further,

$$\begin{aligned}
&\mu_{f(2x)-2f(x)-C(x)}(t)\\
&\geq \min\left(\frac{(1-2^{3-p})t}{(1-2^{3-p})t + 5.2^{-p}\theta(2\|x\|^p)},\right.\\
&\qquad\left.\frac{(1-2^{3-p})t}{(1-2^{3-p})t + 5.2^{-p}\theta(\|2x\|^p + \|x\|^p)}\right)\\
&\geq \frac{(1-2^{3-p})t}{(1-2^{3-p})t + 5.2^{-p}\theta(\|2x\|^p + \|x\|^p)}\\
&= \frac{(2^p - 8)t}{(2^p - 8)t + 5.(2^p + 1)\theta\|x\|^p},
\end{aligned}$$

which is the desired result. This completes the proof. □

Theorem 5.4.3 *Let X be a linear space, $(Y, \mu, \mathcal{T}_\wedge)$ be a complete LRN-space and Φ be a mapping from X^2 to D_L^+, where $\Phi(x, y)$ is denoted by $\Phi_{x,y}$, such that, for some $0 < \alpha < 8$,*

$$\Phi_{\frac{x}{2},\frac{y}{2}}(t) \leq_L \Phi_{x,y}(\alpha t) \tag{5.4.11}$$

for all $x, y \in X$ and $t > 0$. Let $f : X \to Y$ be an odd mapping satisfying (5.4.3). *Then*

$$C(x) := \lim_{n\to\infty} \frac{1}{8^n}\big(f(2^{n+1}x) - 2f(2^n x)\big)$$

exists for all $x \in X$ and defines a cubic mapping $C : X \to Y$ such that

$$\begin{aligned} &\mu_{f(2x)-2f(x)-C(x)}(t) \\ &\quad \geq_L \mathcal{T}_\wedge\left(\Phi_{x,x}\left(\frac{8-\alpha}{5}t\right), \Phi_{2x,x}\left(\frac{8-\alpha}{5}t\right)\right) \end{aligned} \tag{5.4.12}$$

for all $x \in X$ and $t > 0$.

Proof Let (S, d) be the generalized metric space defined in the proof of Theorem 5.4.1. Consider the linear mapping $J : S \to S$ such that

$$Jh(x) := \frac{1}{8}h(2x)$$

for all $x \in X$ and we prove that J is a strictly contractive mapping with the Lipschitz constant $\alpha/8$. Let $h, k \in S$ be given such that $d(h, k) < \varepsilon$. Then we have

$$\mu_{h(x)-k(x)}(\varepsilon t) \geq_L \mathcal{T}_\wedge\big(\Phi_{x,x}(t), \Phi_{2x,x}(t)\big)$$

for all $x \in X$ and $t > 0$ and hence

$$\begin{aligned} \mu_{Jh(x)-Jk(x)}\left(\frac{\alpha}{8}\varepsilon t\right) &= \mu_{\frac{1}{8}h(2x)-\frac{1}{8}k(2x)}\left(\frac{\alpha}{8}\varepsilon t\right) \\ &= \mu_{h(2x)-k(2x)}(\alpha\varepsilon t) \\ &\geq_L \mathcal{T}_\wedge\big(\Phi_{2x,2x}(\alpha t), \Phi_{4x,2x}(\alpha t)\big) \\ &\geq \mathcal{T}_\wedge\big(\Phi_{x,x}(t), \Phi_{2x,x}(t)\big) \end{aligned}$$

for all $x \in X$ and $t > 0$. So $d(h, k) < \varepsilon$ implies that $d(Jh, Jk) \leq \frac{\alpha}{8}\varepsilon$. This means that

$$d(Jh, Jk) \leq \frac{\alpha}{8}d(h, k)$$

for all $g, h \in S$. Letting $g(x) := f(2x) - 2f(x)$ for all $x \in X$, it follows from (5.4.8) that

$$\mu_{g(x)-\frac{1}{8}g(2x)}\left(\frac{5}{8}t\right) \geq_L \mathcal{T}_\wedge\big(\Phi_{x,x}(t), \Phi_{2x,x}(t)\big)$$

for all $x \in X$ and $t > 0$. So $d(g, Jg) \leq \frac{5}{8}$.

Now, there exists a mapping $C : X \to Y$ satisfying the following:

(1) C is a fixed point of J, that is,

$$C(2x) = 8C(x) \tag{5.4.13}$$

for all $x \in X$. Since $g : X \to Y$ is odd, $C : X \to Y$ is an odd mapping. The mapping C is a unique fixed point of J in the set

$$M = \{h \in S : d(h, g) < \infty\}.$$

This implies that C is a unique mapping satisfying (5.4.13) such that there exists $u \in (0, \infty)$ satisfying

$$\mu_{g(x)-C(x)}(ut) \geq_L \mathcal{T}_\wedge(\Phi_{x,x}(t), \Phi_{2x,x}(t))$$

for all $x \in X$ and $t > 0$;

(2) $d(J^n g, C) \to 0$ as $n \to \infty$. This implies the equality

$$\lim_{n\to\infty} \frac{1}{8^n} g(2^n x) = C(x)$$

for all $x \in X$;

(3) $d(h, C) \leq \frac{1}{1-\frac{\alpha}{8}} d(h, Jh)$ for every $h \in M$, which implies the inequality

$$d(g, C) \leq \frac{5}{8-\alpha},$$

from which it follows

$$\mu_{g(x)-C(x)}\left(\frac{5}{8-\alpha} t\right) \geq_L \mathcal{T}_\wedge(\Phi_{x,x}(t), \Phi_{2x,x}(t))$$

for all $x \in X$ and $t > 0$. This implies that the inequality (5.4.12) holds.

From

$$\mu_{Dg(x,y)}(3t) \geq_L \mathcal{T}_\wedge(\Phi_{2x,2y}(t), \Phi_{x,y}(t)) \geq_L \mathcal{T}_\wedge\left(\Phi_{2x,2y}(t), \Phi_{x,y}\left(\frac{t}{8}\right)\right)$$

and (5.4.11), we deduce that

$$\begin{aligned}
\mu_{8^{-n} Dg(2^n x, 2^n y)}(3t) &= \mu_{Dg(2^n x, 2^n y)}(3 \cdot 8^n t) \\
&\geq_L \Phi_{2^n x, 2^n y}(8^{n-1} t) \\
&\geq_L \quad \cdots \\
&\geq \Phi_{x,y}\left(\left(\frac{8}{\alpha}\right)^{n-1} \frac{t}{\alpha}\right)
\end{aligned}$$

for all $x, y \in X$, $t > 0$ and $n \geq 1$. Letting $n \to \infty$, we have

$$\mu_{DC(x,y)}(3t) = 1_{\mathcal{L}}$$

for all $x, y \in X$ and $t > 0$. Thus, the mapping $C : X \to Y$ satisfies (5.4.1). Now, we have

$$\begin{aligned} C(2x) - 8C(x) &= \lim_{n\to\infty}\left[\frac{1}{8^n}g(2^{n+1}x) - \frac{1}{8^{n-1}}g(2^n x)\right] \\ &= 8\lim_{n\to\infty}\left[\frac{1}{8^{n+1}}g(2^{n+1}x) - \frac{1}{8^n}g(2^n x)\right] \\ &= 0 \end{aligned}$$

for all $x \in X$. Since the mapping $x \to C(2x) - 2C(x)$ is cubic (see Lemma 2.2 of [66]), it follows from the equality $C(2x) = 8C(x)$ that the mapping $C : X \to Y$ is cubic. This completes the proof. □

Corollary 5.4.4 *Let $\theta \geq 0$ and p be a real number with $0 < p < 3$. Let X be a normed linear space with norm $\|\cdot\|$. Let $f : X \to Y$ be an odd mapping satisfying* (5.4.10). *Then*

$$C(x) := \lim_{n\to\infty}\frac{1}{8^n}\left(f(2^{n+1}x) - 2f(2^n x)\right)$$

exists for all $x \in X$ and defines a cubic mapping $C : X \to Y$ such that

$$\mu_{f(2x)-2f(x)-C(x)}(t) \geq \frac{(8-2^p)t}{(8-2^p)t + 5(1+2^p)\theta\|x\|^p}$$

for all $x \in X$ and $t > 0$, where (X, μ, T_M) is a complete LRN-space, in which $L =$ [0, 1].

Proof The proof follows from Theorem 5.4.3 by taking

$$\mu_{Df(x,y)}(t) \geq \frac{t}{t + \theta(\|x\|^p + \|y\|^p)}$$

for all $x, y \in X$ and $t > 0$. Then we can choose $\alpha = 2^p$ and so we can get the desired result. □

Theorem 5.4.5 *Let X be a linear space, $(Y, \mu, \mathcal{T}_\wedge)$ be a complete LRN-space and Φ be a mapping from X^2 to D_L^+, where $\Phi(x, y)$ is denoted by $\Phi_{x,y}$, such that, for some* $0 < \alpha < 1/2$,

$$\Phi_{2x,2y}(t) \leq_L \Phi_{x,y}(\alpha t) \tag{5.4.14}$$

for all $x, y \in X$ *and* $t > 0$. *Let* $f : X \to Y$ *be an odd mapping satisfying* (5.4.3). *Then*

$$A(x) := \lim_{n\to\infty} 2^n \left(f\left(\frac{x}{2^{n-1}}\right) - 8f\left(\frac{x}{2^n}\right)\right)$$

exists for all $x \in X$ *and defines an additive mapping* $A : X \to Y$ *such that*

$$\mu_{f(2x)-8f(x)-A(x)}(t) \geq_L \mathcal{T}_\wedge\left(\Phi_{x,x}\left(\frac{1-2\alpha}{5\alpha}t\right), \Phi_{2x,x}\left(\frac{1-2\alpha}{5\alpha}t\right)\right) \quad (5.4.15)$$

for all $x \in X$ *and* $t > 0$.

Proof Let (S, d) be the generalized metric space defined in the proof of Theorem 5.4.1. Letting $y := \frac{x}{2}$ and $g(x) := f(2x) - 8f(x)$ for all $x \in X$ in (5.4.7), we get

$$\mu_{g(x)-2g(\frac{x}{2})}(5t) \geq_L \mathcal{T}_\wedge\left(\Phi_{\frac{x}{2},\frac{x}{2}}(t), \Phi_{x,\frac{x}{2}}(t)\right) \quad (5.4.16)$$

for all $x \in X$ and all $t > 0$. Now, we consider the linear mapping $J : S \to S$ such that

$$Jh(x) := 2h\left(\frac{x}{2}\right)$$

for all $x \in X$. It is easy to see that J is a strictly contractive self-mapping on S with the Lipschitz constant 2α. It follows from (5.4.16) and (5.4.14) that

$$\mu_{g(x)-2g(\frac{x}{2})}(5\alpha t) \geq T_M\left(\Phi_{x,x}(t), \Phi_{2x,x}(t)\right)$$

for all $x \in X$ and all $t > 0$ and so $d(g, Jg) \leq 5\alpha < \infty$.

Now, there exists a mapping $A : X \to Y$ satisfying the following:

(1) A is a fixed point of J, that is,

$$A\left(\frac{x}{2}\right) = \frac{1}{2}A(x) \quad (5.4.17)$$

for all $x \in X$. Since $g : X \to Y$ is odd, $A : X \to Y$ is an odd mapping. The mapping A is a unique fixed point of J in the set $M = \{h \in S : d(h, g) < \infty\}$. This implies that A is a unique mapping satisfying (5.4.17) such that there exists $u \in (0, \infty)$ satisfying

$$\mu_{g(x)-A(x)}(ut) \geq_L \mathcal{T}_\wedge\left(\Phi_{x,x}(t), \Phi_{2x,x}(t)\right)$$

for all $x \in X$ and $t > 0$;

(2) $d(J^n g, A) \to 0$ as $n \to \infty$. This implies the equality

$$\lim_{n\to\infty} 2^n g\left(\frac{x}{2^n}\right) = A(x)$$

for all $x \in X$;

(3) $d(h, A) \le \frac{1}{1-2\alpha} d(h, Jh)$ for all $h \in M$, which implies the inequality

$$d(g, A) \le \frac{5\alpha}{1-2\alpha}.$$

This implies that the inequality (5.4.15) holds.

Since $\mu_{Dg(x,y)}(3t) \ge_L \Phi_{2x,2y}(t)$, it follows that

$$\begin{aligned} \mu_{2^n Dg(\frac{x}{2^n},\frac{y}{2^n})}(3t) &= \mu_{Dg(\frac{x}{2^n},\frac{y}{2^n})}\left(3\frac{t}{2^n}\right) \\ &\ge \Phi_{\frac{x}{2^{n-1}},\frac{y}{2^{n-1}}}\left(\frac{t}{2^n}\right) \\ &\ge_L \cdots \\ &\ge_L \Phi_{x,y}\left(\frac{1}{2}\frac{t}{(2\alpha)^{n-1}}\right) \end{aligned}$$

for all $x, y \in X$, $t > 0$ and $n \ge 1$. Letting $n \to \infty$, we have

$$\mu_{DA(x,y)}(3t) = 1_{\mathcal{L}}$$

for all $x, y \in X$ and $t > 0$. Thus the mapping $A : X \to Y$ satisfies (5.4.1). Now, we have

$$\begin{aligned} A(2x) - 2A(x) &= \lim_{n\to\infty}\left[2^n g\left(\frac{x}{2^{n-1}}\right) - 2^{n+1} g\left(\frac{x}{2^n}\right)\right] \\ &= 2\lim_{n\to\infty}\left[2^{n-1} g\left(\frac{x}{2^{n-1}}\right) - 2^n g\left(\frac{x}{2^n}\right)\right] \\ &= 0 \end{aligned}$$

for all $x \in X$. Since the mapping $x \to A(2x) - 8A(x)$ is additive (see Lemma 2.2 of [66]), from the equality $A(2x) = 2A(x)$, we deduce that the mapping $A : X \to Y$ is additive. This completes the proof. □

Corollary 5.4.6 *Let $\theta \ge 0$ and p be a real number with $p > 1$. Let X be a normed linear space with norm $\|\cdot\|$. Let $f : X \to Y$ be an odd mapping satisfying* (5.4.10). *Then*

$$A(x) := \lim_{n\to\infty} 2^n\left(f\left(\frac{x}{2^{n-1}}\right) - 8f\left(\frac{x}{2^n}\right)\right)$$

exists for all $x \in X$ and defines an additive mapping $A : X \to Y$ such that

$$\mu_{f(2x)-8f(x)-A(x)}(t) \ge \frac{(2^p-2)t}{(2^p-2)t + 5(1+2^p)\theta\|x\|^p}$$

for all $x \in X$ and $t > 0$, where (X, μ, T_M) is a complete LRN-space, in which $L = [0, 1]$.

Proof The proof follows from Theorem 5.4.5 by taking

$$\mu_{Df(x,y)}(t) \geq \frac{t}{t+\theta(\|x\|^p+\|y\|^p)}$$

for all $x, y \in X$ and $t > 0$. Then we can choose $\alpha = 2^{-p}$ and so we can get the desired result. □

Theorem 5.4.7 *Let X be a linear space, $(Y, \mu, \mathcal{T}_\wedge)$ be a complete LRN-space and Φ be a mapping from X^2 to D_L^+, where $\Phi(x, y)$ is denoted by $\Phi_{x,y}$, such that, for some $0 < \alpha < 2$,*

$$\Phi_{x,y}(\alpha t) \geq_L \Phi_{\frac{x}{2},\frac{y}{2}}(t) \tag{5.4.18}$$

for all $x, y \in X$ and $t > 0$. Let $f : X \to Y$ be an odd mapping satisfying (5.4.3). *Then*

$$A(x) := \lim_{n\to\infty} \frac{1}{2^n}\left(f\left(2^{n+1}x\right) - 8f\left(2^n x\right)\right)$$

exists for all $x \in X$ and defines an additive mapping $A : X \to Y$ such that

$$\mu_{f(2x)-8f(x)-A(x)}(t) \geq_L \mathcal{T}_\wedge\left(\Phi_{x,x}\left(\frac{2-\alpha}{5\alpha}t\right), \Phi_{2x,x}\left(\frac{2-\alpha}{5\alpha}t\right)\right) \tag{5.4.19}$$

for all $x \in X$ and $t > 0$.

Proof Let (S, d) be the generalized metric space defined in the proof of Theorem 5.4.1. Consider the linear mapping $J : S \to S$ such that

$$Jh(x) := \frac{1}{2}h(2x)$$

for all $x \in X$. It is easy to see that J is a strictly contractive self-mapping on S with the Lipschitz constant $\alpha/2$. Let $g(x) = f(2x) - 8f(x)$, from (5.4.16), it follows that

$$\mu_{g(x)-\frac{1}{2}g(2x)}\left(\frac{5}{2}t\right) \geq_L \mathcal{T}_\wedge\left(\Phi_{x,x}(t), \Phi_{2x,x}(t)\right)$$

for all $x \in X$ and $t > 0$ and so $d(g, Jg) \leq \frac{5}{2}$.

Now, there exists a mapping $A : X \to Y$ satisfying the following:

(1) A is a fixed point of J, that is,

$$A(2x) = 2A(x) \tag{5.4.20}$$

for all $x \in X$. Since $h : X \to Y$ is odd, $A : X \to Y$ is an odd mapping. The mapping A is a unique fixed point of J in the set $M = \{h \in S : d(h, g) < \infty\}$. This implies that A is a unique mapping satisfying (5.4.20) such that there exists $u \in (0, \infty)$ satisfying

$$\mu_{g(x)-A(x)}(ut) \geq_L \mathcal{T}_\wedge\big(\Phi_{x,x}(t), \Phi_{2x,x}(t)\big)$$

for all $x \in X$ and $t > 0$;

(2) $d(J^n g, A) \to 0$ as $n \to \infty$. This implies the equality

$$\lim_{n\to\infty} \frac{1}{2^n} g\big(2^n x\big) = A(x)$$

for all $x \in X$;

(3) $d(h, A) \leq \frac{1}{1-\frac{\alpha}{2}} d(h, Jh)$, which implies the inequality

$$d(g, A) \leq \frac{5}{2-\alpha}.$$

This implies that the inequality (5.4.19) holds.

Proceeding as in the proof of Theorem 5.4.5, we can show that the mapping $A : X \to Y$ satisfies (5.4.1). Now, we have

$$\begin{aligned} A(2x) - 2A(x) &= \lim_{n\to\infty} \left[\frac{1}{2^n} g\big(2^{n+1}x\big) - \frac{1}{2^{n-1}} g\big(2^n x\big)\right] \\ &= 2 \lim_{n\to\infty} \left[\frac{1}{2^{n+1}} g\big(2^{n+1}x\big) - \frac{1}{2^n} g\big(2^n x\big)\right] \\ &= 0 \end{aligned}$$

for all $x \in X$. Since the mapping $x \to A(2x) - 8A(x)$ is additive (see Lemma 2.2 of [66]), from the equality $A(2x) = 2A(x)$, we deduce that the mapping $A : X \to Y$ is additive. This completes the proof. □

Corollary 5.4.8 *Let $\theta \geq 0$ and p be a real number with $0 < p < 1$. Let X be a normed linear space with norm $\|\cdot\|$. Let $f : X \to Y$ be an odd mapping satisfying (3.3.12). Then*

$$A(x) := \lim_{n\to\infty} \frac{1}{2^n} \big(f\big(2^{n+1}x\big) - 8f\big(2^n x\big)\big)$$

exists for all $x \in X$ and defines an additive mapping $A : X \to Y$ such that

$$\mu_{f(2x)-8f(x)-A(x)}(t) \geq \frac{(2-2^p)t}{(2-2^p)t + 5(1+2^p)\theta\|x\|^p}$$

for all $x \in X$ and $t > 0$, where (X, μ, T_M) is a complete LRN-space, in which $L = [0, 1]$.

Proof The proof follows from Theorem 5.4.7 by taking

$$\mu_{Df(x,y)}(t) \geq \frac{t}{t + \theta(\|x\|^p + \|y\|^p)}$$

for all $x, y \in X$ and $t > 0$. Then we can choose $\alpha = 2^p$ and so we get the desired result. □

5.4.2 The Generalized Hyers–Ulam Stability of the Functional Equation (5.4.1): An Even Case via Fixed Point Method

Using the fixed point method, we prove the generalized Hyers–Ulam stability of the functional equation $Df(x, y) = 0$ in random Banach spaces: an even case.

Theorem 5.4.9 *Let X be a linear space, $(Y, \mu, \mathcal{T}_\wedge)$ be a complete LRN-space and Φ be a mapping from X^2 to D_L^+, where $\Phi(x, y)$ is denoted by $\Phi_{x,y}$, such that, for some $0 < \alpha < \frac{1}{16}$,*

$$\Phi_{x,y}(\alpha t) \geq_L \Phi_{2x,2y}(t) \tag{5.4.21}$$

for all $x \in X$ and $t > 0$. Let $f : X \to Y$ be an even mapping satisfying $f(0) = 0$ and (5.4.3). Then

$$Q(x) := \lim_{n\to\infty} 16^n \left(f\left(\frac{x}{2^{n-1}}\right) - 4f\left(\frac{x}{2^n}\right)\right)$$

exists for all $x \in X$ and defines a quartic mapping $Q : X \to Y$ such that

$$\begin{aligned}
&\mu_{f(2x)-4f(x)-Q(x)}(t) \\
&\quad \geq_L \mathcal{T}_\wedge\left(\Phi_{x,x}\left(\frac{1-16\alpha}{5\alpha}t\right), \Phi_{2x,x}\left(\frac{1-16\alpha}{5\alpha}t\right)\right)
\end{aligned} \tag{5.4.22}$$

for all $x \in X$ and $t > 0$.

Proof Letting $x = y$ in (5.4.3), we get

$$\mu_{f(3y)-6f(2y)+15f(y)}(t) \geq_L \Phi_{y,y}(t) \tag{5.4.23}$$

for all $y \in X$ and $t > 0$. Replacing x by $2y$ in (5.4.3), we get

$$\mu_{f(4y)-4f(3y)+4f(2y)+4f(y)}(t) \geq_L \Phi_{2y,y}(t) \tag{5.4.24}$$

for all $y \in X$ and $t > 0$. By (5.4.23) and (5.4.24), we have

$$\begin{aligned}
&\mu_{f(4x)-20f(2x)+64f(x)}(5t) \\
&\quad \geq_L \mathcal{T}_\wedge\big(\mu_{4(f(3x)-6f(2x)+15f(x))}(4t), \mu_{f(4x)-4f(3x)+4f(2x)+4f(x)}(t)\big) \\
&\quad \geq_L \mathcal{T}_\wedge\big(\Phi_{x,x}(t), \Phi_{2x,x}(t)\big)
\end{aligned} \tag{5.4.25}$$

for all $x \in X$ and $t > 0$. Letting $g(x) := f(2x) - 4f(x)$ for all $x \in X$, we get

$$\mu_{g(x)-16g(\frac{x}{2})}(5t) \geq_L \mathcal{T}_\wedge\big(\Phi_{\frac{x}{2},\frac{x}{2}}(t), \Phi_{x,\frac{x}{2}}(t)\big) \tag{5.4.26}$$

for all $x \in X$ and $t > 0$. Let (S, d) be the generalized metric space defined in the proof of Theorem 5.4.1.

Now, we consider the linear mapping $J : S \to S$ such that

$$Jh(x) := 16h\left(\frac{x}{2}\right)$$

for all $x \in X$. It is easy to see that J is a strictly contractive self-mapping on S with the Lipschitz constant 16α. It follows from (5.4.26) that

$$\mu_{g(x)-16g(\frac{x}{2})}(5\alpha t) \geq_L \mathcal{T}_\wedge\big(\Phi_{x,x}(t), \Phi_{2x,x}(t)\big)$$

for all $x \in X$ and all $t > 0$ and so

$$d(g, Jg) \leq 5\alpha \leq \frac{5}{16} < \infty.$$

Now, there exists a mapping $Q : X \to Y$ satisfying the following:

(1) Q is a fixed point of J, that is,

$$Q\left(\frac{x}{2}\right) = \frac{1}{16}Q(x) \tag{5.4.27}$$

for all $x \in X$. Since $g : X \to Y$ is even with $g(0) = 0$, $Q : X \to Y$ is an even mapping with $Q(0) = 0$. The mapping Q is a unique fixed point of J in the set $M = \{h \in S : d(h, g) < \infty\}$. This implies that Q is a unique mapping satisfying (5.4.27) such that there exists $u \in (0, \infty)$ satisfying

$$\mu_{g(x)-Q(x)}(ut) \geq_L \mathcal{T}_\wedge\big(\Phi_{x,x}(t), \Phi_{2x,x}(t)\big)$$

for all $x \in X$ and $t > 0$;

(2) $d(J^n g, Q) \to 0$ as $n \to \infty$. This implies the equality

$$\lim_{n\to\infty} 16^n g\left(\frac{x}{2^n}\right) = Q(x)$$

for all $x \in X$;

(3) $d(h, Q) \leq \frac{1}{1-16\alpha} d(h, Jh)$ for all $h \in M$, which implies the inequality

$$d(g, Q) \leq \frac{5\alpha}{1 - 16\alpha}.$$

This implies that the inequality (5.4.22) holds.

Proceeding as in the proof of Theorem 5.4.1, we obtain that the mapping $Q : X \to Y$ satisfies (5.4.1). Now, we have

$$Q(2x) - 16Q(x) = \lim_{n\to\infty}\left[16^n g\left(\frac{x}{2^{n-1}}\right) - 16^{n+1} g\left(\frac{x}{2^n}\right)\right]$$

$$= 16 \lim_{n\to\infty}\left[16^{n-1} g\left(\frac{x}{2^{n-1}}\right) - 16^n g\left(\frac{x}{2^n}\right)\right]$$
$$= 0$$

for all $x \in X$. Since the mapping $x \to Q(2x) - 4Q(x)$ is quartic (see Lemma 2.1 of [79]), we get that the mapping $Q : X \to Y$ is quartic. This completes the proof. □

Corollary 5.4.10 *Let $\theta \geq 0$ and p be a real number with $p > 4$. Let X be a normed vector space with norm $\|\cdot\|$. Let $f : X \to Y$ be an even mapping satisfying $f(0) = 0$ and* (5.4.10). *Then*

$$Q(x) := \lim_{n\to\infty} 16^n\left(f\left(\frac{x}{2^{n-1}}\right) - 4f\left(\frac{x}{2^n}\right)\right)$$

exists for all $x \in X$ and defines a quartic mapping $Q : X \to Y$ such that

$$\mu_{f(2x)-4f(x)-Q(x)}(t) \geq \frac{(2^p - 16)t}{(2^p - 16)t + 5(1 + 2^p)\theta\|x\|^p}$$

for all $x \in X$ and $t > 0$, where (X, μ, T_M) is a complete LRN-space, in which $L = [0, 1]$.

Proof The proof follows from Theorem 5.4.9 by taking

$$\mu_{Df(x,y)}(t) \geq \frac{t}{t + \theta(\|x\|^p + \|y\|^p)}$$

for all $x, y \in X$ and $t > 0$. Then we can choose $\alpha = 2^{-p}$ and so we get the desired result. □

Theorem 5.4.11 *Let X be a linear space, $(Y, \mu, \mathcal{T}_\wedge)$ be a complete LRN-space and Φ be a mapping from X^2 to D_L^+, where $\Phi(x, y)$ is denoted by $\Phi_{x,y}$, such that, for some $0 < \alpha < 16$,*

$$\Phi_{x,y}(\alpha t) \geq \Phi_{\frac{x}{2},\frac{y}{2}}(t) \tag{5.4.28}$$

for all $x \in X$ and $t > 0$. Let $f : X \to Y$ be an even mapping satisfying $f(0) = 0$ and (5.4.3). *Then*

$$Q(x) := \lim_{n\to\infty} \frac{1}{16^n}\left(f(2^{n+1}x) - 4f(2^n x)\right)$$

exists for all $x \in X$ and defines a quartic mapping $Q : X \to Y$ such that

$$\mu_{f(2x)-4f(x)-Q(x)}(t) \geq_L \mathcal{T}_\wedge\left(\Phi_{x,x}\left(\frac{16-\alpha}{5}t\right), \Phi_{2x,x}\left(\frac{16-\alpha}{5}t\right)\right) \tag{5.4.29}$$

for all $x \in X$ and $t > 0$.

Proof In the generalized metric space (S, d) defined in the proof of Theorem 5.4.1, we consider the linear mapping $J : S \to S$ such that

$$Jh(x) := \frac{1}{16}h(2x)$$

for all $x \in X$. It is easy to see that J is a strictly contractive self-mapping on S with the Lipschitz constant $\frac{\alpha}{16}$. Letting $g(x) := f(2x) - 4f(x)$ for all $x \in X$, from (5.4.26), it follows that

$$\mu_{g(x)-\frac{1}{16}g(2x)}\left(\frac{5}{16}t\right) \geq_L \mathcal{T}_\wedge\big(\Phi_{x,x}(t), \Phi_{2x,x}(t)\big)$$

for all $x \in X$ and $t > 0$ and so $d(g, Jg) \leq \frac{5}{16}$.

Now, there exists a mapping $Q : X \to Y$ satisfying the following:

(1) Q is a fixed point of J, that is,

$$Q(2x) = 16Q(x) \tag{5.4.30}$$

for all $x \in X$. Since $g : X \to Y$ is even with $g(0) = 0$, $Q : X \to Y$ is an even mapping with $Q(0) = 0$. The mapping Q is a unique fixed point of J in the set $M = \{h \in S : d(h, g) < \infty\}$. This implies that Q is a unique mapping satisfying (5.4.30) such that there exists $u \in (0, \infty)$ satisfying

$$\mu_{g(x)-Q(x)}(ut) \geq_L \mathcal{T}_\wedge\big(\Phi_{x,x}(t), \Phi_{2x,x}(t)\big)$$

for all $x \in X$ and $t > 0$;

(2) $d(J^n g, Q) \to 0$ as $n \to \infty$. This implies the equality

$$\lim_{n\to\infty} \frac{1}{16^n} g(2^n x) = Q(x)$$

for all $x \in X$;

(3) $d(g, Q) \leq \frac{16}{16-\alpha} d(g, Jg)$ for all $h \in M$, which implies the inequality

$$d(g, Q) \leq \frac{5}{16-\alpha}.$$

This implies that the inequality (5.4.29) holds.

Proceeding as in the proof of Theorem 5.4.3, we know that the mapping $Q : X \to Y$ satisfies (5.4.1). Now, we have

$$\begin{aligned}
Q(2x) - 16Q(x) &= \lim_{n\to\infty}\left[\frac{1}{16^n} g\big(2^{n+1}x\big) - \frac{1}{16^{n-1}} g\big(2^n x\big)\right] \\
&= 16 \lim_{n\to\infty}\left[\frac{1}{16^{n+1}} g\big(2^{n+1}x\big) - \frac{1}{16^n} g\big(2^n x\big)\right] \\
&= 0
\end{aligned}$$

for all $x \in X$. Since the mapping $x \to Q(2x) - 4Q(x)$ is quartic (see Lemma 2.1 of [79]), it follows that the mapping $Q : X \to Y$ is quartic. This completes the proof. □

Corollary 5.4.12 *Let $\theta \geq 0$ and p be a real number with $0 < p < 4$. Let X be a normed linear space with norm $\|\cdot\|$. Let $f : X \to Y$ be an even mapping satisfying $f(0) = 0$ and (5.4.10). Then*

$$Q(x) := \lim_{n\to\infty} \frac{1}{16^n}\big(f(2^{n+1}x) - 4f(2^n x)\big)$$

exists for all $x \in X$ and defines a quartic mapping $Q : X \to Y$ such that

$$\mu_{f(2x)-4f(x)-Q(x)}(t) \geq \frac{(16-2^p)t}{(16-2^p)t + 5(1+2^p)\theta\|x\|^p}$$

for all $x \in X$ and $t > 0$, where (X, μ, T_M) is a complete LRN-space, in which $L = [0, 1]$.

Proof The proof follows from Theorem 5.4.11 by taking

$$\mu_{Df(x,y)}(t) \geq \frac{t}{t + \theta(\|x\|^p + \|y\|^p)}$$

for all $x, y \in X$ and $t > 0$. Then we can choose $\alpha = 2^p$ and so we get the desired result. □

Theorem 5.4.13 *Let X be a linear space, $(Y, \mu, \mathcal{T}_\wedge)$ be a complete LRN-space and Φ be a mapping from X^2 to D_L^+, where $\Phi(x, y)$ is denoted by $\Phi_{x,y}$, such that, for some $0 < \alpha < \frac{1}{4}$,*

$$\Phi_{x,y}(\alpha t) \geq_L \Phi_{2x,2y}(t) \tag{5.4.31}$$

for all $x \in X$ and $t > 0$. Let $f : X \to Y$ be an even mapping satisfying $f(0) = 0$ and (5.4.3). Then

$$T(x) := \lim_{n\to\infty} 4^n\left(f\left(\frac{x}{2^{n-1}}\right) - 16f\left(\frac{x}{2^n}\right)\right)$$

exists for all $x \in X$ and defines a quadratic mapping $T : X \to Y$ such that

$$\mu_{f(2x)-16f(x)-T(x)}(t) \geq_L \mathcal{T}_\wedge\left(\Phi_{x,x}\left(\frac{1-4\alpha}{5\alpha}t\right), \Phi_{2x,x}\left(\frac{1-4\alpha}{5\alpha}t\right)\right) \tag{5.4.32}$$

for all $x \in X$ and $t > 0$.

Proof Let (S, d) be the generalized metric space defined in the proof of Theorem 5.4.1. Letting $g(x) := f(2x) - 16f(x)$ for all $x \in X$ in (5.4.25), we get

$$\mu_{g(x)-4g(\frac{x}{2})}(5t) \geq_L \mathcal{T}_\wedge\big(\Phi_{\frac{x}{2},\frac{x}{2}}(t), \Phi_{x,\frac{x}{2}}(t)\big) \tag{5.4.33}$$

for all $x \in X$ and $t > 0$. It is easy to see that the linear mapping $J : S \to S$ is such that

$$Jh(x) := 4h\left(\frac{x}{2}\right)$$

for all $x \in X$ and so J is a strictly contractive self-mapping with the Lipschitz constant 4α. It follows from (5.4.33) that

$$\mu_{g(x)-4g(\frac{x}{2})}(5\alpha t) \geq_L \mathcal{T}_\wedge\big(\Phi_{x,x}(t), \Phi_{2x,x}(t)\big)$$

for all $x \in X$ and $t > 0$ and so $d(g, Jg) \leq 5\alpha < \infty$.

Now, there exists a mapping $T : X \to Y$ satisfying the following:

(1) T is a fixed point of J, that is,

$$T\left(\frac{x}{2}\right) = \frac{1}{4}T(x) \tag{5.4.34}$$

for all $x \in X$. Since $g : X \to Y$ is even with $g(0) = 0$, $T : X \to Y$ is an even mapping with $T(0) = 0$. The mapping T is a unique fixed point of J in the set $M = \{h \in S : d(h, g) < \infty\}$. This implies that T is a unique mapping satisfying (5.4.34) such that there exists $u \in (0, \infty)$ satisfying

$$\mu_{g(x)-T(x)}(ut) \geq_L \mathcal{T}_\wedge\big(\Phi_{x,x}(t), \Phi_{2x,x}(t)\big)$$

for all $x \in X$ and $t > 0$;

(2) $d(J^n g, T) \to 0$ as $n \to \infty$. This implies the equality

$$\lim_{n\to\infty} 4^n g\left(\frac{x}{2^n}\right) = T(x)$$

for all $x \in X$;

(3) $d(h, T) \leq \frac{1}{1-4\alpha} d(h, Jh)$ for all $h \in M$, which implies the inequality

$$d(g, T) \leq \frac{5\alpha}{1-4\alpha}.$$

This implies that the inequality (5.4.32) holds.

Proceeding as in the proof of Theorem 5.4.1, we show that the mapping $T : X \to Y$ satisfies (5.4.1). Now, we have

$$\begin{aligned} T(2x) - 4T(x) &= \lim_{n\to\infty}\left[4^n g\left(\frac{x}{2^{n-1}}\right) - 4^{n+1} g\left(\frac{x}{2^n}\right)\right] \\ &= 4\lim_{n\to\infty}\left[4^{n-1} g\left(\frac{x}{2^{n-1}}\right) - 4^n g\left(\frac{x}{2^n}\right)\right] \\ &= 0 \end{aligned}$$

for all $x \in X$. Since the mapping $x \to T(2x) - 16T(x)$ is quadratic (see Lemma 2.1 of [79]), we get that the mapping $T : X \to Y$ is quadratic. This completes the proof. □

Corollary 5.4.14 *Let $\theta \geq 0$ and p be a real number with $p > 2$. Let X be a normed vector space with norm $\|\cdot\|$. Let $f : X \to Y$ be an even mapping satisfying $f(0) = 0$ and (3.3.12). Then*

$$T(x) := \lim_{n\to\infty} 4^n \left(f\left(\frac{x}{2^{n-1}}\right) - 16 f\left(\frac{x}{2^n}\right)\right)$$

exists for all $x \in X$ and defines a quadratic mapping $T : X \to Y$ such that

$$\mu_{f(2x)-16f(x)-T(x)}(t) \geq \frac{(2^p - 4)t}{(2^p - 4)t + 5(1+2^p)\theta\|x\|^p}$$

for all $x \in X$ and $t > 0$.

Proof The proof follows from Theorem 5.4.13 by taking

$$\Phi_{x,y}(t) := \frac{t}{t + \theta(\|x\|^p + \|y\|^p)}$$

for all $x, y \in X$ and $t > 0$. Then we can choose $\alpha = 2^{-p}$ and so we get the desired result. □

Theorem 5.4.15 *Let X be a linear space, (Y, μ, T_M) be a complete RN-space and Φ be a mapping from X^2 to D^+, where $\Phi(x, y)$ is denoted by $\Phi_{x,y}$, such that, for some $0 < \alpha < 4$,*

$$\Phi_{x,y}(\alpha t) \geq \Phi_{\frac{x}{2},\frac{y}{2}}(t) \tag{5.4.35}$$

for all $x \in X$ and $t > 0$. Let $f : X \to Y$ be an even mapping satisfying $f(0) = 0$ and (5.4.3). Then

$$T(x) := \lim_{n\to\infty} \frac{1}{4^n}\left(f(2^{n+1}x) - 16f(2^n x)\right)$$

exists for all $x \in X$ and defines a quadratic mapping $T : X \to Y$ such that

$$\mu_{f(2x)-16f(x)-T(x)}(t) \geq T_M\left(\Phi_{x,x}\left(\frac{4-\alpha}{5}t\right), \Phi_{2x,x}\left(\frac{4-\alpha}{5}t\right)\right) \tag{5.4.36}$$

for all $x \in X$ and $t > 0$.

Proof It is easy to see that the linear mapping $J : S \to S$ is such that

$$Jh(x) := \frac{1}{4}h(2x)$$

for all $x \in X$ and so J is a strictly contractive self-mapping with the Lipschitz constant $\frac{\alpha}{4}$. Letting $g(x) := f(2x) - 16f(x)$ for all $x \in X$, it follows from (5.4.36) that

$$\mu_{g(x)-\frac{1}{4}g(2x)}\left(\frac{5}{4}t\right) \geq T_M\left(\Phi_{x,x}(t), \Phi_{2x,x}(t)\right)$$

for all $x \in X$ and $t > 0$ and so $d(g, Jg) \leq \frac{5}{4}$.

Now, there exists a mapping $T : X \to Y$ satisfying the following:

(1) T is a fixed point of J, that is,

$$T(2x) = 4T(x) \tag{5.4.37}$$

for all $x \in X$. Since $g : X \to Y$ is even with $g(0) = 0$, $T : X \to Y$ is an even mapping with $T(0) = 0$. The mapping T is a unique fixed point of J in the set $M = \{h \in S : d(h, g) < \infty\}$. This implies that T is a unique mapping satisfying (5.4.37) such that there exists $u \in (0, \infty)$ satisfying

$$\mu_{g(x)-T(x)}(ut) \geq T_M\left(\Phi_{x,x}(t), \Phi_{2x,x}(t)\right)$$

for all $x \in X$ and $t > 0$;

(2) $d(J^n g, T) \to 0$ as $n \to \infty$. This implies the equality

$$\lim_{n\to\infty} \frac{1}{4^n} g(2^n x) = T(x)$$

for all $x \in X$;

(3) $d(h, T) \leq \frac{1}{1-\frac{\alpha}{4}} d(h, Jh)$ for all $h \in M$, which implies the inequality

$$d(g, T) \leq \frac{5}{4 - \alpha}.$$

This implies that the inequality (5.4.36) holds.

We know that the mapping $Q : X \to Y$ satisfies (5.4.1). Now, we have

$$\begin{aligned} T(2x) - 4T(x) &= \lim_{n\to\infty}\left[\frac{1}{4^n} g(2^{n+1}x) - \frac{1}{4^{n-1}} g(2^n x)\right] \\ &= 4 \lim_{n\to\infty}\left[\frac{1}{4^{n+1}} g(2^{n+1}x) - \frac{1}{4^n} g(2^n x)\right] \\ &= 0 \end{aligned}$$

for all $x \in X$. Since the mapping $x \to T(2x) - 16T(x)$ is quadratic (see Lemma 2.1 of [79]), we get that the mapping $T : X \to Y$ is quadratic. This completes the proof. □

Corollary 5.4.16 *Let $\theta \geq 0$ and p be a real number with $0 < p < 2$. Let X be a normed vector space with norm $\|\cdot\|$. Let $f : X \to Y$ be an even mapping satisfying*

$f(0)=0$ and (5.4.10). Then

$$T(x) := \lim_{n\to\infty} \frac{1}{4^n}\left(f\left(2^{n+1}x\right) - 16f\left(2^n x\right)\right)$$

exists for all $x \in X$ and defines a quadratic mapping $T: X \to Y$ such that

$$\mu_{f(2x)-16f(x)-T(x)}(t) \geq \frac{(4-2^p)t}{(4-2^p)t + 5(1+2^p)\theta\|x\|^p}$$

for all $x \in X$ and $t > 0$, where (X, μ, T_M) is a complete LRN-space, in which $L = [0, 1]$.

Proof The proof follows from Theorem 5.4.13 by taking

$$\Phi_{x,y}(t) := \frac{t}{t + \theta(\|x\|^p + \|y\|^p)}$$

for all $x, y \in X$ and t. Then we can choose $\alpha = 2^p$ and so we get the desired result. □

Chapter 6
Stability of Function Equations in Non-Archimedean Random Spaces

Throughout this chapter, we assume that X is a vector space and Y is a complete non-Archimedean normed space.

6.1 Cubic Functional Equations

The functional equation

$$f(2x+y)+f(2x-y)=2f(x+y)+2f(x-y)+12f(x) \qquad (6.1.1)$$

is called the *cubic functional equation*. In particular, every solution of a cubic functional equation is called a *cubic mapping*.

Throughout this section, let $\mathcal{K}$ be a non-Archimedean field, X be a vector space over $\mathcal{K}$ and (Y, μ, T) be a non-Archimedean random Banach space over $\mathcal{K}$.

We investigate the stability of the cubic functional equation (6.1.1), where f is a mapping from X to Y and $f(0)=0$. It is well known that a mapping f satisfies the above functional equation if and only if it is cubic.

Now, we define a random approximately cubic mapping. Let Ψ be a distribution function on $X \times X \times [0, \infty)$ such that $\Psi(x, y, \cdot)$ is nondecreasing and

$$\Psi(cx, cx, t) \geq \Psi\left(x, x, \frac{t}{|c|}\right)$$

for all $x \in X$ and $c \neq 0$.

Definition 6.1.1 A mapping $f : X \to Y$ is said to be Ψ*-approximately cubic* if

$$\mu_{f(2x+y)+f(2x-y)-2f(x+y)-2f(x-y)-12f(x)}(t) \geq \Psi(x, y, t) \qquad (6.1.2)$$

for all $x, y \in X$ and $t > 0$.

In this section, we assume that $2 \neq 0$ in $\mathcal{K}$ (i.e., the characteristic of $\mathcal{K}$ is not 2).

Y.J. Cho et al., *Stability of Functional Equations in Random Normed Spaces*, Springer Optimization and Its Applications 86, DOI 10.1007/978-1-4614-8477-6_6,

Theorem 6.1.2 *Let $\mathcal{K}$ be a non-Archimedean field, X be a linear space over $\mathcal{K}$ and (Y, μ, T) be a non-Archimedean random Banach space over $\mathcal{K}$. Let $f : X \to Y$ be a Ψ-approximately cubic function. If $|2| \neq 1$ and, for some $\alpha \in \mathbb{R}$ with $\alpha > 0$ and $k \geq 2$ with $|2^k| < \alpha$,*

$$\Psi\left(2^{-k}x, 2^{-k}y, t\right) \geq \Psi(x, y, \alpha t) \tag{6.1.3}$$

and

$$\lim_{n\to\infty} \mathrm{T}_{j=n}^{\infty} M\left(x, \frac{\alpha^j t}{|2|^{kj}}\right) = 1 \tag{6.1.4}$$

for all $x, y \in X$ and $t > 0$, then there exists a unique cubic mapping $C : X \to Y$ such that

$$\mu_{f(x)-C(x)}(t) \geq \mathrm{T}_{i=1}^{\infty} M\left(x, \frac{\alpha^{i+1} t}{|2|^{ki}}\right) \tag{6.1.5}$$

for all $x \in X$ and $t > 0$, where

$$M(x, t) := T\left(\Psi(x, 0, t), \Psi(2x, 0, t), \ldots, \Psi\left(2^{k-1}x, 0, t\right)\right)$$

for all $x \in X$ and $t > 0$.

Proof First, we show, by induction on j, that, for all $x \in X$, $t > 0$ and $j \geq 1$,

$$\begin{aligned} &\mu_{f(2^j x)-8^j f(x)}(t) \\ &\quad \geq M_j(x, t) := T\left(\Psi(x, 0, t), \ldots, \Psi\left(2^{j-1}x, 0, t\right)\right). \end{aligned} \tag{6.1.6}$$

Putting $y = 0$ in (6.1.2), we obtain

$$\begin{aligned} \mu_{2f(2x)-16f(x)}(t) &\geq \Psi(x, 0, t), \\ \mu_{f(2x)-8f(x)}(t) &\geq \Psi(x, 0, 2t) \geq \Psi(x, 0, t) \end{aligned}$$

for all $x \in X$ and $t > 0$. This proves (6.1.6) for $j = 1$. Assume that (6.1.6) hold for some $j > 1$. Replacing y by 0 and x by $2^j x$ in (6.1.2), we get

$$\mu_{f(2^{j+1}x)-8f(2^j x)}(t) \geq \Psi\left(2^j x, 0, t\right)$$

for all $x \in X$ and $t > 0$. Since $|8| \leq 1$, we have

$$\begin{aligned} \mu_{f(2^{j+1}x)-8^{j+1}f(x)}(t) &\geq T\left(\mu_{f(2^{j+1}x)-8f(2^j x)}(t), \mu_{8f(2^j x)-8^{j+1}f(x)}(t)\right) \\ &= T\left(\mu_{f(2^{j+1}x)-8f(2^j x)}(t), \mu_{f(2^j x)-8^j f(x)}\left(\frac{t}{|8|}\right)\right) \\ &\geq T\left(\mu_{f(2^{j+1}x)-8f(2^j x)}(t), \mu_{f(2^j x)-8^j f(x)}(t)\right) \\ &\geq T\left(\Psi\left(2^j x, 0, t\right), M_j(x, t)\right) \end{aligned}$$

$$= M_{j+1}(x,t)$$

for all $x \in X$. Thus, (6.1.6) holds for all $j \geq 1$. In particular, we have

$$\mu_{f(2^k x)-8^k f(x)}(t) \geq M(x,t) \tag{6.1.7}$$

for all $x \in X$ and $t > 0$. Replacing x by $2^{-(kn+k)}x$ in (6.1.7) and using the inequality (6.1.3), we obtain

$$\begin{aligned}\mu_{f(\frac{x}{2^{kn}})-8^k f(\frac{x}{2^{kn+k}})}(t) &\geq M\left(\frac{x}{2^{kn+k}}, t\right)\\ &\geq M\left(x, \alpha^{n+1}t\right)\end{aligned} \tag{6.1.8}$$

for all $x \in X$, $t > 0$ and $n \geq 1$. Then we have

$$\begin{aligned}\mu_{(2^{3k})^n f(\frac{x}{(2^k)^n})-(2^{3k})^{n+1} f(\frac{x}{(2^k)^{n+1}})}(t) &\geq M\left(x, \frac{\alpha^{n+1}}{|(2^{3k})^n|}t\right)\\ &\geq M\left(x, \frac{\alpha^{n+1}}{|(2^k)^n|}t\right)\end{aligned} \tag{6.1.9}$$

for all $x \in X$, $t > 0$ and $n \geq 1$ and so

$$\begin{aligned}&\mu_{(2^{3k})^n f(\frac{x}{(2^k)^n})-(2^{3k})^{n+p} f(\frac{x}{(2^k)^{n+p}})}(t)\\ &\geq \mathrm{T}_{j=n}^{n+p}\left(\mu_{(2^{3k})^j f(\frac{x}{(2^k)^j})-(2^{3k})^{j+p} f(\frac{x}{(2^k)^{j+p}})}(t)\right)\\ &\geq \mathrm{T}_{j=n}^{n+p} M\left(x, \frac{\alpha^{j+1}}{|(2^k)^j|}t\right)\end{aligned}$$

for all $x \in X$, $t > 0$ and $n \geq 1$. Since $\lim_{n\to\infty} \mathrm{T}_{j=n}^{\infty} M(x, \frac{\alpha^{j+1}}{|(2^k)^j|}t) = 1$ for all $x \in X$ and $t > 0$, the sequence $\{(2^{3k})^n f(\frac{x}{(2^k)^n})\}$ is a Cauchy sequence in the non-Archimedean random Banach space (Y, μ, T). Hence, we can define a mapping $C : X \to Y$ such that

$$\lim_{n\to\infty} \mu_{(2^{3k})^n f(\frac{x}{(2^k)^n})-C(x)}(t) = 1 \tag{6.1.10}$$

for all $x \in X$ and $t > 0$. It follows that, for all $n \geq 1$, $x \in X$ and $t > 0$,

$$\begin{aligned}\mu_{f(x)-(2^{3k})^n f(\frac{x}{(2^k)^n})}(t) &= \mu_{\sum_{i=0}^{n-1}(2^{3k})^i f(\frac{x}{(2^k)^i})-(2^{3k})^{i+1} f(\frac{x}{(2^k)^{i+1}})}(t)\\ &\geq \mathrm{T}_{i=0}^{n-1}\left(\mu_{(2^{3k})^i f(\frac{x}{(2^k)^i})-(2^{3k})^{i+1} f(\frac{x}{(2^k)^{i+1}})}(t)\right)\\ &\geq \mathrm{T}_{i=0}^{n-1} M\left(x, \frac{\alpha^{i+1}t}{|2^{3k}|^i}\right)\end{aligned}$$

$$\geq \mathrm{T}_{i=0}^{n-1} M\left(x, \frac{\alpha^{i+1}t}{|2^k|^i}\right).$$

Therefore, we have

$$\begin{aligned}\mu_{f(x)-C(x)}(t) &\geq T\left(\mu_{f(x)-(2^{3k})^n f(\frac{x}{(2^k)^n})}(t), \mu_{(2^{3k})^n f(\frac{x}{(2^k)^n})-C(x)}(t)\right)\\ &\geq T\left(\mathrm{T}_{i=0}^{n-1} M\left(x, \frac{\alpha^{i+1}t}{|2^k|^i}\right), \mu_{(2^{3k})^n f(\frac{x}{(2^k)^n})-C(x)}(t)\right).\end{aligned}$$

By letting $n \to \infty$, we obtain

$$\mu_{f(x)-C(x)}(t) \geq \mathrm{T}_{i=1}^{\infty} M\left(x, \frac{\alpha^{i+1}t}{|2^k|^i}\right).$$

This proves (6.1.5). Since T is continuous, from a well-known result in probabilistic metric space (see, for example, [241, Chap. 12]), it follows that

$$\lim_{n\to\infty} \mu_{Df(x,y)}(t) = \mu_{C(2x+y)+C(2x-y)-2C(x+y)-2C(x-y)-12C(x)}(t)$$

for all $x, y \in X$ and $t > 0$, where

$$\begin{aligned}Df(x, y) &= (2^{3k})^n f(2^{-kn}(2x + y)) + (2^{3k})^n f(2^{-kn}(2x - y))\\ &\quad - 2(2^{3k})^n f(2^{-kn}(x + y)) - 2(2^{3k})^n f(2^{-kn}(x - y))\\ &\quad - 12(2^{3k})^n f(3^{-kn}x).\end{aligned}$$

On the other hand, replacing x, y by $2^{-kn}x, 2^{-kn}y$ in (6.1.2) and using (NA-RN2) and (6.1.3), we have

$$\begin{aligned}\mu_{Df(x,y)}(t) &\geq \Psi\left(2^{-kn}x, 2^{-kn}y, \frac{t}{|2^k|^n}\right)\\ &\geq \Psi\left(x, y, \frac{\alpha^n t}{|2^k|^n}\right)\end{aligned}$$

for all $x, y \in X$ and $t > 0$. Since $\lim_{n\to\infty} \Psi(x, y, \frac{\alpha^n t}{|2^k|^n}) = 1$, we show that C is a cubic mapping.

Finally, if $C' : X \to Y$ is another cubic mapping such that

$$\mu_{C'(x)-f(x)}(t) \geq M(x, t)$$

for all $x \in X$ and $t > 0$, then, for all $n \geq 1$, $x \in X$ and $t > 0$,

$$\mu_{C(x)-C'(x)}(t) \geq T\left(\mu_{C(x)-(2^{3k})^n f(\frac{x}{(2^k)^n})}(t), \mu_{(2^{3k})^n f(\frac{x}{(2^k)^n})-C'(x)}(t)\right).$$

Therefore, from (6.1.10), it follows that $C = C'$. This completes the proof. □

Corollary 6.1.3 *Let $\mathcal{K}$ be a non-Archimedean field, X be a linear space over $\mathcal{K}$ and (Y, μ, T) be a non-Archimedean random Banach space over $\mathcal{K}$ under the t-norm $T \in \mathcal{H}$. Let $f : X \to Y$ be a Ψ-approximately cubic function. If, for some $\alpha \in \mathbb{R}$ with $\alpha > 0$, $|2| \neq 1$ and $k \geq 2$ with $|2^k| < \alpha$,*

$$\Psi\left(2^{-k}x, 2^{-k}y, t\right) \geq \Psi(x, y, \alpha t)$$

for all $x \in X$ and $t > 0$, then there exists a unique cubic mapping $C : X \to Y$ such that

$$\mu_{f(x)-C(x)}(t) \geq \mathrm{T}_{i=1}^{\infty} M\left(x, \frac{\alpha^{i+1}t}{|2|^{ki}}\right)$$

for all $x \in X$ and $t > 0$, where

$$M(x, t) := T\left(\Psi(x, 0, t), \Psi(2x, 0, t), \ldots, \Psi\left(2^{k-1}x, 0, t\right)\right)$$

for all $x \in X$ and $t > 0$.

Proof Since

$$\lim_{n\to\infty} M\left(x, \frac{\alpha^j t}{|2|^{kj}}\right) = 1$$

for all $x \in X$ and $t > 0$ and T is of Hadžić type, then it follows that

$$\lim_{n\to\infty} \mathrm{T}_{j=n}^{\infty} M\left(x, \frac{\alpha^j t}{|2|^{kj}}\right) = 1$$

for all $x \in X$ and $t > 0$. Now, if we can apply Theorem 6.1.2, then we get the conclusion. □

6.2 Quartic Functional Equations

Let $\mathcal{K}$ be a non-Archimedean field, X be a vector space over $\mathcal{K}$ and (Y, μ, T) be a non-Archimedean random Banach space over $\mathcal{K}$. The functional equation

$$\begin{aligned} &f(2x+y) + f(2x-y) \\ &= 4f(x+y) + 4f(x-y) + 24f(x) - 6f(y) \end{aligned} \tag{6.2.1}$$

is called the *quartic functional equation* since the function $f(x) = cx^4$ is a solution.

We investigate the stability of the quartic functional equation (6.2.1), where f is a mapping from X to Y and $f(0) = 0$. It is well known that a function f satisfies the above functional equation if and only if it is quartic.

Next, we define a random approximately quartic mapping. Let Ψ be a distribution function on $X \times X \times [0, \infty)$ such that $\Psi(x, y, \cdot)$ is nondecreasing and

$$\Psi(cx, cx, t) \geq \Psi\left(x, x, \frac{t}{|c|}\right)$$

for all $x \in X$ and $c \neq 0$.

Definition 6.2.1 A mapping $f : X \to Y$ is said to be *Ψ-approximately quartic* if

$$\mu_{f(2x+y)+f(2x-y)-4f(x+y)-4f(x-y)-24f(x)+6f(y)}(t) \geq \Psi(x, y, t) \tag{6.2.2}$$

for all $x, y \in X$ and $t > 0$.

Theorem 6.2.2 *Let $\mathcal{K}$ be a non-Archimedean field, X be a linear space over $\mathcal{K}$ and (Y, μ, T) be a non-Archimedean random Banach space over $\mathcal{K}$. Let $f : X \to Y$ be a Ψ-approximately quartic function. If, for some $\alpha \in \mathbb{R}$ with $\alpha > 0$ and $k \geq 3$ with $|2^k| < \alpha$,*

$$\Psi\left(2^{-k}x, 2^{-k}y, t\right) \geq \Psi(x, y, \alpha t) \tag{6.2.3}$$

and

$$\lim_{n\to\infty} \mathrm{T}_{j=n}^{\infty} M\left(x, \frac{\alpha^j t}{|2|^{kj}}\right) = 1 \tag{6.2.4}$$

for all $x, y \in X$ and $t > 0$, then there exists a unique quartic mapping $Q : X \to Y$ such that

$$\mu_{f(x)-Q(x)}(t) \geq \mathrm{T}_{i=1}^{\infty} M\left(x, \frac{\alpha^{i+1} t}{|2|^{ki}}\right) \tag{6.2.5}$$

for all $x \in X$ and $t > 0$, where

$$M(x, t) := T\left(\Psi(x, 0, t), \Psi(2x, 0, t), \ldots, \Psi\left(2^{k-1}x, 0, t\right)\right)$$

for all $x \in X$ and $t > 0$.

Proof First, we show, by induction on j, that, for all $x \in X$, $t > 0$ and $j \geq 1$,

$$\begin{aligned}&\mu_{f(2^j x)-16^j f(x)}(t)\\ &\quad\geq M_j(x, t) := T\left(\Psi(x, 0, t), \ldots, \Psi\left(2^{j-1}x, 0, t\right)\right).\end{aligned} \tag{6.2.6}$$

Putting $y = 0$ in (6.2.2), we have

$$\mu_{2f(2x)-32f(x)}(t) \geq \Psi(x, 0, t)$$

and

$$\mu_{f(2x)-16f(x)}(t) \geq \Psi(x, 0, 2t) \geq \Psi(x, 0, t)$$

for all $x \in X$ and $t > 0$. This proves (6.2.6) for $j = 1$. Assume that (6.2.6) hold for some $j > 1$. Replacing y by 0 and x by $2^j x$ in (6.2.2), we get

$$\mu_{f(2^{j+1}x)-16f(2^jx)}(t) \geq \Psi(2^jx, 0, t)$$

for all $x \in X$ and $t > 0$. Since $|16| \leq 1$, it follows that

$$\begin{aligned}
\mu_{f(2^{j+1}x)-16^{j+1}f(x)}(t) &\geq T\big(\mu_{f(2^{j+1}x)-16f(2^jx)}(t), \mu_{16f(2^jx)-16^{j+1}f(x)}(t)\big) \\
&= T\left(\mu_{f(2^{j+1}x)-16f(2^jx)}(t), \mu_{f(2^jx)-16^jf(x)}\left(\frac{t}{|16|}\right)\right) \\
&\geq T\big(\mu_{f(2^{j+1}x)-16f(2^jx)}(t), \mu_{f(2^jx)-16^jf(x)}(t)\big) \\
&\geq T\big(\Psi(2^jx, 0, t), M_j(x, t)\big) \\
&= M_{j+1}(x, t)
\end{aligned}$$

for all $x \in X$ and $t > 0$ and so (6.2.6) holds for all $j \geq 1$. In particular, we have

$$\mu_{f(2^kx)-16^kf(x)}(t) \geq M(x, t) \tag{6.2.7}$$

for all $x \in X$ and $t > 0$. Replacing x by $2^{-(kn+k)}x$ in (6.2.7) and using the inequality (6.2.3), we obtain

$$\mu_{f(\frac{x}{2^{kn}})-16^kf(\frac{x}{2^{kn+k}})}(t) \geq M\left(\frac{x}{2^{kn+k}}, t\right) \geq M(x, \alpha^{n+1}t) \tag{6.2.8}$$

for all $x \in X$, $t > 0$ and $n \geq 0$. Then we have

$$\begin{aligned}
\mu_{(2^{4k})^n f(\frac{x}{(2^k)^n})-(2^{4k})^{n+1}f(\frac{x}{(2^k)^{n+1}})}(t) &\geq M\left(x, \frac{\alpha^{n+1}}{|(2^{4k})^n|}t\right) \\
&\geq M\left(x, \frac{\alpha^{n+1}}{|(2^k)^n|}t\right)
\end{aligned} \tag{6.2.9}$$

for all $x \in X$, $t > 0$ and $n \geq 0$ and so

$$\begin{aligned}
&\mu_{(2^{4k})^n f(\frac{x}{(2^k)^n})-(2^{4k})^{n+p}f(\frac{x}{(2^k)^{n+p}})}(t) \\
&\geq T_{j=n}^{n+p}\big(\mu_{(2^{4k})^j f(\frac{x}{(2^k)^j})-(2^{4k})^{j+p}f(\frac{x}{(2^k)^{j+p}})}(t)\big) \\
&\geq T_{j=n}^{n+p} M\left(x, \frac{\alpha^{j+1}}{|(2^k)^j|}t\right)
\end{aligned}$$

for all $x \in X$, $t > 0$ and $n \geq 0$. Since $\lim_{n\to\infty} \mathrm{T}_{j=n}^{\infty} M(x, \frac{\alpha^{j+1}}{|(2^k)^j|}t) = 1$ for all $x \in X$ and $t > 0$, it follows that $\{(2^{4k})^n f(\frac{x}{(2^k)^n})\}$ is a Cauchy sequence in the non-Archimedean random Banach space (Y, μ, T). Hence, we can define a mapping

$Q : X \to Y$ such that

$$\lim_{n\to\infty} \mu_{(2^{4k})^n f(\frac{x}{(2^k)^n}) - Q(x)}(t) = 1 \tag{6.2.10}$$

for all $x \in X$ and $t > 0$. It follows that, for all $n \geq 1$, $x \in X$ and $t > 0$,

$$\begin{aligned}
\mu_{f(x)-(2^{4k})^n f(\frac{x}{(2^k)^n})}(t) &= \mu_{\sum_{i=0}^{n-1} (2^{4k})^i f(\frac{x}{(2^k)^i}) - (2^{4k})^{i+1} f(\frac{x}{(2^k)^{i+1}})}(t) \\
&\geq \mathrm{T}_{i=0}^{n-1} \big(\mu_{(2^{4k})^i f(\frac{x}{(2^k)^i}) - (2^{4k})^{i+1} f(\frac{x}{(2^k)^{i+1}})}(t)\big) \\
&\geq \mathrm{T}_{i=0}^{n-1} M\left(x, \frac{\alpha^{i+1} t}{|2^k|^i}\right)
\end{aligned}$$

and so

$$\begin{aligned}
&\mu_{f(x)-Q(x)}(t) \\
&\quad \geq T\big(\mu_{f(x)-(2^{4k})^n f(\frac{x}{(2^k)^n})}(t), \mu_{(2^{4k})^n f(\frac{x}{(2^k)^n}) - Q(x)}(t)\big) \\
&\quad \geq T\left(\mathrm{T}_{i=0}^{n-1} M\left(x, \frac{\alpha^{i+1} t}{|2^k|^i}\right), \mu_{(2^{4k})^n f(\frac{x}{(2^k)^n}) - Q(x)}(t)\right).
\end{aligned} \tag{6.2.11}$$

Taking $n \to \infty$ in (6.2.11), we have

$$\mu_{f(x)-Q(x)}(t) \geq \mathrm{T}_{i=1}^{\infty} M\left(x, \frac{\alpha^{i+1} t}{|2^k|^i}\right),$$

which proves (6.2.5). Since T is continuous, from a well-known result in probabilistic metric space (see, for example, [241, Chap. 12]), it follows that

$$\begin{aligned}
&\lim_{n\to\infty} \mu_{Df(x,y)}(t) \\
&\quad = \mu_{Q(2x+y)+Q(2x-y)-4Q(x+y)-4Q(x-y)-24Q(x)+6Q(y)}(t)
\end{aligned}$$

for all $x, y \in X$ and $t > 0$, where

$$\begin{aligned}
Df(x,y) = {} & (2^{4k})^n f(2^{-kn}(2x+y)) + (2^{4k})^n f(2^{-kn}(2x-y)) \\
& - 4(2^{4k})^n f(2^{-kn}(x+y)) - 4(2^{4k})^n f(2^{-kn}(x-y)) \\
& - 24(2^{4k})^n f(2^{-kn}x) + 6(2^{4k})^n f(2^{-kn}(y)).
\end{aligned}$$

On the other hand, replacing x, y by $2^{-kn}x, 2^{-kn}y$ in (6.2.2) and (6.2.3), we get

$$\begin{aligned}
\mu_{Df(x,y)}(t) &\geq \Psi\left(2^{-kn}x, 2^{-kn}y, \frac{t}{|2^k|^n}\right) \\
&\geq \Psi\left(x, y, \frac{\alpha^n t}{|2^k|^n}\right)
\end{aligned}$$

for all $x, y \in X$ and $t > 0$. Since $\lim_{n\to\infty} \Psi(x, y, \frac{\alpha^n t}{|2^k|^n}) = 1$, we show that Q is a quartic mapping.

Finally, if $Q' : X \to Y$ is another quartic mapping such that

$$\mu_{Q'(x)-f(x)}(t) \geq M(x, t)$$

for all $x \in X$ and $t > 0$, then, for all $n \in \mathbb{N}$, $x \in X$ and $t > 0$,

$$\begin{aligned}&\mu_{Q(x)-Q'(x)}(t)\\ &\quad \geq T\big(\mu_{Q(x)-(2^{4k})^n f(\frac{x}{(2^k)^n})}(t), \mu_{(2^{4k})^n f(\frac{x}{(2^k)^n})-Q'(x)}(t)\big).\end{aligned}$$

Therefore, we conclude that $Q = Q'$. This completes the proof. □

Corollary 6.2.3 *Let $\mathcal{K}$ be a non-Archimedean field, X be a linear space over $\mathcal{K}$ and (Y, μ, T) be a non-Archimedean random Banach space over $\mathcal{K}$ under the t-norm $T \in \mathcal{H}$. Let $f : X \to Y$ be a Ψ-approximately quartic function. If, for some $\alpha \in \mathbb{R}$ with $\alpha > 0$ and $k \geq 2$ with $|2^k| < \alpha$,*

$$\Psi\big(2^{-k}x, 2^{-k}y, t\big) \geq \Psi(x, y, \alpha t)$$

for all $x, y \in X$ and $t > 0$, then there exists a unique quartic mapping $Q : X \to Y$ such that

$$\mu_{f(x)-Q(x)}(t) \geq \mathrm{T}_{i=1}^{\infty} M\left(x, \frac{\alpha^{i+1}t}{|2|^{ki}}\right)$$

for all $x \in X$ and $t > 0$, where

$$M(x, t) := T\big(\Psi(x, 0, t), \Psi(2x, 0, t), \ldots, \Psi\big(2^{k-1}x, 0, t\big)\big)$$

for all $x \in X$ and $t > 0$.

Proof Since

$$\lim_{n\to\infty} M\left(x, \frac{\alpha^j t}{|2|^{kj}}\right) = 1$$

for all $x \in X$ and $t > 0$ and T is of Hadžić type, it follows that

$$\lim_{n\to\infty} \mathrm{T}_{j=n}^{\infty} M\left(x, \frac{\alpha^j t}{|2|^{kj}}\right) = 1$$

for all $x \in X$ and $t > 0$. Now, if we can apply Theorem 6.2.2, then we can get the conclusion. □

6.3 Another Quartic Functional Equations

Let $\mathcal{K}$ be a non-Archimedean field, X be a vector space over $\mathcal{K}$ and (Y, μ, T) be a non-Archimedean random Banach space over $\mathcal{K}$.

We investigate the stability of the quartic functional equation

$$\begin{aligned} &16f(x+4y) + f(4x-y) \\ &= 306\left[9f\left(x+\frac{y}{3}\right) + f(x+2y)\right] \\ &\quad + 136f(x-y) - 1394f(x+y) + 425f(y) - 1530f(x) \end{aligned}$$

for all $x, y \in X$, where f is a mapping from X to Y and $f(0) = 0$.

Now, we define a random approximately quartic mapping. Let Ψ be a distribution function on $X \times X \times [0, \infty)$ such that $\Psi(x, y, \cdot)$ is symmetric, nondecreasing and

$$\Psi(cx, cx, t) \geq \Psi\left(x, x, \frac{t}{|c|}\right)$$

for all $x \in X$ and $c \neq 0$.

Definition 6.3.1 A mapping $f : X \to Y$ is said to be Ψ*-approximately quartic* if

$$\mu_{Df(x,y)}(t) \geq \Psi(x, y, t) \tag{6.3.1}$$

for all $x, y \in X$ and $t > 0$, where

$$\begin{aligned} Df(x,y) = 16f(x+4y) + f(4x-y) - 306\left[9f\left(x+\frac{y}{3}\right) + f(x+2y)\right] \\ - 136f(x-y) + 1394f(x+y) - 425f(y) + 1530f(x). \end{aligned}$$

In this section, we assume that $4 \neq 0$ in $\mathcal{K}$ (i.e., the characteristic of $\mathcal{K}$ is not 4).

Theorem 6.3.2 *Let $\mathcal{K}$ be a non-Archimedean field, X be a linear space over $\mathcal{K}$ and (Y, μ, T) be a non-Archimedean random Banach space over $\mathcal{K}$. Let $f : X \to Y$ be a Ψ-approximately quartic mapping. If, for some $\alpha \in \mathbb{R}$ with $\alpha > 0$ and $k > 3$ with $|4^k| < \alpha$,*

$$\Psi\left(4^{-k}x, 4^{-k}y, t\right) \geq \Psi(x, y, \alpha t) \tag{6.3.2}$$

and

$$\lim_{n\to\infty} \mathrm{T}_{j=n}^{\infty} M\left(x, \frac{\alpha^j t}{|4|^{kj}}\right) = 1 \tag{6.3.3}$$

for all $x \in X$ and $t > 0$, then there exists a unique quartic mapping $Q : X \to Y$ such that

$$\mu_{f(x)-Q(x)}(t) \geq \mathrm{T}_{i=1}^{\infty} M\left(x, \frac{\alpha^{i+1} t}{|4|^{ki}}\right) \tag{6.3.4}$$

for all $x \in X$ and $t > 0$, where

$$M(x,t) := T\big(\Psi(x,0,t), \Psi(4x,0,t), \dots, \Psi\big(4^{k-1}x,0,t\big)\big)$$

for all $x \in X$ and $t > 0$.

Proof First, we show, by induction on j, that, for all $x \in X$, $t > 0$ and $j \geq 1$,

$$\begin{aligned} &\mu_{f(4^j x)-256^j f(x)}(t) \\ &\quad \geq M_j(x,t) := T\big(\Psi(x,0,t), \dots, \Psi\big(4^{j-1}x,0,t\big)\big). \end{aligned} \tag{6.3.5}$$

Putting $y = 0$ in (6.3.1), we have

$$\mu_{f(4x)-256f(x)}(t) \geq \Psi(x,0,t)$$

for all $x \in X$ and $t > 0$. This proves (6.3.5) for $j = 1$. Assume that (6.3.5) holds for some $j \geq 1$. Replacing y by 0 and x by $4^j x$ in (6.3.1), respectively, we get

$$\mu_{f(4^{j+1}x)-256f(4^j x)}(t) \geq \Psi\big(4^j x,0,t\big)$$

for all $x \in X$ and $t > 0$. Since $|256| \leq 1$, we have

$$\begin{aligned} &\mu_{f(4^{j+1}x)-256^{j+1}f(x)}(t) \\ &\quad \geq T\big(\mu_{f(4^{j+1}x)-256f(4^j x)}(t), \mu_{256f(4^j x)-256^{j+1}f(x)}(t)\big) \\ &\quad = T\left(\mu_{f(4^{j+1}x)-256f(4^j x)}(t), \mu_{f(4^j x)-256^j f(x)}\left(\frac{t}{|256|}\right)\right) \\ &\quad \geq T\big(\mu_{f(4^{j+1}x)-256f(4^j x)}(t), \mu_{f(4^j x)-256^j f(x)}(t)\big) \\ &\quad \geq T\big(\Psi\big(4^j x,0,t\big), M_j(x,t)\big) \\ &\quad = M_{j+1}(x,t) \end{aligned}$$

for all $x \in X$ and $t > 0$. Thus, (6.3.5) holds for all $j \geq 1$. In particular, we have

$$\mu_{f(4^k x)-256^k f(x)}(t) \geq M(x,t) \tag{6.3.6}$$

for all $x \in X$ and $t > 0$. Replacing x by $4^{-(kn+k)}x$ in (6.3.6) and using the inequality (6.3.2), we have

$$\mu_{f(\frac{x}{4^{kn}})-256^k f(\frac{x}{4^{kn+k}})}(t) \geq M\left(\frac{x}{4^{kn+k}}, t\right) \geq M\big(x, \alpha^{n+1}t\big) \tag{6.3.7}$$

for all $x \in X$, $t > 0$ and $n \geq 0$. Then it follows that

$$\mu_{(4^{4k})^n f(\frac{x}{(4^k)^n})-(4^{4k})^{n+1} f(\frac{x}{(4^k)^{n+1}})}(t) \geq M\left(x, \frac{\alpha^{n+1}}{|(4^{4k})^n|}t\right)$$

for all $x \in X$, $t > 0$ and $n \geq 0$ and so

$$\begin{aligned}
&\mu_{(4^{4k})^n f(\frac{x}{(4^k)^n})-(4^{4k})^{n+p} f(\frac{x}{(4^k)^{n+p}})}(t) \\
&\geq T_{j=n}^{n+p}\big(\mu_{(4^{4k})^j f(\frac{x}{(4^k)^j})-(4^{4k})^{j+p} f(\frac{x}{(4^k)^{j+p}})}(t)\big) \\
&\geq T_{j=n}^{n+p} M\left(x, \frac{\alpha^{j+1}}{|(4^{4k})^j|}t\right) \\
&\geq T_{j=n}^{n+p} M\left(x, \frac{\alpha^{j+1}}{|(4^k)^j|}t\right)
\end{aligned}$$

for all $x \in X$, $t > 0$ and $n \geq 0$. Since $\lim_{n\to\infty} \mathrm{T}_{j=n}^{\infty} M(x, \frac{\alpha^{j+1}}{|(4^k)^j|}t) = 1$ for all $x \in X$ and $t > 0$, it follows that $\{(4^{4k})^n f(\frac{x}{(4^k)^n})\}$ is a Cauchy sequence in a non-Archimedean random Banach space (Y, μ, T). Hence, we can define a mapping $Q : X \to Y$ such that

$$\lim_{n\to\infty} \mu_{(4^{4k})^n f(\frac{x}{(4^k)^n})-Q(x)}(t) = 1 \tag{6.3.8}$$

for all $x \in X$ and $t > 0$. It follows that, for all $n \geq 1$, $x \in X$ and $t > 0$,

$$\begin{aligned}
\mu_{f(x)-(4^{4k})^n f(\frac{x}{(4^k)^n})}(t) &= \mu_{\sum_{i=0}^{n-1}(4^{4k})^i f(\frac{x}{(4^k)^i})-(4^{4k})^{i+1} f(\frac{x}{(4^k)^{i+1}})}(t) \\
&\geq \mathrm{T}_{i=0}^{n-1}\big(\mu_{(4^{4k})^i f(\frac{x}{(4^k)^i})-(4^{4k})^{i+1} f(\frac{x}{(4^k)^{i+1}})}(t)\big) \\
&\geq \mathrm{T}_{i=0}^{n-1} M\left(x, \frac{\alpha^{i+1}t}{|4^{4k}|^i}\right).
\end{aligned}$$

Therefore, we have

$$\begin{aligned}
\mu_{f(x)-Q(x)}(t) &\geq T\big(\mu_{f(x)-(4^{4k})^n f(\frac{x}{(4^k)^n})}(t), \mu_{(4^{4k})^n f(\frac{x}{(4^k)^n})-Q(x)}(t)\big) \\
&\geq T\left(\mathrm{T}_{i=0}^{n-1} M\left(x, \frac{\alpha^{i+1}t}{|4^{4k}|^i}\right), \mu_{(4^{4k})^n f(\frac{x}{(4^k)^n})-Q(x)}(t)\right).
\end{aligned}$$

By letting $n \to \infty$, we obtain

$$\mu_{f(x)-Q(x)}(t) \geq \mathrm{T}_{i=1}^{\infty} M\left(x, \frac{\alpha^{i+1}t}{|4^k|^i}\right),$$

which proves (6.3.4). Since T is continuous, from a well-known result in probabilistic metric space (see, for example, [241], Chap. 12), it follows that

$$\lim_{n\to\infty} \mu_{D_1 f(x,y)}(t) = \mu_{D_2 f(x,y)}(t)$$

for all $x, y \in X$ and $t > 0$, where

$$\begin{aligned} D_1 f(x,y) &= (4^k)^n 16 f(4^{-kn}(x+4y)) + (4^k)^n f(4^{-kn}(4x-y)) \\ &\quad - 306\left[(4^k)^n 9 f\left(4^{-kn}\left(x+\frac{y}{3}\right)\right) + (4^k)^n f(4^{-kn}(x+2y))\right] \\ &\quad - 136(4^k)^n f(4^{-kn}(x-y)) + 1394(4^k)^n f(4^{-kn}(x+y)) \\ &\quad - 425(4^k)^n f(4^{-kn}y) + 1530(4^k)^n f(4^{-kn}x) \end{aligned}$$

and

$$\begin{aligned} D_2 f(x,y) = 16Q(x+4y) + Q(4x-y) - 306\left[9Q\left(x+\frac{y}{3}\right) + Q(x+2y)\right] \\ - 136Q(x-y) + 1394Q(x+y) - 425Q(y) + 1530Q(x). \end{aligned}$$

On the other hand, replacing x, y by $4^{-kn}x, 4^{-kn}y$, respectively, in (6.3.1) and using (NA-RN2) and (6.3.2), we get

$$\begin{aligned} \mu_{D_1 f(x,y)}(t) &\geq \Psi\left(4^{-kn}x, 4^{-kn}y, \frac{t}{|4^k|^n}\right) \\ &\geq \Psi\left(x, y, \frac{\alpha^n t}{|4^k|^n}\right) \end{aligned}$$

for all $x, y \in X$ and all $t > 0$. Since $\lim_{n\to\infty} \Psi(x, y, \frac{\alpha^n t}{|4^k|^n}) = 1$, we show that Q is a quartic mapping.

Finally, if $Q' : X \to Y$ is another quartic mapping such that

$$\mu_{Q'(x)-f(x)}(t) \geq M(x,t)$$

for all $x \in X$ and $t > 0$, then we have

$$\mu_{Q(x)-Q'(x)}(t) \geq T\left(\mu_{Q(x)-(4^{4k})^n f(\frac{x}{(4^k)^n})}(t), \mu_{(4^{4k})^n f(\frac{x}{(4^k)^n})-Q'(x)}(t)\right)$$

for all $n \in N$, $x \in X$ and $t > 0$. Therefore, we conclude that $Q = Q'$. This completes the proof. □

Corollary 6.3.3 *Let $\mathcal{K}$ be a non-Archimedean field, X be a vector space over $\mathcal{K}$ and (Y, μ, T) be a non-Archimedean random Banach space over $\mathcal{K}$ under the t-norm $T \in \mathcal{H}$. Let $f : X \to Y$ be a Ψ-approximately quartic mapping. If, for some $\alpha \in \mathbb{R}$ with $\alpha > 0$ and $k > 3$ with $|4^k| < \alpha$,*

$$\Psi(4^{-k}x, 4^{-k}y, t) \geq \Psi(x, y, \alpha t)$$

for all $x \in X$ and $t > 0$, then there exists a unique quartic mapping $Q : X \to Y$ such that

$$\mu_{f(x)-Q(x)}(t) \geq \mathrm{T}_{i=1}^{\infty} M\left(x, \frac{\alpha^{i+1}t}{|4|^{ki}}\right)$$

for all $x \in X$ and $t > 0$, where

$$M(x,t) := T\big(\Psi(x,0,t), \Psi(4x,0,t), \ldots, \Psi(4^{k-1}x,0,t)\big)$$

for all $x \in X$ and $t > 0$.

Proof Since

$$\lim_{n\to\infty} M\left(x, \frac{\alpha^j t}{|4|^{kj}}\right) = 1$$

for all $x \in X$ and $t > 0$ and T is of Hadžić type, it follows that

$$\lim_{n\to\infty} \mathrm{T}_{j=n}^{\infty} M\left(x, \frac{\alpha^j t}{|4|^{kj}}\right) = 1$$

for all $x \in X$ and $t > 0$. Now, if we can apply Theorem 6.3.2, then we get the conclusion. □

Example 6.3.4 Let (X, μ, T_M) be a non-Archimedean random normed space in which

$$\mu_x(t) = \frac{t}{t + \|x\|}$$

for all $x \in X$ and $t > 0$ and (Y, μ, T_M) be a complete non-Archimedean random normed space. Define

$$\Psi(x,y,t) = \frac{t}{1+t}.$$

It is easy to see that (6.3.2) holds for $\alpha = 1$. Also, since $M(x,t) = \frac{t}{1+t}$, we have

$$\begin{aligned}\lim_{n\to\infty} \mathrm{T}_{M,j=n}^{\infty} M\left(x, \frac{\alpha^j t}{|4|^{kj}}\right) &= \lim_{n\to\infty}\left(\lim_{m\to\infty} \mathrm{T}_{M,j=n}^{m} M\left(x, \frac{t}{|4|^{kj}}\right)\right)\\ &= \lim_{n\to\infty}\lim_{m\to\infty}\left(\frac{t}{t+|4^k|^n}\right)\\ &= 1,\end{aligned}$$

for all $x \in X$ and $t > 0$. Let $f : X \to Y$ be a Ψ-approximately quartic mapping. Thus all the conditions of Theorem 6.3.2 hold and so there exists a unique quartic mapping $Q : X \to Y$ such that

$$\mu_{f(x)-Q(x)}(t) \geq \frac{t}{t+|4^k|}.$$

6.4 Mixed AQCQ Functional Equations

In this section, we prove the generalized Hyers–Ulam stability of the following additive-quadratic-cubic-quartic functional equation:

$$f(x+2y)+f(x-2y)=4f(x+y)+4f(x-y)-6f(x)+f(2y)$$
$$+f(-2y)-4f(y)-4f(-y) \tag{6.4.1}$$

in various complete random normed spaces.

6.4.1 The Generalized Hyers–Ulam Stability of the Quartic Functional Equations (6.4.1) in Non-Archimedean RN-Spaces: An Odd Case

Let $\mathcal{K}$ be a non-Archimedean field, X be a vector space over $\mathcal{K}$ and let (Y, μ, T) be a non-Archimedean random Banach space over $\mathcal{K}$.

Now, we define a random approximately AQCQ mapping. Let Ψ be a distribution function on $\mathcal{X} \times \mathcal{X} \times [0, \infty)$ such that $\Psi(x, y, \cdot)$ is nondecreasing and

$$\Psi(cx, cx, t) \geq \Psi\left(x, x, \frac{t}{|c|}\right)$$

for all $x \in X$ and $c \neq 0$.

Definition 6.4.1 A mapping $f : X \to Y$ is said to be *Ψ-approximately AQCQ* if

$$\mu_{Df(x,y)}(t) \geq \Psi(x, y, t) \tag{6.4.2}$$

for all $x, y \in X$ and $t > 0$.

In this section, we assume that $2 \neq 0$ in $\mathcal{K}$ (i.e., the characteristic of $\mathcal{K}$ is not 2).

First, we prove the generalized Hyers–Ulam stability of the functional equation $Df(x, y) = 0$ in non-Archimedean random spaces: an odd case.

Theorem 6.4.2 *Let $\mathcal{K}$ be a non-Archimedean field, X be a vector space over $\mathcal{K}$ and (Y, μ, T) be a non-Archimedean random Banach space over $\mathcal{K}$. Let $f : X \to Y$ be an odd mapping and Ψ-approximately AQCQ mapping. If, for some $\alpha \in \mathbb{R}$ with $\alpha > 0$ and $k > 3$ with $|2^k| < \alpha$,*

$$\Psi(2^{-k}x, 2^{-k}y, t) \geq \Psi(x, y, \alpha t) \tag{6.4.3}$$

and

$$\lim_{n\to\infty} \mathrm{T}_{j=n}^{\infty} M\left(2x, \frac{\alpha^j t}{|8|^{kj}}\right) = 1 \tag{6.4.4}$$

for all $x, y \in X$ and $t > 0$, then there exists a unique cubic mapping $C : X \to Y$ such that

$$\mu_{f(x)-2f(\frac{x}{2})-C(\frac{x}{2})}(t) \geq \mathrm{T}_{i=1}^{\infty} M\left(x, \frac{\alpha^{i+1}t}{|8|^{ki}}\right) \tag{6.4.5}$$

for all $x \in X$ and $t > 0$, where

$$M(x,t) := T^{k-1}\left[\Psi\left(\frac{x}{2}, \frac{x}{2}, \frac{t}{|4|}\right), \Psi\left(x, \frac{x}{2}, t\right), \ldots, \right.$$
$$\left. \Psi\left(\frac{2^{k-1}x}{2}, \frac{2^{k-1}x}{2}, \frac{t}{|4|}\right), \Psi\left(2^{k-1}x, \frac{2^{k-1}x}{2}, t\right)\right].$$

Proof Letting $x = y$ in (6.4.2), we get

$$\mu_{f(3y)-4f(2y)+5f(y)}(t) \geq \Psi(y, y, t) \tag{6.4.6}$$

for all $y \in X$ and $t > 0$. Replacing x by $2y$ in (6.4.2), we get

$$\mu_{f(4y)-4f(3y)+6f(2y)-4f(y)}(t) \geq \Psi(2y, y, t) \tag{6.4.7}$$

for all $y \in X$ and $t > 0$. By (6.4.6) and (6.4.7), we have

$$\begin{aligned}
&\mu_{f(4y)-10f(2y)+16f(y)}(t) \\
&\geq T\left(\mu_{4(f(3y)-4f(2y)+5f(y))}(t), \mu_{f(4y)-4f(3y)+6f(2y)-4f(y)}(t)\right) \\
&= T\left(\mu_{f(3y)-4f(2y)+5f(y)}\left(\frac{t}{|4|}\right), \mu_{f(4y)-4f(3y)+6f(2y)-4f(y)}(t)\right) \\
&\geq T\left(\Psi\left(y, y, \frac{t}{|4|}\right), \Psi(2y, y, t)\right)
\end{aligned} \tag{6.4.8}$$

for all $y \in X$ and $t > 0$. Letting $y := \frac{x}{2}$ and $g(x) := f(2x) - 2f(x)$ for all $x \in X$ in (6.4.8), we get

$$\mu_{g(x)-8g(\frac{x}{2})}(t) \geq T\left(\Psi\left(\frac{x}{2}, \frac{x}{2}, \frac{t}{|4|}\right), \Psi\left(x, \frac{x}{2}, t\right)\right) \tag{6.4.9}$$

for all $x \in X$ and $t > 0$.

Now, we show, by induction on j, that, for all $x \in X$, $t > 0$ and $j \geq 1$,

$$\begin{aligned}
\mu_{g(2^{j-1}x)-8^j g(\frac{x}{2})}(t) &\geq M_j(x,t) \\
&:= T^{2j-1}\left[\Psi\left(\frac{x}{2}, \frac{x}{2}, \frac{t}{|4|}\right), \Psi\left(x, \frac{x}{2}, t\right), \ldots, \right. \\
&\qquad \left. \Psi\left(\frac{2^{j-1}x}{2}, \frac{2^{j-1}x}{2}, \frac{t}{|4|}\right), \Psi\left(2^{j-1}x, \frac{2^{j-1}x}{2}, t\right)\right].
\end{aligned} \tag{6.4.10}$$

Putting $j = 1$ in (6.4.10), we obtain (6.4.9). Assume that (6.4.10) holds for some $j \geq 1$. Replacing x by $2^j x$ in (6.4.9), we get

$$\mu_{g(2^j x)-8g(2^{j-1}x)}(t) \geq T\left(\Psi\left(2^{j-1}x, 2^{j-1}x, \frac{t}{|4|}\right), \Psi(2^j x, 2^{j-1}x, t)\right). \tag{6.4.11}$$

Since $|8| \leq 1$,

$$\begin{aligned}
&\mu_{g(2^j x)-8^{j+1}g(\frac{x}{2})}(t) \\
&\geq T\left(\mu_{g(2^j x)-8g(2^{j-1}x)}(t), \mu_{8g(2^{j-1}x)-8^{j+1}g(\frac{x}{2})}(t)\right) \\
&= T\left(\mu_{g(2^j x)-8g(2^{j-1}x)}(t), \mu_{g(2^{j-1}x)-8^j g(\frac{x}{2})}\left(\frac{t}{|8|}\right)\right) \\
&\geq T^2\left(\Psi\left(2^{j-1}x, 2^{j-1}x, \frac{t}{|4|}\right), \Psi(2^j x, 2^{j-1}x, t), M_j(x,t)\right) \\
&= M_{j+1}(x,t)
\end{aligned}$$

for all $x \in X$ and $t > 0$. Thus, (6.4.10) holds for all $j \geq 2$. In particular, we have

$$\mu_{g(2^{k-1}x)-8^k g(\frac{x}{2})}(t) \geq M(x,t) \tag{6.4.12}$$

for all $x \in X$ and $t > 0$. Replacing x by $2^{-(kn+k-1)}x$ in (6.4.12) and using the inequality (6.4.3), we have

$$\mu_{g(\frac{x}{2^{kn}})-8^k g(\frac{x}{2^{k(n+1)}})}(t) \geq M\left(\frac{2x}{2^{k(n+1)}}, t\right) \tag{6.4.13}$$

for all $x \in X$, $t > 0$ and $n \geq 0$. Then we have

$$\mu_{8^{kn}g(\frac{x}{2^{kn}})-8^{k(n+1)}g(\frac{x}{2^{k(n+1)}})}(t) \geq M\left(2x, \frac{\alpha^{n+1}}{|8^{k(n+1)}|}t\right) \tag{6.4.14}$$

for all $x \in X$, $t > 0$ and $n \geq 0$ and so

$$\begin{aligned}
\mu_{8^{kn}g(\frac{x}{2^{kn}})-8^{k(n+p)}g(\frac{x}{2^{k(n+p)}})}(t) &\geq T_{j=n}^{n+p}\left(\mu_{8^{kj}g(\frac{x}{2^{kj}})-8^{k(j+p)}g(\frac{x}{2^{k(j+p)}})}(t)\right) \\
&\geq T_{j=n}^{n+p} M\left(2x, \frac{\alpha^{j+1}}{|(8^k)^{j+1}|}t\right) \\
&\geq T_{j=n}^{n+p} M\left(2x, \frac{\alpha^{j+1}}{|(8^k)^{j+1}|}t\right)
\end{aligned}$$

for all $x \in X$, $t > 0$ and $n \geq 0$. Since

$$\lim_{n\to\infty} \mathrm{T}_{j=n}^{\infty} M\left(2x, \frac{\alpha^{j+1}}{|(8^k)^{j+1}|}t\right) = 1$$

for all $x \in X$ and $t > 0$. then it follows that $\{8^{kn} g(\frac{x}{2^{kn}})\}$ is a Cauchy sequence in the non-Archimedean random Banach space (Y, μ, T). Hence, we can define a mapping $C : X \to Y$ such that

$$\lim_{n\to\infty} \mu_{(8^{8k})^n g(\frac{x}{2^{kn}})-C(x)}(t) = 1 \tag{6.4.15}$$

for all $x \in X$ and $t > 0$. It follows that, for all $n \geq 1$, $x \in X$ and $t > 0$,

$$\begin{aligned}
\mu_{g(x)-(8^{8k})^n g(\frac{x}{2^{kn}})}(t) &= \mu_{\sum_{i=0}^{n-1}(8^{8k})^i g(\frac{x}{2^{ki}})-(8^{8k})^{i+1} g(\frac{x}{2^{k(i+1)}})}(t) \\
&\geq \mathrm{T}_{i=0}^{n-1}\left(\mu_{8^{8k})^i g(\frac{x}{2^{ki}})-(8^{8k})^{i+1} g(\frac{x}{2^{k(i+1)}})}(t)\right) \\
&\geq \mathrm{T}_{i=0}^{n-1} M\left(2x, \frac{\alpha^{i+1} t}{|8^k|^{i+1}}\right).
\end{aligned}$$

Therefore, we have

$$\begin{aligned}
\mu_{g(x)-C(x)}(t) &\geq T\left(\mu_{g(x)-(8^{8k})^n g(\frac{x}{2^{kn}})}(t), \mu_{(8^{8k})^n g(\frac{x}{2^{kn}})-C(x)}(t)\right) \\
&\geq T\left(\mathrm{T}_{i=0}^{n-1} M\left(2x, \frac{\alpha^{i+1} t}{|8^k|^{i+1}}\right), \mu_{(8^{8k})^n g(\frac{x}{2^{kn}})-C(x)}(t)\right).
\end{aligned}$$

By letting $n \to \infty$, we obtain

$$\mu_{g(x)-C(x)}(t) \geq \mathrm{T}_{i=1}^{\infty} M\left(2x, \frac{\alpha^{i+1} t}{|8^k|^{i+1}}\right)$$

and so

$$\mu_{f(x)-2f(\frac{x}{2})-C(\frac{x}{2})}(t) \geq \mathrm{T}_{i=1}^{\infty} M\left(x, \frac{\alpha^{i+1} t}{|8^k|^{i+1}}\right),$$

which proves (6.4.5). From $Dg(x, y) = Df(2x, 2y) - 2Df(x, y)$ and (6.4.2), it follows that

$$\mu_{Df(2x,2y)}(t) \geq \Psi(2x, 2y, t)$$

and

$$\mu_{-2Df(x,y)}(t) = \mu_{Df(x,y)}\left(\frac{t}{|2|}\right) \geq \mu_{Df(x,y)}(t) \geq \Psi(x, y, t)$$

and so, by (NA-RN3) and (6.4.2),

$$\begin{aligned}
\mu_{Dg(x,y)}(t) &\geq T\left(\mu_{Df(2x,2y)}(t), \mu_{-2Df(x,y)}(t)\right) \\
&\geq T\left(\Psi(2x, 2y, t), \Psi(x, y, t)\right) \\
&:= N(x, y, t).
\end{aligned}$$

Thus, it follows that

$$\begin{aligned}\mu_{8^{kn}Dg(\frac{x}{2^{kn}},\frac{y}{2^{kn}})}(t) &= \mu_{Dg(\frac{x}{2^{kn}},\frac{y}{2^{kn}})}\left(\frac{t}{|8|^{kn}}\right)\\ &\geq N\left(\frac{x}{2^{kn}},\frac{y}{2^{kn}},\frac{t}{|8|^{kn}}\right)\\ &\geq \cdots\\ &\geq N\left(x,y,\frac{\alpha^{n-1}t}{|8|^{k(n-1)}}\right)\end{aligned}$$

for all $x, y \in X$, $t > 0$ and $n \geq 1$. Since

$$\lim_{n\to\infty} N\left(x,y,\frac{\alpha^{n-1}t}{|8|^{k(n-1)}}\right) = 1$$

for all $x, y \in X$ and $t > 0$, then, it follows that

$$\mu_{DC(x,y)}(t) = 1$$

for all $x, y \in X$ and $t > 0$. Thus, the mapping $C : X \to Y$ satisfies (6.4.1).

Now, we have

$$\begin{aligned}C(2x) - 8C(x) &= \lim_{n\to\infty}\left[8^n g\left(\frac{x}{2^{n-1}}\right) - 8^{n+1} g\left(\frac{x}{2^n}\right)\right]\\ &= 8\lim_{n\to\infty}\left[8^{n-1} g\left(\frac{x}{2^{n-1}}\right) - 8^n g\left(\frac{x}{2^n}\right)\right]\\ &= 0\end{aligned}$$

for all $x \in X$. Since the mapping $x \to C(2x) - 2C(x)$ is cubic (see Lemma 2.2 of [66]), from the equality $C(2x) = 8C(x)$, we deduce that the mapping $C : X \to Y$ is cubic. This completes the proof. □

Corollary 6.4.3 *Let $\mathcal{K}$ be a non-Archimedean field, X be a vector space over $\mathcal{K}$ and (Y, μ, T) be a non-Archimedean random Banach space over $\mathcal{K}$ under the t-norm $T \in \mathcal{H}$. Let $f : X \to Y$ be an odd and Ψ-approximately AQCQ mapping. If, for some $\alpha \in \mathbb{R}$ with $\alpha > 0$ and $k > 3$ with $|2^k| < \alpha$,*

$$\Psi(2^{-k}x, 2^{-k}y, t) \geq \Psi(x, y, \alpha t)$$

for all $x, y \in X$ and $t > 0$, then there exists a unique cubic mapping $C : X \to Y$ such that

$$\mu_{f(x)-2f(\frac{x}{2})-C(\frac{x}{2})}(t) \geq \mathrm{T}_{i=1}^{\infty} M\left(x, \frac{\alpha^{i+1}t}{|8|^{ki}}\right)$$

for all $x \in X$ and $t > 0$.

Proof Since

$$\lim_{n\to\infty} M\left(x, \frac{\alpha^j t}{|8|^{kj}}\right) = 1$$

for all $x \in X$ and $t > 0$ and T is of Hadžić type, from Proposition 1.1.10, it follows that

$$\lim_{n\to\infty} \mathrm{T}_{j=n}^{\infty} M\left(x, \frac{\alpha^j t}{|8|^{kj}}\right) = 1$$

for all $x \in X$ and $t > 0$. Now, if we can apply Theorem 6.4.2, we can get the result. □

Example 6.4.4 Let (X, μ, T_M) be a non-Archimedean random normed space in which

$$\mu_x(t) = \frac{t}{t + \|x\|}$$

for all $x \in X$ and $t > 0$ and (Y, μ, T_M) be a complete non-Archimedean random normed space (see Example 2.4.3). Define

$$\Psi(x, y, t) = \frac{t}{1+t}.$$

It is easy to see that (6.4.3) holds for $\alpha = 1$. Also, since

$$M(x, t) = \frac{t}{1+t},$$

we have

$$\begin{aligned}
\lim_{n\to\infty} \mathrm{T}_{M,j=n}^{\infty} M\left(x, \frac{\alpha^j t}{|8|^{kj}}\right) &= \lim_{n\to\infty}\left(\lim_{m\to\infty} \mathrm{T}_{M,j=n}^{m} M\left(x, \frac{t}{|8|^{kj}}\right)\right) \\
&= \lim_{n\to\infty} \lim_{m\to\infty}\left(\frac{t}{t + |8^k|^n}\right) \\
&= 1
\end{aligned}$$

for all $x \in X$ and $t > 0$. Let $f : X \to Y$ be an odd and Ψ-approximately AQCQ mapping. Thus, all the conditions of Theorem 6.4.2 hold and so there exists a unique cubic mapping $C : X \to Y$ such that

$$\mu_{f(x)-2f(\frac{x}{2})-C(\frac{x}{2})}(t) \geq \frac{t}{t + |8^k|}$$

for all $x \in X$ and $t > 0$.

Theorem 6.4.5 *Let $\mathcal{K}$ be a non-Archimedean field, X be a vector space over $\mathcal{K}$ and (Y, μ, T) be a non-Archimedean random Banach space over $\mathcal{K}$. Let $f : X \to Y$ be* *an odd mapping and Ψ-approximately AQCQ mapping. If, for some $\alpha \in \mathbb{R}$ with $\alpha > 0$ and $k > 1$ with $|2^k| < \alpha$,*

$$\Psi\left(2^{-k}x, 2^{-k}y, t\right) \geq \Psi(x, y, \alpha t)$$

and

$$\lim_{n\to\infty} \mathrm{T}_{j=n}^{\infty} M\left(2x, \frac{\alpha^j t}{|2|^{kj}}\right) = 1 \tag{6.4.16}$$

for all $x, y \in X$ and $t > 0$, then there exists a unique additive mapping $A : X \to Y$ such that

$$\mu_{f(x)-8f(\frac{x}{2})-A(\frac{x}{2})}(t) \geq \mathrm{T}_{i=1}^{\infty} M\left(x, \frac{\alpha^{i+1}t}{|2|^{ki}}\right) \tag{6.4.17}$$

for all $x \in X$ and $t > 0$, where

$$M(x,t) := T^{k-1}\left[\Psi\left(\frac{x}{2}, \frac{x}{2}, \frac{t}{|4|}\right), \Psi\left(x, \frac{x}{2}, t\right), \dots, \right.$$
$$\left. \Psi\left(\frac{2^{k-1}x}{2}, \frac{2^{k-1}x}{2}, \frac{t}{|4|}\right), \Psi\left(2^{k-1}x, \frac{2^{k-1}x}{2}, t\right)\right]$$

for all $x \in X$ and $t > 0$.

Proof Letting $y := \frac{x}{2}$ and $g(x) := f(2x) - 8f(x)$ for all $x \in X$ in (6.4.8), we have

$$\mu_{g(x)-2g(\frac{x}{2})}(t) \geq T\left(\Psi\left(\frac{x}{2}, \frac{x}{2}, \frac{t}{|4|}\right), \Psi\left(x, \frac{x}{2}, t\right)\right)$$

for all $x \in X$ and $t > 0$. The rest of the proof is similar to the proof of Theorem 6.4.2. □

Corollary 6.4.6 *Let $\mathcal{K}$ be a non-Archimedean field, X be a vector space over $\mathcal{K}$ and (Y, μ, T) be a non-Archimedean random Banach space over $\mathcal{K}$ under the t-norm $T \in \mathcal{H}$. Let $f : X \to Y$ be an odd and Ψ-approximately AQCQ mapping. If, for some $\alpha \in \mathbb{R}$ with $\alpha > 0$ and $k > 1$ with $|2^k| < \alpha$,*

$$\Psi\left(2^{-k}x, 2^{-k}y, t\right) \geq \Psi(x, y, \alpha t)$$

for all $x, y \in X$ and $t > 0$, then there exists a unique additive mapping $A : X \to Y$ such that

$$\mu_{f(x)-8f(\frac{x}{2})-A(\frac{x}{2})}(t) \geq \mathrm{T}_{i=1}^{\infty} M\left(x, \frac{\alpha^{i+1}t}{|2|^{ki}}\right)$$

for all $x \in X$ and $t > 0$.

Proof Since

$$\lim_{n\to\infty} M\left(x, \frac{\alpha^j t}{|2|^{kj}}\right) = 1$$

for all $x \in X$ and $t > 0$ and T is of Hadžić type, from Proposition 1.1.10, it follows that

$$\lim_{n\to\infty} \mathrm{T}_{j=n}^{\infty} M\left(x, \frac{\alpha^j t}{|2|^{kj}}\right) = 1$$

for all $x \in X$ and $t > 0$. Now, if we can apply Theorem 6.4.5, then we get the result. □

Example 6.4.7 Let (X, μ, T_M) be a non-Archimedean random normed space in which

$$\mu_x(t) = \frac{t}{t + \|x\|}$$

for all $x \in X$ and $t > 0$ and (Y, μ, T_M) be a complete non-Archimedean random normed space (see Example 2.4.3). Define

$$\Psi(x, y, t) = \frac{t}{1+t}.$$

It is easy to see that (6.4.3) holds for $\alpha = 1$. Also, since

$$M(x, t) = \frac{t}{1+t},$$

we have

$$\begin{aligned}
\lim_{n\to\infty} \mathrm{T}_{M,j=n}^{\infty} M\left(x, \frac{\alpha^j t}{|2|^{kj}}\right) &= \lim_{n\to\infty}\left(\lim_{m\to\infty} \mathrm{T}_{M,j=n}^{m} M\left(x, \frac{t}{|2|^{kj}}\right)\right) \\
&= \lim_{n\to\infty}\lim_{m\to\infty}\left(\frac{t}{t+|2^k|^n}\right) \\
&= 1
\end{aligned}$$

for all $x \in X$ and $t > 0$. Let $f : X \to Y$ be an odd and Ψ-approximately AQCQ mapping. Thus, all the conditions of Theorem 6.4.2 hold and so there exists a unique additive mapping $A : X \to Y$ such that

$$\mu_{f(x)-8f(\frac{x}{2})-A(\frac{x}{2})}(t) \geq \frac{t}{t+|2^k|}$$

for all $x \in X$ and $t > 0$.

6.4.2 The Generalized Hyers–Ulam Stability of the Functional Equation (6.4.1) in Non-Archimedean RN-Spaces: An Even Case

Now, we prove the generalized Hyers–Ulam stability of the functional equation $Df(x, y) = 0$ in non-Archimedean Banach spaces: an even case.

Theorem 6.4.8 *Let $\mathcal{K}$ be a non-Archimedean field, X be a vector space over $\mathcal{K}$ and (Y, μ, T) be a non-Archimedean random Banach space over $\mathcal{K}$. Let $f : X \to Y$ be an even mapping, $f(0) = 0$ and Ψ-approximately AQCQ mapping. If, for some $\alpha \in \mathbb{R}$ with $\alpha > 0$ and $k > 4$ with $|2^k| < \alpha$,*

$$\Psi\left(2^{-k}x, 2^{-k}y, t\right) \geq \Psi(x, y, \alpha t)$$

and

$$\lim_{n\to\infty} \mathrm{T}_{j=n}^{\infty} M\left(2x, \frac{\alpha^j t}{|16|^{kj}}\right) = 1 \tag{6.4.18}$$

for all $x, y \in X$ and $t > 0$, then there exists a unique quartic mapping $Q : X \to Y$ such that

$$\mu_{f(x)-4f(\frac{x}{2})-Q(\frac{x}{2})}(t) \geq \mathrm{T}_{i=1}^{\infty} M\left(x, \frac{\alpha^{i+1}t}{|16|^{ki}}\right) \tag{6.4.19}$$

for all $x \in X$ and $t > 0$, where

$$M(x, t) := T^{k-1}\left[\Psi\left(\frac{x}{2}, \frac{x}{2}, \frac{t}{|4|}\right), \Psi\left(x, \frac{x}{2}, t\right), \dots, \Psi\left(\frac{2^{k-1}x}{2}, \frac{2^{k-1}x}{2}, \frac{t}{|4|}\right), \Psi\left(2^{k-1}x, \frac{2^{k-1}x}{2}, t\right)\right]$$

for all $x \in X$ and $t > 0$.

Proof Letting $x = y$ in (6.4.2), we get

$$\mu_{f(3y)-6f(2y)+15f(y)}(t) \geq \Psi(y, y, t) \tag{6.4.20}$$

for all $y \in X$ and $t > 0$. Replacing x by $2y$ in (6.4.2), we get

$$\mu_{f(4y)-4f(3y)+4f(2y)+4f(y)}(t) \geq \Psi(2y, y, t) \tag{6.4.21}$$

for all $y \in X$ and $t > 0$. By (6.4.20) and (6.4.21), we have

$$\begin{aligned}&\mu_{f(4y)-20f(2y)+64f(y)}(t)\\&\quad \geq T\left(\mu_{4(f(3y)-4f(2y)+5f(y))}(t), \mu_{f(4y)-4f(3y)+6f(2y)-4f(y)}(t)\right)\end{aligned}$$

$$= T\left(\mu_{f(3y)-4f(2y)+5f(y)}\left(\frac{t}{|4|}\right), \mu_{f(4y)-4f(3y)+6f(2y)-4f(y)}(t)\right)$$

$$\geq T\left(\Psi\left(y, y, \frac{t}{|4|}\right), \Psi(2y, y, t)\right) \tag{6.4.22}$$

for all $y \in X$ and $t > 0$. Letting $y := \frac{x}{2}$ and $g(x) := f(2x) - 4f(x)$ for all $x \in X$ in (6.4.22), respectively, we get

$$\mu_{g(x)-16g(\frac{x}{2})}(t) \geq T\left(\Psi\left(\frac{x}{2}, \frac{x}{2}, \frac{t}{|4|}\right), \Psi\left(x, \frac{x}{2}, t\right)\right) \tag{6.4.23}$$

for all $x \in X$ and $t > 0$. The rest of the proof is similar to the proof of Theorem 6.4.2. This completes the proof. □

Corollary 6.4.9 *Let $\mathcal{K}$ be a non-Archimedean field, X be a vector space over $\mathcal{K}$ and (Y, μ, T) be a non-Archimedean random Banach space over $\mathcal{K}$ under the t-norm $T \in \mathcal{H}$. Let $f : X \to Y$ be an even, $f(0) = 0$ and Ψ-approximately AQCQ mapping. If, for some $\alpha \in \mathbb{R}$ with $\alpha > 0$ and $k > 4$ with $|2^k| < \alpha$,*

$$\Psi\left(2^{-k}x, 2^{-k}y, t\right) \geq \Psi(x, y, \alpha t)$$

for all $x \in X$ and $t > 0$, then there exists a unique quartic mapping $Q : X \to Y$ such that

$$\mu_{f(x)-4f(\frac{x}{2})-Q(\frac{x}{2})}(t) \geq \mathrm{T}_{i=1}^{\infty} M\left(x, \frac{\alpha^{i+1}t}{|16|^{ki}}\right)$$

for all $x \in X$ and $t > 0$.

Proof Since

$$\lim_{n\to\infty} M\left(x, \frac{\alpha^j t}{|16|^{kj}}\right) = 1$$

for all $x \in X$ and $t > 0$ and T is of Hadžić type, from Proposition 1.1.10, it follows that

$$\lim_{n\to\infty} \mathrm{T}_{j=n}^{\infty} M\left(x, \frac{\alpha^j t}{|16|^{kj}}\right) = 1$$

for all $x \in X$ and $t > 0$. Now, if we apply Theorem 6.4.8, then we get the result. □

Example 6.4.10 Let (X, μ, T_M) be a non-Archimedean random normed space in which

$$\mu_x(t) = \frac{t}{t + \|x\|}$$

for all $x \in X$ and $t > 0$ and (Y, μ, T_M) be a complete non-Archimedean random normed space (see Example 2.4.3). Define

$$\Psi(x, y, t) = \frac{t}{1+t}.$$

It is easy to see that (6.4.3) holds for $\alpha = 1$. Also, since

$$M(x, t) = \frac{t}{1+t},$$

we have

$$\begin{aligned}\lim_{n\to\infty} \mathrm{T}_{M,j=n}^{\infty} M\left(x, \frac{\alpha^j t}{|16|^{kj}}\right) &= \lim_{n\to\infty}\left(\lim_{m\to\infty} \mathrm{T}_{M,j=n}^{m} M\left(x, \frac{t}{|16|^{kj}}\right)\right)\\ &= \lim_{n\to\infty}\lim_{m\to\infty}\left(\frac{t}{t+|16^k|^n}\right)\\ &= 1,\end{aligned}$$

for all $x \in X$ and $t > 0$. Let $f : X \to Y$ be an even, $f(0) = 0$ and Ψ-approximately AQCQ mapping. Thus all the conditions of Theorem 6.4.8 hold and so there exists a unique quartic mapping $Q : X \to Y$ such that

$$\mu_{f(x)-4f(\frac{x}{2})-Q(\frac{x}{2})}(t) \geq \frac{t}{t+|16^k|}$$

for all $x \in X$ and $t > 0$.

Theorem 6.4.11 *Let $\mathcal{K}$ be a non-Archimedean field, X be a vector space over $\mathcal{K}$ and (Y, μ, T) be a non-Archimedean random Banach space over $\mathcal{K}$. Let $f : X \to Y$ be an even mapping, $f(0) = 0$ and Ψ-approximately AQCQ mapping. If, for some $\alpha \in \mathbb{R}$ with $\alpha > 0$ and $k > 2$ with $|2^k| < \alpha$,*

$$\Psi\left(2^{-k}x, 2^{-k}y, t\right) \geq \Psi(x, y, \alpha t)$$

and

$$\lim_{n\to\infty} \mathrm{T}_{j=n}^{\infty} M\left(2x, \frac{\alpha^j t}{|4|^{kj}}\right) = 1 \tag{6.4.24}$$

for all $x \in X$ and $t > 0$, then there exists a unique quadratic mapping $Q : X \to Y$ such that

$$\mu_{f(x)-16f(\frac{x}{2})-Q(\frac{x}{2})}(t) \geq \mathrm{T}_{i=1}^{\infty} M\left(x, \frac{\alpha^{i+1} t}{|4|^{ki}}\right) \tag{6.4.25}$$

for all $x \in X$ and $t > 0$, where

$$M(x,t) := T^{k-1}\Big[\Psi\Big(\frac{x}{2},\frac{x}{2},\frac{t}{|4|}\Big),\Psi\Big(x,\frac{x}{2},t\Big),\dots,$$
$$\Psi\Big(\frac{2^{k-1}x}{2},\frac{2^{k-1}x}{2},\frac{t}{|4|}\Big),\Psi\Big(2^{k-1}x,\frac{2^{k-1}x}{2},t\Big)\Big]$$

for all $x \in X$ and $t > 0$.

Proof Letting $y := \frac{x}{2}$ and $g(x) := f(2x) - 16f(x)$ for all $x \in X$ in (6.4.22), respectively, we get

$$\mu_{g(x)-4g(\frac{x}{2})}(t) \geq T\Big(\Psi\Big(\frac{x}{2},\frac{x}{2},\frac{t}{|4|}\Big),\Psi\Big(x,\frac{x}{2},t\Big)\Big)$$

for all $x \in X$ and $t > 0$. The rest of the proof is similar to the proof of Theorem 6.4.8. □

Corollary 6.4.12 *Let $\mathcal{K}$ be a non-Archimedean field, X be a vector space over $\mathcal{K}$ and (Y,μ,T) be a non-Archimedean random Banach space over $\mathcal{K}$ under the t-norm $T \in \mathcal{H}$. Let $f : X \to Y$ be an even, $f(0) = 0$ and Ψ-approximately AQCQ mapping. If, for some $\alpha \in \mathbb{R}$ with $\alpha > 0$ and $k > 2$ with $|2^k| < \alpha$,*

$$\Psi(2^{-k}x, 2^{-k}y, t) \geq \Psi(x, y, \alpha t)$$

for all $x \in X$ and $t > 0$, then there exists a unique quadratic mapping $Q : X \to Y$ such that

$$\mu_{f(x)-16f(\frac{x}{2})-Q(\frac{x}{2})}(t) \geq \mathrm{T}_{i=1}^{\infty} M\Big(x, \frac{\alpha^{i+1}t}{|4|^{ki}}\Big)$$

for all $x \in X$ and $t > 0$.

Proof Since

$$\lim_{n\to\infty} M\Big(x, \frac{\alpha^j t}{|4|^{kj}}\Big) = 1$$

for all $x \in X$ and $t > 0$ and T is of Hadžić type, from Proposition 1.1.10, it follows that

$$\lim_{n\to\infty} \mathrm{T}_{j=n}^{\infty} M\Big(x, \frac{\alpha^j t}{|4|^{kj}}\Big) = 1$$

for all $x \in X$ and $t > 0$. Now, if we apply Theorem 6.4.11, then we obtain the result. □

Example 6.4.13 Let (X, μ, T_M) be a non-Archimedean random normed space in which

$$\mu_x(t) = \frac{t}{t + \|x\|}$$

for all $x \in X$ and $t > 0$ and (Y, μ, T_M) be a complete non-Archimedean random normed space (see Example 2.4.3). Define

$$\Psi(x, y, t) = \frac{t}{1+t}.$$

It is easy to see that (6.4.3) holds for $\alpha = 1$. Also, since

$$M(x, t) = \frac{t}{1+t},$$

we have

$$\begin{aligned}\lim_{n\to\infty} \mathrm{T}_{M,j=n}^{\infty} M\left(x, \frac{\alpha^j t}{|4|^{kj}}\right) &= \lim_{n\to\infty}\left(\lim_{m\to\infty} \mathrm{T}_{M,j=n}^{m} M\left(x, \frac{t}{|4|^{kj}}\right)\right) \\ &= \lim_{n\to\infty}\lim_{m\to\infty}\left(\frac{t}{t+|4^k|^n}\right) \\ &= 1\end{aligned}$$

for all $x \in X$ and $t > 0$. Let $f : X \to Y$ be an even, $f(0) = 0$ and Ψ-approximately AQCQ mapping. Thus, all the conditions of Theorem 6.4.11 hold and so there exists a unique quadratic mapping $Q : X \to Y$ such that

$$\mu_{f(x)-16f(\frac{x}{2})-Q(\frac{x}{2})}(t) \geq \frac{t}{t+|4^k|}$$

for all $x \in X$ and $t > 0$.

Chapter 7
Stability of Functional Equations Related to Inner Product Spaces

A square norm on an inner product space satisfies the parallelogram equality:

$$\|x+y\|^2+\|x-y\|^2=2\|x\|^2+2\|y\|^2.$$

From the above equation, we consider the following functional equation:

$$f(x+y)+f(x-y)=2f(x)+2f(y)$$

related to an inner product space. A square norm on an inner product space also satisfies the following:

$$\sum_{i,j=1}^{3}\|x_i-x_j\|^2=6\sum_{i=1}^{3}\|x_i\|^2$$

for all $x_1,x_2,x_3\in\mathbb{R}$ with $x_1+x_2+x_3=0$. From the above equality, we can define the following functional equation:

$$f(x-y)+f(2x+y)+f(x+2y)=3f(x)+3f(y)+3f(x+y),$$

which is called the *quadratic functional equation*. In fact, $f(x)=ax^2$ in $\mathbb{R}$ satisfies the quadratic functional equation.

In this chapter, we investigate the generalized Hyers–Ulam stability of functional equations in random normed spaces related to inner product spaces.

7.1 AQ Functional Equations

Th.M. Rassias [217] introduced the following equality:

$$\sum_{i,j=1}^{n}\|x_i-x_j\|^2=2n\sum_{i=1}^{n}\|x_i\|^2,\qquad \sum_{i=1}^{n}x_i=0,$$

Y.J. Cho et al., *Stability of Functional Equations in Random Normed Spaces*, Springer Optimization and Its Applications 86, DOI 10.1007/978-1-4614-8477-6_7,

for a fixed integer $n \geq 3$. For any mapping $f : X \to Y$, where X is a vector space and Y is a complete random normed space, we consider the following functional equation:

$$\sum_{i,j=1}^{n} f(x_i - x_j) = 2n \sum_{i=1}^{n} f(x_i) \tag{7.1.1}$$

for all $x_1, \ldots, x_n \in X$ with $\sum_{i=1}^{n} x_i = 0$.

In this section, we prove the generalized Hyers–Ulam stability of the functional equation (7.1.1) related to inner product spaces.

7.1.1 The Generalized Hyers–Ulam Stability of the Functional Equation (7.1.1): An Odd Case

Now, we investigate the functional equation (7.1.1) for an odd mapping in RN-spaces.

For any mapping $f : X \to Y$, we define

$$Df(x_1, \ldots, x_n) := \sum_{i,j=1}^{n} f(x_i - x_j) - 2n \sum_{i=1}^{n} f(x_i)$$

for all $x_1, \ldots, x_n \in X$ with $\sum_{i=1}^{n} x_i = 0$. For any odd mapping $f : X \to Y$, we note that, if f satisfies the following:

$$Df(x_1, x_2, \ldots, x_n) = 0$$

for all $x_1, \ldots, x_n \in X$ with $\sum_{i=1}^{n} x_i = 0$, then the mapping f is additive.

Now, we prove the generalized Hyers–Ulam stability of the functional equation (7.1.1) of an odd mapping in RN-spaces.

Theorem 7.1.1 *Let $f : X \to Y$ be an odd mapping for which there exists a $\rho : X^n \to D^+$, where $\rho(x_1, x_2, \ldots, x_n)$ is denoted by $\rho_{(x_1,x_2,\ldots,x_n)}$, such that*

$$\mu_{Df(x_1,x_2,\ldots,x_n)}(t) \geq \rho_{(x_1,x_2,\ldots,x_n)}(t) \tag{7.1.2}$$

for all $(x_1, x_2, \ldots, x_n) \in X^n$ and $t > 0$. If

$$T_{k=1}^{\infty} \rho_{(\frac{x}{2^{k+l}}, \frac{x}{2^{k+l}}, -\frac{x}{2^{k+l-1}}, 0, \ldots, 0)}\left(\frac{nt}{2^{2k+l-2}}\right) = 1 \tag{7.1.3}$$

and

$$\lim_{m\to\infty} \rho_{(\frac{x}{2^m}, \frac{y}{2^m}, -\frac{x+y}{2^m}, 0, \ldots, 0)}\left(\frac{nt}{2^{m-1}}\right) = 1 \tag{7.1.4}$$

for all $x, y \in X$, $t > 0$ and $l \geq 0$, then there exists a unique additive mapping $A : X \to Y$ such that

$$\mu_{f(x)-A(x)}(t) \geq T_{k=1}^{\infty} \rho_{(\frac{x}{2^k}, \frac{x}{2^k}, -\frac{x}{2^{k-1}}, 0, \ldots, 0)}\left(\frac{nt}{2^{2k-2}}\right) \tag{7.1.5}$$

for all $x \in X$ and $t > 0$.

Proof Putting $x_1 = x_2 = \frac{x}{2}$, $x_3 = -x$ and $x_4 = \cdots = x_n = 0$ in (7.1.2), respectively, we get

$$\mu_{2n(f(x)-2f(\frac{x}{2}))}(t) \geq \rho_{(\frac{x}{2}, \frac{x}{2}, -x, 0, \ldots, 0)}(t),$$

which is equivalent to the following:

$$\mu_{f(x)-2f(\frac{x}{2})}(t) \geq \rho_{(\frac{x}{2}, \frac{x}{2}, -x, 0, \ldots, 0)}(2nt)$$

for all $x \in X$ and $t > 0$. Replacing x and t by $\frac{x}{2^{k-1}}$ and $\frac{t}{2^{2k-1}}$, respectively, in the above inequality, we get

$$\mu_{2^{k-1} f(\frac{x}{2^{k-1}}) - 2^k f(\frac{x}{2^k})}\left(\frac{t}{2^k}\right) \geq \rho_{(\frac{x}{2^k}, \frac{x}{2^k}, -\frac{x}{2^{k-1}}, 0, \ldots, 0)}\left(\frac{nt}{2^{2k-2}}\right)$$

for all $x \in X$ and $t > 0$. Since $\mu_x(s) \leq \mu_x(t)$ for all s and t with $0 < s \leq t$, we obtain

$$\begin{aligned}
\mu_{f(x)-2^m f(\frac{x}{2^m})}(t) &= \mu_{\sum_{k=1}^{m} (2^{k-1} f(\frac{x}{2^{k-1}}) - 2^k f(\frac{x}{2^k}))}(t) \\
&\geq \mu_{\sum_{k=1}^{m} (2^{k-1} f(\frac{x}{2^{k-1}}) - 2^k f(\frac{x}{2^k}))}\left(\sum_{k=1}^{m} \frac{t}{2^k}\right) \\
&\geq T_{k=1}^{m} \rho_{(\frac{x}{2^k}, \frac{x}{2^k}, -\frac{x}{2^{k-1}}, 0, \ldots, 0)}\left(\frac{nt}{2^{2k-2}}\right).
\end{aligned}$$

Replacing x by $\frac{x}{2^l}$ in the above inequality, we get

$$\mu_{f(\frac{x}{2^l}) - 2^m f(\frac{x}{2^{m+l}})}(t) \geq T_{k=1}^{m} \rho_{(\frac{x}{2^{k+l}}, \frac{x}{2^{k+l}}, -\frac{x}{2^{k+l-1}}, 0, \ldots, 0)}\left(\frac{nt}{2^{2k-2}}\right),$$

which is equivalent to the following:

$$\begin{aligned}
&\mu_{2^l f(\frac{x}{2^l}) - 2^{m+l} f(\frac{x}{2^{m+l}})}(t) \\
&\quad \geq T_{k=1}^{m} \rho_{(\frac{x}{2^{k+l}}, \frac{x}{2^{k+l}}, -\frac{x}{2^{k+l-1}}, 0, \ldots, 0)}\left(\frac{nt}{2^{2k+l-2}}\right)
\end{aligned} \tag{7.1.6}$$

for all $x \in X$, $t > 0$ and $l \geq 0$. Since the right-hand side of the inequality (7.1.6) tends to 1 as $m \to \infty$ by (7.1.3), the sequence $\{2^m f(\frac{x}{2^m})\}$ is a Cauchy sequence. Thus, we define $A(x) := \lim_{m\to\infty} 2^m f(\frac{x}{2^m})$ for all $x \in X$, which is an odd mapping.

Now, we show that A is an additive mapping. From (7.1.2), it follows that

$$\mu_{2^m(f(\frac{x+y}{2^m})-f(\frac{x}{2^m})-f(\frac{y}{2^m}))}(t) \geq \rho_{(\frac{x}{2^m},\frac{y}{2^m},-(\frac{x+y}{2^m}),0,\ldots,0)}\left(\frac{nt}{2^{m-1}}\right).$$

Taking $m \to \infty$ in the above inequality, by (7.1.4), the mapping A is additive. By letting $l = 0$ and taking $m \to \infty$ in (7.1.6), we get (7.1.5).

Finally, to prove the uniqueness of the additive mapping A subject to (7.1.4), let us assume that there exists another additive mapping B which satisfies (7.1.5). Since

$$\begin{aligned}\mu_{A(x)-B(x)}(2t) &= \mu_{A(x)-2^m f(\frac{x}{2^m})+2^m f(\frac{x}{2^m})-B(x)}(2t)\\ &\geq T\big(\mu_{A(x)-2^m f(\frac{x}{2^m})}(t), \mu_{2^m f(\frac{x}{2^m})-B(x)}(t)\big)\end{aligned}$$

and

$$\lim_{m\to\infty} \mu_{A(x)-2^m f(\frac{x}{2^m})} = \lim_{m\to\infty} \mu_{B(x)-2^m f(\frac{x}{2^m})} = 1$$

for all $x \in X$ and $t > 0$, we get

$$\lim_{m\to\infty} T\big(\mu_{A(x)-2^m f(\frac{x}{2^m})}(t), \mu_{2^m f(\frac{x}{2^m})-B(x)}(t)\big) = 1.$$

Thus, we have $A = B$. This completes the proof. □

Corollary 7.1.2 *Let $\theta \geq 0$ and p be a constant with $p > 1$. Let X be a normed vector space and Y be a complete RN-space. Let $f : X \to Y$ be an odd mapping satisfying*

$$\mu_{Df(x_1,x_2,\ldots,x_n)}(t) \geq \frac{t}{t+\theta\sum_{i=1}^n \|x_i\|^p}$$

for all $(x_1, x_2, \ldots, x_n) \in X$ with $\sum_{i=1}^n x_i = 0$ and $t > 0$. If

$$T_{k=1}^{\infty}\left(\frac{2^{(k+l)p}nt}{2^{(k+l)p}nt + 2^{2k+l-2}(2+2^p)\theta\|x\|^p}\right) = 1$$

for all $x \in X$, $t > 0$ and $l \geq 0$, then there exists a unique additive mapping $A : X \to Y$ such that

$$\mu_{f(x)-A(x)}(t) \geq T_{k=1}^{\infty}\left(\frac{2^{kp}nt}{2^{kp}nt + 2^{2k-2}(2+2^p)\theta\|x\|^p}\right)$$

for all $x \in X$ and $t > 0$.

Proof If we define

$$\rho_{(x_1,x_2,\ldots,x_n)}(t) = \frac{t}{t+\theta\sum_{i=1}^n \|x_i\|^p}$$

for all $(x_1, x_2, \ldots, x_n) \in X$ with $\sum_{i=1}^{n} x_i = 0$ and $t > 0$, then, from Theorem 7.1.1, we get the desired result. □

Theorem 7.1.3 *Let $f : X \to Y$ be an odd mapping for which there exists a $\rho : X^n \to D^+$ satisfying* (7.1.2). *If*

$$T_{k=1}^{\infty}\rho_{(2^{k+l-2}x,2^{k+l-2}x,-2^{k+l-1}x,0,\ldots,0)}\left(2^{l+1}nt\right) = 1 \tag{7.1.7}$$

and

$$\lim_{m\to\infty}\rho_{(2^m x,2^m y,-2^m(x+y),0,\ldots,0)}\left(2^{m+1}nt\right) = 1 \tag{7.1.8}$$

for all $x, y \in X$, $t > 0$ and $l \geq 0$, then there exists a unique additive mapping $A : X \to Y$ such that

$$\mu_{f(x)-A(x)}(t) \geq T_{k=1}^{\infty}\rho_{(2^{k-2}x,2^{k-2}x,-2^{k-1}x,0,\ldots,0)}(2nt) \tag{7.1.9}$$

for all $x \in X$ and $t > 0$.

Proof Putting $x_1 = x_2 = x$, $x_3 = -2x$ and $x_4 = \cdots = x_n = 0$ in (7.1.2), respectively, we get

$$\mu_{2n(f(2x)-2f(x))}(t) \geq \rho_{(x,x,-2x,0,\ldots,0)}(t),$$

which is equivalent to the following:

$$\mu_{f(x)-\frac{1}{2}f(2x)}(t) \geq \rho_{(\frac{x}{2},\frac{x}{2},-x,0,\ldots,0)}(4nt)$$

for all $x \in X$ and $t > 0$. Replacing x and t by $2^{k-1}x$ and $2t$, respectively, in the above inequality, we get

$$\mu_{\frac{1}{2^{k-1}}f(2^{k-1}x)-\frac{1}{2^k}f(2^k x)}\left(\frac{t}{2^k}\right) \geq \rho_{(2^{k-2}x,2^{k-2}x,-2^{k-1}x,0,\ldots,0)}(2nt)$$

for all $x \in X$ and $t > 0$. Since $\mu_x(s) \leq \mu_x(t)$ for all s and t with $0 < s \leq t$, we obtain

$$\begin{aligned}
\mu_{f(x)-\frac{1}{2^m}f(2^m x)}(t) &= \mu_{\sum_{k=1}^{m}(\frac{1}{2^{k-1}}f(2^{k-1}x)-\frac{1}{2^k}f(2^k x))}(t)\\
&\geq \mu_{\sum_{k=1}^{m}(\frac{1}{2^{k-1}}f(2^{k-1}x)-\frac{1}{2^k}f(2^k x))}\left(\sum_{k=1}^{m}\frac{t}{2^k}\right)\\
&\geq T_{k=1}^{m}\rho_{(2^{k-2}x,2^{k-2}x,-2^{k-1}x,0,\ldots,0)}(2nt).
\end{aligned}$$

Replacing x by $2^l x$ in the above inequality, we get

$$\mu_{f(2^l x)-\frac{1}{2^m}f(2^{m+l}x)}(t) \geq T_{k=1}^{m}\rho_{(2^{k+l-2}x,2^{k+l-2}x,-2^{k+l-1}x,0,\ldots,0)}(2nt),$$

which is equivalent to the following:

$$\mu_{\frac{1}{2^l}f(2^l x)-\frac{1}{2^{m+l}}f(2^{m+l}x)}(t) \geq T_{k=1}^{m}\rho_{(2^{k+l-2}x,2^{k+l-2}x,-2^{k+l-1}x,0,\dots,0)}\left(2^{l+1}nt\right) \tag{7.1.10}$$

for all $x \in X$, $t > 0$ and $l \geq 0$. Since the right-hand side of the inequality (7.1.10) tends to 1 as $m \to \infty$ by (7.1.7), the sequence $\{\frac{1}{2^m}f(2^m x)\}$ is a Cauchy sequence. Thus, we define $A(x) := \lim_{m\to\infty}\frac{1}{2^m}f(2^m x)$ for all $x \in X$, which is an odd mapping.

Now, we show that A is an additive mapping. From (7.1.2), it follows that

$$\mu_{\frac{1}{2^m}(f(2^m(x+y))-f(2^m x)-f(2^m y))}(t) \geq \rho_{(2^m x,2^m y,-2^m(x+y),0,\dots,0)}\left(2^{m+1}nt\right).$$

Taking $m \to \infty$ in the above inequality, by (7.1.8), the mapping A is additive. Thus, by letting $l = 0$ and taking $m \to \infty$ in (7.1.10), we get (7.1.9).

The rest of the proof is the same as in the proof of Theorem 7.1.1. This completes the proof. □

Corollary 7.1.4 *Let $\theta \geq 0$ and p be a constant with $0 < p < 1$. Let X be a normed vector space and Y be a complete RN-space. Let $f : X \to Y$ be an odd mapping satisfying*

$$\mu_{Df(x_1,x_2,\dots,x_n)}(t) \geq \frac{t}{t+\theta\sum_{i=1}^{n}\|x_i\|^p}$$

for all $(x_1, x_2, \dots, x_n) \in X$ with $\sum_{i=1}^{n} x_i = 0$ and $t > 0$. If

$$T_{k=1}^{\infty}\left(\frac{2^{l+1}nt}{2^{l+1}nt+2^{(k+l-1)p}(2^{1-p}+1)\theta\|x\|^p}\right) = 1$$

for all $x \in X$, $t > 0$ and $l \geq 0$, then there exists a unique additive mapping $A : X \to Y$ such that

$$\mu_{f(x)-A(x)}(t) \geq T_{k=1}^{\infty}\left(\frac{2nt}{2nt+2^{(k-1)p}(2^{1-p}+1)\theta\|x\|^p}\right)$$

for all $x \in X$ and $t > 0$.

Proof If we define

$$\rho_{(x_1,x_2,\dots,x_n)}(t) = \frac{t}{t+\theta\sum_{i=1}^{n}\|x_i\|^p}$$

for all $(x_1, x_2, \dots, x_n) \in X$ with $\sum_{i=1}^{n} x_i = 0$ and $t > 0$, then, from Theorem 7.1.3, we get the desired result. □

7.1.2 The Generalized Hyers–Ulam Stability of the Functional Equation (7.1.1): An Even Case

Now, we are going to prove the generalized Hyers–Ulam stability of the functional equation (7.1.1) of an even mapping in RN-spaces.

For any even mapping $f : X \to Y$ with $f(0) = 0$, we note that, if f satisfies the following:

$$Df(x_1, x_2, \ldots, x_n) = 0$$

for all $x_1, \ldots, x_n \in X$ with $\sum_{i=1}^n x_i = 0$, then the mapping f is quadratic.

Theorem 7.1.5 *Let $f : X \to Y$ be an even mapping with $f(0) = 0$ for which there exists a $\rho : X^n \to D^+$ satisfying* (7.1.2). *If*

$$T_{k=1}^{\infty} \rho_{(\frac{x}{2^{k+l}}, -\frac{x}{2^{k+l}}, 0, \ldots, 0)}\left(\frac{t}{2^{3k+2l-3}}\right) = 1 \tag{7.1.11}$$

and

$$\lim_{m\to\infty} \rho_{(\frac{x}{2^m}, \frac{y}{2^m}, -\frac{x+y}{2^m}, 0, \ldots, 0)}\left(\frac{t}{2^{2m-1}}\right) = 1 \tag{7.1.12}$$

for all $x, y \in X$, $t > 0$ and $l \geq 0$, then there exists a unique quadratic mapping $Q : X \to Y$ such that

$$\mu_{f(x)-Q(x)}(t) \geq T_{k=1}^{\infty} \rho_{(\frac{x}{2^k}, -\frac{x}{2^k}, 0, \ldots, 0)}\left(\frac{t}{2^{3k-3}}\right) \tag{7.1.13}$$

for all $x \in X$ and $t > 0$.

Proof Putting $x_1 = x$, $x_2 = -x$ and $x_3 = \cdots = x_n = 0$ in (7.1.2), respectively, we get

$$\mu_{2(f(2x)-4f(x))}(t) \geq \rho_{(x,-x,0,\ldots,0)}(t)$$

for all $x \in X$ and $t > 0$, which is equivalent to the following:

$$\mu_{f(x)-4f(\frac{x}{2})}(t) \geq \rho_{(\frac{x}{2},-\frac{x}{2},0,\ldots,0)}(2t)$$

for all $x \in X$ and $t > 0$. Replacing x and t by $\frac{x}{2^{k-1}}$ and $\frac{t}{2^{3k-2}}$, respectively, in the above inequality, we get

$$\mu_{4^{k-1}f(\frac{x}{2^{k-1}})-4^k f(\frac{x}{2^k})}\left(\frac{t}{2^k}\right) \geq \rho_{(\frac{x}{2^k}, -\frac{x}{2^k} 0, \ldots, 0)}\left(\frac{t}{2^{3k-3}}\right)$$

for all $x \in X$ and $t > 0$. Since $\mu_x(s) \le \mu_x(t)$ for all s and t with $0 < s \le t$, we obtain

$$\begin{aligned}
\mu_{f(x)-4^m f(\frac{x}{2^m})}(t) &= \mu_{\sum_{k=1}^m (4^{k-1} f(\frac{x}{2^{k-1}}) - 4^k f(\frac{x}{2^k}))}(t) \\
&\ge \mu_{\sum_{k=1}^m (4^{k-1} f(\frac{x}{2^{k-1}}) - 4^k f(\frac{x}{2^k}))}\left(\sum_{k=1}^m \frac{t}{2^k}\right) \\
&\ge T_{k=1}^m \rho_{(\frac{x}{2^k}, -\frac{x}{2^k}, 0, \dots, 0)}\left(\frac{t}{2^{3k-3}}\right).
\end{aligned}$$

Replacing x by $\frac{x}{2^l}$ in the above inequality, we get

$$\mu_{f(\frac{x}{2^l})-4^m f(\frac{x}{2^{m+l}})}(t) \ge T_{k=1}^m \rho_{(\frac{x}{2^{k+l}}, -\frac{x}{2^{k+l}}, 0, \dots, 0)}\left(\frac{t}{2^{3k-3}}\right),$$

which is equivalent to the following:

$$\begin{aligned}
&\mu_{4^l f(\frac{x}{2^l})-4^{m+l} f(\frac{x}{2^{m+l}})}(t) \\
&\quad \ge T_{k=1}^m \rho_{(\frac{x}{2^{k+l}}, -\frac{x}{2^{k+l}}, 0, \dots, 0)}\left(\frac{t}{2^{3k+2l-3}}\right)
\end{aligned} \tag{7.1.14}$$

for all $x \in X$, $t > 0$ and $l \ge 0$. Since the right-hand side of the inequality (7.1.14) tends to 1 as $m \to \infty$ by (7.1.11), the sequence $\{4^m f(\frac{x}{2^m})\}$ is a Cauchy sequence. Thus, we define $Q(x) := \lim_{m\to\infty} 4^m f(\frac{x}{2^m})$ for all $x \in X$, which is an even mapping.

Now, we show that Q is an quadratic mapping. From (7.1.2), it follows that

$$\begin{aligned}
&\mu_{4^m(f(\frac{x-y}{2^m})+f(\frac{2x+y}{2^m})+f(\frac{x+2y}{2^m})-3f(\frac{x+y}{2^m})-3f(\frac{x}{2^m})-3f(\frac{y}{2^m}))}(t) \\
&\ge \rho_{(\frac{x}{2^m}, \frac{y}{2^m}, -\frac{x+y}{2^m}, 0, \dots, 0)}\left(\frac{t}{2^{2m-1}}\right).
\end{aligned}$$

Taking $m \to \infty$ in the above inequality, by (7.1.12), the mapping Q is quadratic. Moreover, letting $l = 0$ and taking $m \to \infty$ in (7.1.14), we get (7.1.13).

The rest of the proof is the same as in the proof of Theorem 7.1.1. This completes the proof. □

Corollary 7.1.6 *Let $\theta \ge 0$ and p be a constant with $p > 2$. Let X be a normed vector space and Y be a complete RN-space. Let $f : X \to Y$ be an even mapping satisfying*

$$\mu_{Df(x_1, x_2, \dots, x_n)}(t) \ge \frac{t}{t + \theta \sum_{i=1}^n \|x_i\|^p}$$

for all $(x_1, x_2, \ldots, x_n) \in X$ *with* $\sum_{i=1}^{n} x_i = 0$ *and* $t > 0$. *If*

$$T_{k=1}^{\infty}\left(\frac{2^{(k+l)p}t}{2^{(k+l)p}t + 2^{3k+2l-2}\theta\|x\|^p}\right) = 1$$

for all $x \in X$, $t > 0$ *and all* $l \geq 0$, *then there exists a unique quadratic mapping* $Q : X \to Y$ *such that*

$$\mu_{f(x)-Q(x)}(t) \geq T_{k=1}^{\infty}\left(\frac{2^{kp}t}{2^{kp}t + 2^{3k-2}\theta\|x\|^p}\right)$$

for all $x \in X$ *and* $t > 0$.

Proof If we define

$$\rho_{(x_1,x_2,\ldots,x_n)}(t) = \frac{t}{t + \theta\sum_{i=1}^{n}\|x_i\|^p}$$

for all $(x_1, x_2, \ldots, x_n) \in X$ with $\sum_{i=1}^{n} x_i = 0$ and $t > 0$, then, from Theorem 7.1.5, we get the desired result. □

Theorem 7.1.7 *Let* $f : X \to Y$ *be an even mapping with* $f(0) = 0$ *for which there exits a* $\rho : X^n \to D^+$ *satisfying* (7.1.2). *If*

$$T_{k=1}^{\infty}\rho_{(2^{k+l-1}x,-2^{k+l-1}x,0,\ldots,0)}\left(2^{k+2l-1}t\right) = 1 \tag{7.1.15}$$

and

$$\lim_{m\to\infty}\rho_{(2^m x,2^m y,-2^m(x+y),0,\ldots,0)}\left(2^{m+1}t\right) = 1 \tag{7.1.16}$$

for all $x, y \in X$, $t > 0$ *and* $l \geq 0$, *then there exists a unique quadratic mapping* $Q : X \to Y$ *such that*

$$\mu_{f(x)-Q(x)}(t) \geq T_{k=1}^{\infty}\rho_{(2^k x,-2^k x,0,\ldots,0)}\left(2^{k-1}t\right) \tag{7.1.17}$$

for all $x \in X$ *and* $t > 0$.

Proof Letting $x_1 = x$, $x_2 = -x$ and $x_3 = \cdots = x_n = 0$ in (7.1.2), respectively, we get

$$\mu_{2(f(2x)-4f(x))}(t) \geq \rho_{(x,-x,0,\ldots,0)}(t)$$

for all $x \in X$ and $t > 0$, which is equivalent to the following:

$$\mu_{f(x)-\frac{1}{4}f(2x)}\left(\frac{t}{4}\right) \geq \rho_{(x,-x,0,\ldots,0)}(2t)$$

for all $x \in X$ and $t > 0$. Replacing x and t by $2^{k-1}x$ and $2^{k-2}t$, respectively, in the above inequality, we get

$$\mu_{\frac{1}{4^{k-1}}f(2^{k-1}x)-\frac{1}{4^k}f(2^kx)}\left(\frac{t}{2^k}\right) \geq \rho_{(2^{k-1}x,-2^{k-1}x,0,\ldots,0)}\left(2^{k-1}t\right)$$

for all $x \in X$ and $t > 0$. Since $\mu_x(s) \leq \mu_x(t)$ for all s and t with $0 < s \leq t$, we obtain

$$\begin{aligned}\mu_{f(x)-\frac{1}{4^m}f(2^mx)}(t) &= \mu_{\sum_{k=1}^m(\frac{1}{4^{k-1}}f(2^{k-1}x)-\frac{1}{4^k}f(2^kx))}(t)\\ &\geq \mu_{\sum_{k=1}^m(\frac{1}{4^{k-1}}f(2^{k-1}x)-\frac{1}{4^k}f(2^kx))}\left(\sum_{k=1}^m\frac{t}{2^k}\right)\\ &\geq T_{k=1}^m\rho_{(2^{k-1}x,-2^{k-1}x,0,\ldots,0)}\left(2^{k-1}t\right).\end{aligned}$$

Replacing x by 2^lx in the above inequality, we get

$$\mu_{f(2^lx)-\frac{1}{4^m}f(2^{m+l}x)}(t) \geq T_{k=1}^m\rho_{(2^{k+l-1}x,-2^{k+l-1}x,0,\ldots,0)}\left(2^{k-1}t\right),$$

which is equivalent to the following:

$$\begin{aligned}&\mu_{\frac{1}{4^l}f(2^lx)-\frac{1}{4^{m+l}}f(2^{m+l}x)}(t)\\ &\quad\geq T_{k=1}^m\rho_{(2^{k+l-1}x,-2^{k+l-1}x,0,\ldots,0)}\left(2^{k+2l-1}t\right)\end{aligned} \tag{7.1.18}$$

for all $x \in X$, $t > 0$ and $l \geq 0$. Since the right-hand side of the inequality (7.1.18) tends to 1 as $m \to \infty$ by (7.1.15), the sequence $\{\frac{1}{4^m}f(2^mx)\}$ is a Cauchy sequence. Thus we define $Q(x) := \lim_{m\to\infty}\frac{1}{4^m}f(2^mx)$ for all $x \in X$, which is an even mapping.

Now, we show that Q is a quadratic mapping. From (7.1.2), it follows that

$$\begin{aligned}&\mu_{\frac{1}{4^m}(f(2^m(x-y))+f(2^m(2x+y))+f(2^m(x+2y))-3f(2^m(x+y))-3f(2^mx)-3f(2^my))}(t)\\ &\geq \rho_{(2^mx,2^my,-2^m(x+y),0,\ldots,0)}\left(2^{m+1}t\right).\end{aligned}$$

Taking $m \to \infty$ in the above inequality, by (7.1.16), the mapping Q is quadratic. Moreover, letting $l = 0$ and taking $m \to \infty$ in (7.1.18), we get (7.1.17).

The rest of the proof is the same as in the proof of Theorem 7.1.5. This completes the proof. □

Corollary 7.1.8 *Let $\theta \geq 0$ and p be a constant with $0 < p < 2$. Let X be a normed vector space X and Y be a complete RN-space. Let $f : X \to Y$ be an even mapping satisfying*

$$\mu_{Df(x_1,x_2,\ldots,x_n)}(t) \geq \frac{t}{t+\theta\sum_{i=1}^n\|x_i\|^p}$$

for all $(x_1, x_2, \ldots, x_n) \in X$ *with* $\sum_{i=1}^n x_i = 0$ *and* $t > 0$. *If*

$$T_{k=1}^{\infty}\left(\frac{2^{k+2l-2}t}{2^{k+2l-2}t + 2^{(k+l)p}\theta\|x\|^p}\right) = 1$$

for all $x \in X$, $t > 0$ *and* $l \geq 0$, *then there exists a unique quadratic mapping* $Q : X \to Y$ *such that*

$$\mu_{f(x)-Q(x)}(t) \geq \lim_{m\to\infty} T_{k=1}^{m}\left(\frac{2^{k-2}t}{2^{k-2}t + 2^{kp}\theta\|x\|^p}\right)$$

for all $x \in X$ *and* $t > 0$.

Proof If we define

$$\rho_{(x_1,x_2,\ldots,x_n)}(t) = \frac{t}{t + \theta\sum_{i=1}^n \|x_i\|^p}$$

for all $(x_1, x_2, \ldots, x_n) \in X$ with $\sum_{i=1}^n x_i = 0$ and $t > 0$, then, from Theorem 7.1.7, then we get the desired result. □

7.2 Non-Archimedean Lattice RN-Spaces

The main objective of this section is to prove the Hyers–Ulam stability of the following functional equation related to inner product spaces:

$$\sum_{i=1}^n f\left(x_i - \frac{1}{n}\sum_{j=1}^n x_j\right) = \sum_{i=1}^n f(x_i) - nf\left(\frac{1}{n}\sum_{i=1}^n x_i\right) \tag{7.2.1}$$

for all $n \geq 2$ in non-Archimedean lattice random normed spaces.

7.2.1 The Hyers–Ulam Stability of Functional Equations in Non-Archimedean Lattice Random Spaces

In the rest of this section, unless otherwise explicitly stated, we assume that G is an additive group and X is a complete non-Archimedean lattice random space. For convenience, we use the following abbreviation: for any mapping $f : G \to X$,

$$\Delta f(x_1, \ldots, x_n) = \sum_{i=1}^n f\left(x_i - \frac{1}{n}\sum_{j=1}^n x_j\right) - \sum_{i=1}^n f(x_i) + nf\left(\frac{1}{n}\sum_{i=1}^n x_i\right)$$

for all $x_1, \ldots, x_n \in G$, where $n \geq 2$ is a fixed integer.

Lemma 7.2.1 [185] *Let V_1 and V_2 be real vector spaces. If an odd mapping $f : V_1 \to V_2$ satisfies the functional equation* (7.2.1), *then f is additive.*

Let $\mathcal{K}$ be a non-Archimedean field, X be a vector space over $\mathcal{K}$ and $(Y, \mu, \mathcal{T}_\wedge)$ be a non-Archimedean complete LRN-space over $\mathcal{K}$.

In the following theorem, we prove the Hyers–Ulam stability of the functional equation (7.2.1) in non-Archimedean lattice random spaces for an odd case.

Theorem 7.2.2 *Let $\mathcal{K}$ be a non-Archimedean field and $(\mathcal{X}, \mu, \mathcal{T}_\wedge)$ be a non-Archimedean complete LRN-space over $\mathcal{K}$. Let $\varphi : G^n \to D_L^+$ be a distribution function such that*

$$\lim_{m\to\infty} \varphi_{2^m x_1, 2^m x_2, \ldots, 2^m x_n}\left(|2|^m t\right) = 1_{\mathcal{L}} = \lim_{m\to\infty} \Phi_{2^{m-1}x}\left(|2|^m t\right) \tag{7.2.2}$$

for all $x, x_1, x_2, \ldots, x_n \in G$ and

$$\tilde{\varphi}_x(t) = \lim_{m\to\infty} \min\left\{\Phi_{2^k x}\left(|2|^k t\right) : 0 \le k < m\right\} \tag{7.2.3}$$

exists for all $x \in G$, where

$$\begin{aligned}\Phi_x(t) := \min\Bigg\{\varphi_{2x,0,\ldots,0}(t), \min\Bigg\{\varphi_{x,x,0,\ldots,0}\left(\frac{|2|t}{n}\right),\\ \varphi_{x,-x,\ldots,-x}\left(|2|t\right), \varphi(-x, x, \ldots, x)\Bigg\}\Bigg\}\end{aligned} \tag{7.2.4}$$

for all $x \in G$. Suppose that an odd mapping $f : G \to X$ satisfies the inequality

$$\mu_{\Delta f(x_1,\ldots,x_n)}(t) \ge_L \varphi_{x_1,x_2,\ldots,x_n}(t) \tag{7.2.5}$$

for all $x_1, x_2, \ldots, x_n \in G$ and $t > 0$. Then there exists an additive mapping $A : G \to X$ such that

$$\mu_{f(x)-A(x)}(t) \ge_L \tilde{\varphi}_x\left(|2|t\right) \tag{7.2.6}$$

for all $x \in G$, $t > 0$ and, if

$$\lim_{\ell\to\infty} \lim_{m\to\infty} \min\left\{\Phi_{2^k x}\left(|2|^k t\right) : \ell \le k < m + \ell\right\} = 1_{\mathcal{L}}, \tag{7.2.7}$$

then A is a unique additive mapping satisfying (7.2.6).

Proof Letting $x_1 = nx_1$ and $x_i = nx_1'$ for each $i = 2, \ldots, n$ in (7.2.5), respectively, since f is an odd mapping, we obtain

$$\begin{aligned}&\mu_{nf(x_1+(n-1)x_1')+f((n-1)(x_1-x_1'))-(n-1)f(x_1-x_1')-f(nx_1)-(n-1)f(nx_1')}(t)\\ &\ge_L \varphi_{nx_1,nx_1',\ldots,nx_1'}(t)\end{aligned} \tag{7.2.8}$$

for all $x_1, x_1' \in G$ and $t > 0$. Interchanging x_1 with x_1' in (7.2.8) and using the oddness of f, we get

$$\begin{aligned}&\mu_{nf((n-1)x_1+x_1')-f((n-1)(x_1-x_1'))+(n-1)f(x_1-x_1')-(n-1)f(nx_1)-f(nx_1')}(t)\\ &\geq_L \varphi_{nx_1',nx_1,\dots,nx_1}(t)\end{aligned} \tag{7.2.9}$$

for all $x_1, x_1' \in G$ and $t > 0$. It follows from (7.2.8) and (7.2.9) that

$$\mu_{Df(x_1,x_1')}(t) \geq_L \min\{\varphi_{nx_1,nx_1',\dots,nx_1'}(t), \varphi_{nx_1',nx_1,\dots,nx_1}(t)\} \tag{7.2.10}$$

for all $x_1, x_1' \in G$ and $t > 0$, where

$$\begin{aligned}Df(x_1,x_1') &= nf(x_1+(n-1)x_1') - nf((n-1)x_1+x_1') + 2f((n-1)(x_1-x_1'))\\ &\quad - 2(n-1)f(x_1-x_1') + (n-2)f(nx_1) - (n-2)f(nx_1').\end{aligned}$$

Setting $x_1 = nx_1$, $x_2 = -nx_1'$ and $x_i = 0$ for each $i = 3, \dots, n$ in (7.2.5), respectively, since f is an odd mapping, we get

$$\begin{aligned}&\mu_{f((n-1)x_1+x_1')-f(x_1+(n-1)x_1')+2f(x_1-x_1')-f(nx_1)+f(nx_1')}(t)\\ &\geq_L \varphi_{nx_1,-nx_1',0,\dots,0}(t)\end{aligned} \tag{7.2.11}$$

for all $x_1, x_1' \in G$ and $t > 0$. It follows from (7.2.10) and (7.2.11) that

$$\begin{aligned}&\mu_{f((n-1)(x_1-x_1'))+f(x_1-x_1')-f(nx_1)+f(nx_1')}(t)\\ &\geq_L \min\left\{\varphi_{nx_1,-nx_1',0,\dots,0}\left(\frac{|2|}{n}\right), \varphi_{nx_1,nx_1',\dots,nx_1'}\left(\frac{|2|}{n}\right),\right.\\ &\qquad\left.\varphi_{nx_1',nx_1,\dots,nx_1}\left(\frac{|2|}{n}\right)\right\}\end{aligned} \tag{7.2.12}$$

for all $x_1, x_1' \in G$ and $t > 0$. Putting $x_1 = n(x_1 - x_1')$ and $x_i = 0$ for each $i = 2, \dots, n$ in (7.2.5), respectively, we obtain

$$\mu_{f(n(x_1-x_1'))-f((n-1)(x_1-x_1'))-f((x_1-x_1'))}(t) \geq_L \varphi_{n(x_1-x_1'),0,\dots,0}(t) \tag{7.2.13}$$

for all $x_1, x_1' \in G$ and $t > 0$. It follows from (7.2.12) and (7.2.13) that

$$\begin{aligned}&\mu_{f(n(x_1-x_1'))-f(nx_1)+f(nx_1')}(t)\\ &\geq_L \min\left\{\varphi_{n(x_1-x_1'),0,\dots,0}(t), \varphi_{nx_1,-nx_1',0,\dots,0}\left(\frac{|2|}{n}t\right),\right.\\ &\qquad\left.\min\left\{\varphi_{nx_1,nx_1',\dots,nx_1'}\left(\frac{|2|}{n}t\right), \varphi_{nx_1',nx_1,\dots,nx_1}\left(\frac{|2|}{n}t\right)\right\}\right\}\end{aligned} \tag{7.2.14}$$

for all $x_1, x_1' \in G$ and $t > 0$. Replacing x_1 and x_1' by $\frac{x}{n}$ and $\frac{-x}{n}$ in (7.2.14), respectively, we obtain

$$\mu_{f(2x)-2f(x)}(t) \geq_L \min\left\{\varphi_{2x,0,\ldots,0}(t), \min\left\{\varphi_{x,x,0,\ldots,0}\left(\frac{|2|}{n}t\right), \varphi_{x,-x,\ldots,-x}(t), \varphi_{-x,x,\ldots,x}(t)\right\}\right\}$$

for all $x \in G$ and $t > 0$. Hence, we have

$$\mu_{\frac{f(2x)}{2}-f(x)}(t) \geq_L \Phi_x\big(|2|t\big) \tag{7.2.15}$$

for all $x \in G$ and $t > 0$. Replacing x by $2^{m-1}x$ in (7.2.15), we have

$$\mu_{\frac{f(2^{m-1}x)}{2^{m-1}}-\frac{f(2^m x)}{2^m}}(t) \geq_L \Phi_{2^{m-1}x}\big(|2|^m t\big) \tag{7.2.16}$$

for all $x \in G$ and $t > 0$. It follows from (7.2.2) and (7.2.16) that the sequence $\{\frac{f(2^m x)}{2^m}\}$ is a Cauchy sequence. Since X is complete, we conclude that $\{\frac{f(2^m x)}{2^m}\}$ is convergent. So one can define the mapping $A : G \to X$ by $A(x) := \lim_{m\to\infty} \frac{f(2^m x)}{2^m}$ for all $x \in G$. It follows from (7.2.15) and (7.2.16) that

$$\mu_{f(x)-\frac{f(2^m x)}{2^m}}(t) \geq_L \min\big\{\Phi_{2^k x}\big(|2|^{k+1}t\big) : 0 \leq k < m\big\} \tag{7.2.17}$$

for all $m \geq 1$, $x \in G$ and $t > 0$. By taking $m \to \infty$ in (7.2.17) and using (7.2.3), one gets (7.2.6). By (7.2.2) and (7.2.5), we obtain

$$\begin{aligned}
\mu_{\Delta A(x_1,x_2,\ldots,x_n)}(t) &= \lim_{m\to\infty} \mu_{\Delta f(2^m x_1, 2^m x_2,\ldots,2^m x_n)}\big(|2|^m t\big)\\
&\geq_L \lim_{m\to\infty} \varphi_{2^m x_1, 2^m x_2,\ldots,2^m x_n}\big(|2|^m t\big)\\
&= 1_{\mathcal{L}}
\end{aligned}$$

for all $x_1, x_2, \ldots, x_n \in G$ and $t > 0$. Thus, the mapping A satisfies (7.2.1) and so, by Lemma 7.2.1, A is additive.

Finally, if A' is another additive mapping satisfying (7.2.6), then we have

$$\begin{aligned}
\mu_{A(x)-A'(x)}(t) &= \lim_{\ell\to\infty} \mu_{A(2^\ell x)-A'(2^\ell x)}\big(|2|^\ell t\big)\\
&\geq_L \lim_{\ell\to\infty} \min\big\{\mu_{A(2^\ell x)-f(2^\ell x)}\big(|2|^\ell t\big), \mu_{f(2^\ell x)-Q'(2^\ell x)}\big(|2|^\ell t\big)\big\}\\
&\geq_L \lim_{\ell\to\infty}\lim_{m\to\infty} \min\big\{\tilde{\varphi}_{2^k x}\big(|2|^{k+1}\big) : \ell \leq k < m+\ell\big\}\\
&= 0
\end{aligned}$$

for all $x \in G$ and so $A = A'$. This completes the proof. □

Corollary 7.2.3 *Let $\rho : [0, \infty) \to [0, \infty)$ be a function satisfying the following conditions:*

(a) $\rho(|2|t) \le \rho(|2|)\rho(t)$ *for all* $t \ge 0$;
(b) $\rho(|2|) < |2|$.

Let $\varepsilon > 0$ and $(G, \mu, \mathcal{T}_\wedge)$ be a LRN-space in which $L = D^+$. Suppose that an odd mapping $f : G \to X$ satisfies the following inequality:

$$\mu_{\Delta f(x_1,\dots,x_n)}(t) \ge_L \frac{t}{t + \varepsilon \sum_{i=1}^n \rho(\|x_i\|)}$$

for all $x_1, \dots, x_n \in G$ and $t > 0$. Then there exists a unique additive mapping $A : G \to X$ such that

$$\mu_{f(x)-A(x)}(t) \ge_L \frac{t}{t + \frac{2n}{|2|^2} \varepsilon\rho(\|x\|)}$$

for all $x \in G$ and $t > 0$.

Proof Define a mapping $\varphi : G^n \to D^+$ by $\varphi_{x_1,\dots,x_n}(t) := \frac{t}{t+\varepsilon \sum_{i=1}^n \rho(\|x_i\|)}$ for all $t > 0$. Then we have

$$\lim_{m\to\infty} \varphi_{2^m x_1,\dots,2^m x_n}\left(|2|^m t\right) \ge_L \lim_{m\to\infty} \varphi_{x_1,\dots,x_n}\left(\left(\frac{|2|}{\rho(|2|)}\right)^m\right) = 1_{\mathcal{L}}$$

for all $x_1, \dots, x_n \in G$ and $t > 0$. So we have

$$\tilde{\varphi}_x(t) := \lim_{m\to\infty} \min\left\{\Phi_{2^k x}\left(|2|^k\right) : 0 \le k < m\right\} = \Phi_x(t)$$

and

$$\lim_{\ell\to\infty} \lim_{m\to\infty} \min\left\{\Phi_{2^k x}\left(|2|^k\right) : \ell \le k < m + \ell\right\} = \lim_{\ell\to\infty} \Phi_{2^\ell x}\left(|2|^\ell\right) = 1_{\mathcal{L}}$$

for all $x \in G$ and $t > 0$. It follows from (7.2.4) that

$$\Phi_x(t) = \min\left\{\frac{t}{t + \varepsilon\rho(\|2x\|)}, \frac{t}{t + \frac{1}{|2|}2n\varepsilon\rho(\|x\|)}\right\} = \frac{|2|t}{|2|t + 2n\ \varepsilon\rho(\|x\|)}.$$

Applying Theorem 7.2.2, we conclude that

$$\mu_{f(x)-A(x)}(t) \ge_L \tilde{\varphi}_x\left(|2|t\right) = \Phi_x\left(|2|t\right) = \frac{t}{t + \frac{2n}{|2|^2} \varepsilon\rho(\|x\|)}$$

for all $x \in G$ and $t > 0$. This completes the proof. □

Lemma 7.2.4 [185] *Let V_1 and V_2 be real vector spaces. If an even mapping $f : V_1 \to V_2$ satisfies the functional equation* (7.2.1), *then f is quadratic.*

In the following theorem, we prove the Hyers–Ulam stability of the functional equation (7.2.1) in non-Archimedean LRN-spaces for an even case.

Theorem 7.2.5 *Let $\varphi : G^n \to D_L^+$ be a function such that*

$$\lim_{m\to\infty} \varphi_{2^m x_1, 2^m x_2, \ldots, 2^m x_n}\big(|2|^{2m} t\big) = 1_{\mathcal{L}} = \lim_{m\to\infty} \tilde{\varphi}'_{2^{m-1}x}\big(|2|^{2m} t\big) \tag{7.2.18}$$

for all $x, x_1, x_2, \ldots, x_n \in G$ and $t > 0$ and

$$\tilde{\varphi}'_x(t) = \lim_{m\to\infty} \min\big\{\tilde{\varphi}'_{2^k x}\big(|2|^{2k} t\big) : 0 \le k < m\big\} \tag{7.2.19}$$

exists for all $x \in G$ and $t > 0$, where

$$\begin{aligned}\tilde{\varphi}'_x(t) := \min\big\{&\varphi_{nx,nx,0,\ldots,0}\big(|2n-2|t\big), \varphi_{nx,0,\ldots,0}\big(|n-1|t\big),\\ &\varphi_{x,(n-1)x,0,\ldots,0}\big(|n-1|t\big), \Psi_x\big(|n-1|t\big)\big\}\end{aligned} \tag{7.2.20}$$

and

$$\Psi_x(t) := \min\left\{n\varphi_{nx,0,\ldots,0}\left(\frac{|2|}{n}t\right), \varphi_{nx,0,\ldots,0}\big(|2|t\big), \varphi_{0,nx,\ldots,nx}\big(|2|t\big)\right\} \tag{7.2.21}$$

for all $x \in G$ and $t > 0$. Suppose that an even mapping $f : G \to X$ with $f(0) = 0$ satisfies the inequality (7.2.5) *for all $x_1, x_2, \ldots, x_n \in G$ and $t > 0$. Then there exists a quadratic mapping $Q : G \to X$ such that*

$$\mu_{f(x)-Q(x)}(t) \ge_L \tilde{\varphi}'_x\big(|2|^2 t\big) \tag{7.2.22}$$

for all $x \in G$, $t > 0$ and, if

$$\lim_{\ell\to\infty}\lim_{m\to\infty} \min\big\{\tilde{\varphi}'_{2^k x}\big(|2|^{2k} t\big) : \ell \le k < m + \ell\big\} = 1_{\mathcal{L}} \tag{7.2.23}$$

then Q is a unique quadratic mapping satisfying (7.2.22).

Proof Letting $x_1 = nx_1$ and $x_i = nx_2$ for each $i = 2, \ldots, n$ in (7.2.5), respectively, since f is an even mapping, we obtain

$$\begin{aligned}&\mu_{nf(x_1+(n-1)x_2)+f((n-1)(x_1-x_2))+(n-1)f(x_1-x_2)-f(nx_1)-(n-1)f(nx_2)}(t)\\ &\ge_L \varphi_{nx_1,nx_2,\ldots,nx_2}(t)\end{aligned} \tag{7.2.24}$$

for all $x_1, x_2 \in G$ and $t > 0$. Interchanging x_1 with x_2 in (7.2.24), since f is an even mapping, we obtain

$$\begin{aligned}&\mu_{nf((n-1)x_1+x_2)+f((n-1)(x_1-x_2))+(n-1)f(x_1-x_2)-(n-1)f(nx_1)-f(nx_2)}(t)\\ &\ge_L \varphi_{nx_2,nx_1,\ldots,nx_1}(t)\end{aligned} \tag{7.2.25}$$

for all $x_1, x_2 \in G$ and $t > 0$. It follows from (7.2.24) and (7.2.25) that

$$\mu_{Df(x_1,x_2)}(t) \geq_L \min\{\varphi_{nx_1,nx_2,\ldots,nx_2}(t), \varphi_{nx_2,nx_1,\ldots,nx_1}(t)\} \qquad (7.2.26)$$

for all $x_1, x_2 \in G$ and $t > 0$, where

$$\begin{aligned} &Df(x_1,x_2) \\ &\quad = nf\big((n-1)x_1 + x_2\big) + nf\big(x_1 + (n-1)x_2\big) + 2f\big((n-1)(x_1 - x_2)\big) \\ &\qquad + 2(n-1)f(x_1 - x_2) - nf(nx_1) - nf(nx_2). \end{aligned}$$

Setting $x_1 = nx_1$, $x_2 = -nx_2$ and $x_i = 0$ for each $i = 3, \ldots, n$ in (7.2.5), respectively, since f is an even mapping, we obtain

$$\begin{aligned} &\mu_{f((n-1)x_1+x_2)+f(x_1+(n-1)x_2)+2(n-1)f(x_1-x_2)-f(nx_1)-f(nx_2)}(t) \\ &\quad \geq_L \varphi_{nx_1,-nx_2,0,\ldots,0}(t) \end{aligned} \qquad (7.2.27)$$

for all $x_1, x_2 \in G$ and $t > 0$. So it follows from (7.2.26) and (7.2.27) that

$$\begin{aligned} &\mu_{f((n-1)(x_1-x_2))-(n-1)^2 f(x_1-x_2)}(t) \\ &\quad \geq_L \min\bigg\{\varphi_{nx_1,-nx_2,0,\ldots,0}\bigg(\frac{|2|}{n}\bigg), \varphi_{nx_1,nx_2,\ldots,nx_2}\big(|2|t\big), \\ &\qquad\qquad \varphi_{nx_2,nx_1,\ldots,nx_1}\big(|2|t\big)\bigg\} \end{aligned} \qquad (7.2.28)$$

for all $x_1, x_2 \in G$ and $t > 0$. Setting $x_1 = x$ and $x_2 = 0$ in (7.2.28), we obtain

$$\begin{aligned} &\mu_{f((n-1)x)-(n-1)^2 f(x)}(t) \\ &\quad \geq_L \min\bigg\{\varphi_{nx,0,\ldots,0}\bigg(\frac{|2|}{n}t\bigg), \varphi_{nx,0,\ldots,0}\big(|2|t\big), \varphi_{0,nx,\ldots,nx}\big(|2|t\big)\bigg\} \end{aligned} \qquad (7.2.29)$$

for all $x \in G$ and $t > 0$. Putting $x_1 = nx$ and $x_i = 0$ for each $i = 2, \ldots, n$ in (7.2.5), respectively, one obtains

$$\mu_{f(nx)-f((n-1)x)-(2n-1)f(x)}(t) \geq_L \varphi_{nx,0,\ldots,0}(t) \qquad (7.2.30)$$

for all $x \in G$ and $t > 0$. It follows from (7.2.29) and (7.2.30) that

$$\begin{aligned} &\mu_{f(nx)-n^2 f(x)}(t) \\ &\quad \geq_L \min\bigg\{\varphi_{nx,0,\ldots,0}(t), \varphi_{nx,0,\ldots,0}\bigg(\frac{|2|}{n}t\bigg), \varphi_{nx,0,\ldots,0}\big(|2|t\big), \\ &\qquad\qquad \varphi_{0,nx,\ldots,nx}\big(|2|t\big)\bigg\} \end{aligned} \qquad (7.2.31)$$

for all $x \in G$ and $t > 0$. Letting $x_2 = -(n-1)x_1$ and replacing x_1 by $\frac{x}{n}$ in (7.2.27), we get

$$\mu_{f((n-1)x)-f((n-2)x)-(2n-3)f(x)}(t) \geq_L \varphi_{x,(n-1)x,0,\dots,0}(t) \tag{7.2.32}$$

for all $x \in G$ and $t > 0$. It follows from (7.2.29) and (7.2.32) that

$$\begin{aligned}\mu_{f((n-2)x)-(n-2)^2 f(x)}(t) \\ \geq_L \min\Big\{&\varphi_{x,(n-1)x,0,\dots,0}(t), \varphi_{nx,0,\dots,0}\Big(\frac{|2|}{n}t\Big), \\ &\varphi_{nx,0,\dots,0}(|2|t), \varphi_{0,nx,\dots,nx}(|2|t)\Big\}\end{aligned} \tag{7.2.33}$$

for all $x \in G$ and $t > 0$. It follows from (7.2.31) and (7.2.33) that

$$\begin{aligned}&\mu_{f(nx)-f((n-2)x)-4(n-1)f(x)}(t) \\ &\quad\geq_L \min\{\varphi_{nx,0,\dots,0}(t), \varphi_{x,(n-1)x,0,\dots,0}(t), \Psi_x(t)\}\end{aligned} \tag{7.2.34}$$

for all $x \in G$ and $t > 0$. Setting $x_1 = x_2 = nx$ and $x_i = 0$ for each $i = 3, \dots, n$ in (7.2.5), respectively, we obtain

$$\mu_{f((n-2)x)+(n-1)f(2x)-f(nx)}(t) \geq_L \varphi_{nx,nx,0,\dots,0}(|2|t) \tag{7.2.35}$$

for all $x \in G$ and $t > 0$. It follows from (7.2.34) and (7.2.35) that

$$\begin{aligned}&\mu_{f(2x)-4f(x)}(t) \\ &\quad\geq_L \min\{\varphi_{nx,nx,0,\dots,0}(|2n-2|t), \varphi_{nx,0,\dots,0}(|n-1|t), \\ &\qquad\quad \varphi_{x,(n-1)x,0,\dots,0}(|n-1|t), \Psi_x(|n-1|t)\}\end{aligned} \tag{7.2.36}$$

for all $x \in G$ and $t > 0$. Thus, we have

$$\mu_{f(x)-\frac{f(2x)}{2^2}}(t) \geq_L \tilde{\varphi}'_x(|2|^2 t) \tag{7.2.37}$$

for all $x \in G$ and $t > 0$. Replacing x by $2^{m-1}x$ in (7.2.37), we have

$$\mu_{\frac{f(2^{m-1}x)}{2^{2(m-1)}}-\frac{f(2^m x)}{2^{2m}}}(t) \geq_L \tilde{\varphi}'_{2^{m-1}x}(|2|^{2m} t) \tag{7.2.38}$$

for all $x \in G$ and $t > 0$. It follows from (7.2.18) and (7.2.38) that the sequence $\{\frac{f(2^m x)}{2^{2m}}\}$ is a Cauchy sequence. Since X is complete, we conclude that $\{\frac{f(2^m x)}{2^{2m}}\}$ is convergent. So one can define the mapping $Q: G \to X$ by

$$Q(x) := \lim_{m\to\infty} \frac{f(2^m x)}{2^{2m}}$$

for all $x \in G$. By using induction, it follows from (7.2.37) and (7.2.38) that

$$\mu_{f(x)-\frac{f(2^m x)}{2^{2m}}}(t) \geq_L \min\{\tilde{\varphi}'_{2^k x}(|2|^{2k+2}t) : 0 \leq k < m\} \tag{7.2.39}$$

for all $n \geq 1$, $x \in G$ and $t > 0$. By taking $m \to \infty$ in (7.2.39) and using (7.2.19), one gets (7.2.22).

The rest of proof is similar to proof of Theorem 7.2.2. This completes the proof. □

Corollary 7.2.6 *Let $\eta : [0, \infty) \to [0, \infty)$ be a function satisfying the following conditions*:

(a) $\eta(|l|t) \leq \eta(|l|)\eta(t)$ *for all* $t \geq 0$;
(b) $\eta(|l|) < |l|^2$ *for* $l \in \{2, n-1, n\}$.

Let $\varepsilon > 0$ and $(G, \mu, \mathcal{T}_\wedge)$ be a LRN-space in which $L = D^+$. Suppose that an even mapping $f : G \to X$ with $f(0) = 0$ satisfies the following inequality:

$$\mu_{\Delta f(x_1,\dots,x_n)}(t) \geq \frac{t}{t + \varepsilon \sum_{i=1}^n \eta(\|x_i\|)}$$

for all $x_1, \dots, x_n \in G$ and $t > 0$. Then there exists a unique quadratic mapping $Q : G \to X$ such that

$$\mu_{f(x)-Q(x)}(t) \geq \begin{cases} \frac{t}{t+\frac{2}{|2|^2}\varepsilon\eta(\|x\|)}, & n = 2; \\ \frac{t}{t+\frac{n}{|2|^3|n-1|}\varepsilon\eta(\|nx\|)}, & n > 2, \end{cases}$$

for all $x \in G$ and $t > 0$.

Proof Define a mapping $\varphi : G^n \to D^+$ by $\varphi_{x_1,\dots,x_n}(t) := \frac{t}{t+\varepsilon \sum_{i=1}^n \eta(\|x_i\|)}$. Then we have

$$\lim_{m\to\infty} \varphi_{2^m x_1,\dots,2^m x_n}(|2|^{2m}t) \geq \lim_{m\to\infty} \varphi_{x_1,\dots,x_n}\left(\left(\frac{|2|^2}{\eta(|2|)}\right)^m\right) = 1_{\mathcal{L}}$$

for all $x_1, \dots, x_n \in G$ and $t > 0$. We have

$$\tilde{\varphi}'_x(t) := \lim_{m\to\infty} \min\{\tilde{\varphi}'_{2^k x}(|2|^{2k}t) : 0 \leq k < m\}$$

and

$$\lim_{\ell\to\infty}\lim_{m\to\infty} \min\{\tilde{\varphi}'_{2^k x}(|2|^{2k}t) : \ell \leq k < m + \ell\} = \lim_{\ell\to\infty} \tilde{\varphi}'_{2^\ell x}(|2|^{2\ell}t) = 0$$

for all $x \in G$ and $t > 0$. It follows from (7.2.21) that

$$\Psi_x(t) = \min\left\{\frac{|2|t}{|2|t + 2n\varepsilon\eta(\|nx\|)}, \frac{|2|t}{|2|t + 2\varepsilon\eta(\|nx\|)}, \frac{|2|t}{|2|t + 2(n-1)\varepsilon\eta(\|nx\|)}\right\}$$
$$= \frac{|2|t}{|2|t + n\varepsilon\eta(\|nx\|)}.$$

Hence, by using (7.2.20), we obtain

$$\tilde{\varphi}'_x(t) = \min\left\{\frac{|2n-2|t}{|2n-2|t + 2\varepsilon\eta(\|nx\|)}, \frac{|n-1|t}{|n-1|t + \varepsilon\eta(\|nx\|)}, \right.$$
$$\left. \frac{|2n-2|t}{|2n-2|t + n\varepsilon\eta(\|nx\|)}, \frac{|n-1|t}{|n-1|t + \varepsilon(\eta(\|x\|) + \eta(\|(n-1)x\|))}\right\}$$
$$= \begin{cases} \frac{t}{t+2\varepsilon\eta(\|x\|)}, & n = 2; \\ \frac{|2\|n-1|t}{|2\|n-1|t+n\varepsilon\eta(\|nx\|)}, & n > 2, \end{cases}$$

for all $x \in G$ and $t > 0$. Applying Theorem 7.2.5, we conclude the required result. This completes the proof. □

Lemma 7.2.7 [185] *Let V_1 and V_2 be real vector spaces. A mapping $f : V_1 \to V_2$ satisfies* (7.2.1) *if and only if there exist a symmetric bi-additive mapping $B : V_1 \times V_1 \to V_2$ and an additive mapping $A : V_1 \to V_2$ such that $f(x) = B(x,x) + A(x)$ for all $x \in V_1$.*

Now, we are ready to prove the main theorem concerning the Hyers–Ulam stability problem for the functional equation (7.2.1) in non-Archimedean spaces.

Theorem 7.2.8 *Let $\varphi : G^n \to D_L^+$ be a function satisfying* (7.2.2) *and* (7.2.18) *for all $x, x_1, x_2, \ldots, x_n \in G$, and $\tilde{\varphi}_x(t)$ and $\tilde{\varphi}'_x(t)$ exist for all $x \in G$ and $t > 0$, where $\tilde{\varphi}_x(t)$ and $\tilde{\varphi}'_x(t)$ are defined as in Theorems* 7.2.2 *and* 7.2.5. *Suppose that a mapping $f : G \to X$ with $f(0) = 0$ satisfies the inequality* (7.2.5) *for all $x_1, x_2, \ldots, x_n \in G$. Then there exist an additive mapping $A : G \to X$ and a quadratic mapping $Q : G \to X$ such that*

$$\mu_{f(x)-A(x)-Q(x)}(t)$$
$$\geq_L \min\left\{\tilde{\varphi}_x(|2|^2 t), \tilde{\varphi}_{-x}(|2|^2 t), \tilde{\varphi}'_x(|2|t), \frac{1}{|2|}\tilde{\varphi}'_{-x}(|2|t)\right\} \tag{7.2.40}$$

for all $x \in G$ and $t > 0$. If we have

$$\lim_{\ell\to\infty}\lim_{m\to\infty} \min\{\varphi_{2^k x}(|2|^k t) : \ell \leq k < m + \ell\} = 1_{\mathcal{L}}$$
$$= \lim_{\ell\to\infty}\lim_{m\to\infty} \min\{\tilde{\varphi}'_{2^k x}(|2|^{2k} t) : \ell \leq k < m + \ell\},$$

then A is a unique additive mapping and Q is a unique quadratic mapping satisfying (7.2.40).

Proof Let $f_e(x) = \frac{1}{2}(f(x) + f(-x))$ for all $x \in G$. Then we have

$$\begin{aligned}\|\Delta f_e(x_1, \ldots, x_n)\| &= \left\|\frac{1}{2}(\Delta f(x_1, \ldots, x_n) + \Delta f(-x_1, \ldots, -x_n))\right\| \\ &\le \frac{1}{|2|}\max\{\varphi(x_1, \ldots, x_n), \varphi(-x_1, \ldots, -x_n)\}\end{aligned}$$

for all $x_1, x_2, \ldots, x_n \in G$ and $t > 0$. By Theorem 7.2.5, there exists a quadratic mapping $Q : G \to X$ such that

$$\mu_{f_e(x)-Q(x)}(t) \ge_L \min\{\tilde{\varphi}'_x(|2|^3 t), \tilde{\varphi}'_{-x}(|2|^3 t)\} \tag{7.2.41}$$

for all $x \in G$ and $t > 0$. Also, let $f_o(x) = \frac{1}{2}(f(x) - f(-x))$ for all $x \in G$. By Theorem 7.2.2, there exists an additive mapping $A : G \to X$ such that

$$\mu_{f_o(x)-A(x)}(t) \ge_L \min\{\tilde{\varphi}_x(|2|^2 t), \tilde{\varphi}_{-x}(|2|^2 t)\} \tag{7.2.42}$$

for all $x \in G$ and $t > 0$. Hence, (7.2.40) follows from (7.2.41) and (7.2.42).

The rest of proof is trivial. This completes the proof. □

Corollary 7.2.9 *Let $\gamma : [0, \infty) \to [0, \infty)$ be a function satisfying the following conditions*:

(a) $\gamma(|l|t) \le \gamma(|l|)\gamma(t)$ *for all* $t \ge 0$;
(b) $\gamma(|l|) < |l|^2$ *for* $l \in \{2, n-1, n\}$.

Let $\varepsilon > 0$, $(G, \mu, T_\wedge)$ be a LRN-space in which $L = D^+$ and $f : G \to X$ is a mapping satisfying the following:

$$\mu_{\Delta f(x_1,\ldots,x_n)}(t) \ge \frac{t}{t + \varepsilon \sum_{i=1}^n \gamma(\|x_i\|)}$$

for all $x_1, \ldots, x_n \in G$, $t > 0$ and $f(0) = 0$. Then there exist a unique additive mapping $A : G \to X$ and a unique quadratic mapping $Q : G \to X$ such that

$$\mu_{f(x)-A(x)-Q(x)}(t) \ge \frac{|2|^3 t}{|2|^3 t + 2n\varepsilon\gamma(\|x\|)}$$

for all $x \in G$ and $t > 0$.

Proof The result follows by Corollaries 7.2.6 and 7.2.3. □

Chapter 8
Random Banach Algebras and Stability Results

In this chapter, we define random normed algebras, provide some characteristic examples of them and also prove the stability of random homomorphisms, Cauchy–Jensen functional equations and random $*$-derivations in random Banach algebras.

8.1 Random Homomorphisms

Now, we define random homomorphism and random derivation as follows:

A *random normed algebra* is a random normed space with algebraic structure such that

(RN4) $\mu_{xy}(ts) \geq \mu_x(t)\mu_y(s)$ for all $x, y \in X$ and all $t, s > 0$.

For example, every normed algebra $(X, \|\cdot\|)$ defines a random normed algebra (X, μ, T_M), where

$$\mu_x(t) = \frac{t}{t + \|x\|}$$

for all $t > 0$. This space is called the *induced random normed algebra*.

Let (X, μ, T_M) and (Y, μ, T_M) be random normed algebras.

(1) An $\mathbb{R}$-linear mapping $f : X \to Y$ is called a *random homomorphism* if

$$f(xy) = f(x)f(y)$$

for all $x, y \in X$.

(2) An $\mathbb{R}$-linear mapping $f : X \to X$ is called a *random derivation* if

$$f(xy) = f(x)y + xf(y)$$

for all $x, y \in X$.

Gilányi [93] showed that, if f satisfies the functional inequality

$$\left\|2f(x) + 2f(y) - f(x - y)\right\| \leq \left\|f(x + y)\right\|, \tag{8.1.1}$$

Y.J. Cho et al., *Stability of Functional Equations in Random Normed Spaces*, Springer Optimization and Its Applications 86, DOI 10.1007/978-1-4614-8477-6_8,

then f satisfies the *Jordan–von Neumann functional equation*

$$2f(x)+2f(y)=f(x+y)+f(x-y).$$

Also, in [232], we can see more results on this functional equation. Fechner [86] and Gilányi [94] proved the Hyers–Ulam stability of the functional inequality (8.1.1). Park et al. [195] investigated the Cauchy additive functional inequality

$$\|f(x)+f(y)+f(z)\| \le \|f(x+y+z)\| \tag{8.1.2}$$

and the Cauchy–Jensen additive functional inequality

$$\|f(x)+f(y)+f(2z)\| \le \left\|2f\left(\frac{x+y}{2}+z\right)\right\| \tag{8.1.3}$$

and proved the Hyers–Ulam stability of the functional inequalities (8.1.2) and (8.1.3) in Banach spaces.

Throughout this chapter, assume that (X,μ,T_M) is a random normed algebra and (Y,μ,T_M) is a complete random normed algebra.

8.1.1 The Stability of Random Homomorphisms in Random Normed Algebras

In this section, using the fixed point method, we prove the Hyers–Ulam stability of *random homomorphisms* in complete random normed algebras associated with the Cauchy additive functional inequality (8.1.2).

Theorem 8.1.1 *Let $\varphi : X^3 \to [0,\infty)$ be a function such that there exists $L<\frac{1}{2}$ with*

$$\varphi(x,y,z) \le \frac{L}{2}\varphi(2x,2y,2z)$$

for all $x,y,z\in X$. Let $f:X\to Y$ be an odd mapping satisfying

$$\begin{aligned}&\mu_{rf(x)+f(ry)+f(rz)}(t)\\ &\quad \ge \min\left\{\mu_{f(rx+ry+rz)}\left(\frac{t}{2}\right), \frac{t}{t+\varphi(x,y,z)}\right\}\end{aligned} \tag{8.1.4}$$

and

$$\mu_{f(xy)-f(x)f(y)}(t) \ge \frac{t}{t+\varphi(x,y,0)} \tag{8.1.5}$$

for all $r\in\mathbb{R}$, $x,y,z\in X$ and $t>0$. Then $H(x):=\lim_{n\to\infty}2^n f(\frac{x}{2^n})$ exists for any $x\in X$ and defines a random homomorphism $H:X\to Y$ such that

$$\mu_{f(x)-A(x)}(t) \ge \frac{(2-2L)t}{(2-2L)t+L\varphi(x,x,-2x)} \tag{8.1.6}$$

for all $x \in X$ and $t > 0$.

Proof Since f is odd, $f(0) = 0$. So $\mu_{f(0)}(\frac{t}{2}) = 1$. Letting $r = 1$ and $y = x$ and replacing z by $-2x$ in (8.1.4), respectively, we get

$$\mu_{f(2x)-2f(x)}(t) \geq \frac{t}{t + \varphi(x, x, -2x)} \tag{8.1.7}$$

for all $x \in X$. Consider the following set

$$S := \{g : X \to Y\}$$

and introduce the generalized metric on S:

$$d(g, h) = \inf\left\{\nu \in \mathbb{R}_+ : \mu_{g(x)-h(x)}(\nu t) \geq \frac{t}{t + \varphi(x, x, -2x)}, \ \forall x \in X, \ t > 0\right\},$$

where, as usual, $\inf \phi = +\infty$. It is easy to show that (S, d) is complete (see the proof of [161, Lemma 2.1]).

Now, we consider the linear mapping $J : S \to S$ such that

$$Jg(x) := 2g\left(\frac{x}{2}\right)$$

for all $x \in X$. Let $g, h \in S$ be given such that $d(g, h) = \varepsilon$. Then we have

$$\mu_{g(x)-h(x)}(\varepsilon t) \geq \frac{t}{t + \varphi(x, x, -2x)}$$

for all $x \in X$ and all $t > 0$ and hence we have

$$\begin{aligned} \mu_{Jg(x)-Jh(x)}(L\varepsilon t) &= \mu_{2g(\frac{x}{2})-2h(\frac{x}{2})}(L\varepsilon t) \\ &= \mu_{g(\frac{x}{2})-h(\frac{x}{2})}\left(\frac{L}{2}\varepsilon t\right) \\ &\geq \frac{\frac{Lt}{2}}{\frac{Lt}{2} + \varphi(\frac{x}{2}, \frac{x}{2}, -x)} \\ &\geq \frac{\frac{Lt}{2}}{\frac{Lt}{2} + \frac{L}{2}\varphi(x, x, -2x)} \\ &= \frac{t}{t + \varphi(x, x, -2x)} \end{aligned}$$

for all $x \in X$ and $t > 0$. So $d(g, h) = \varepsilon$ implies that $d(Jg, Jh) \leq L\varepsilon$. This means that

$$d(Jg, Jh) \leq Ld(g, h)$$

for all $g, h \in S$. It follows from (8.1.7) that

$$\mu_{f(x)-2f(\frac{x}{2})}\left(\frac{L}{2}t\right) \geq \frac{t}{t+\varphi(x,x,-2x)}$$

for all $x \in X$ and $t > 0$. So $d(f, Jf) \leq \frac{L}{2}$.

Now, there exists a mapping $H : X \to Y$ satisfying the following:

(1) H is a fixed point of J, i.e.,

$$H\left(\frac{x}{2}\right) = \frac{1}{2}H(x) \tag{8.1.8}$$

for all $x \in X$. Since $f : X \to Y$ is odd, $H : X \to Y$ is an odd mapping. The mapping H is a unique fixed point of J in the set

$$M = \{g \in S : d(f, g) < \infty\}.$$

This implies that H is a unique mapping satisfying (8.1.8) such that there exists a $\nu \in (0, \infty)$ satisfying

$$\mu_{f(x)-H(x)}(\nu t) \geq \frac{t}{t+\varphi(x,x,-2x)}$$

for all $x \in X$;

(2) $d(J^n f, H) \to 0$ as $n \to \infty$. This implies the equality

$$\lim_{n\to\infty} 2^n f\left(\frac{x}{2^n}\right) = H(x)$$

for all $x \in X$;

(3) $d(f, H) \leq \frac{1}{1-L}d(f, Jf)$, which implies the inequality

$$d(f, H) \leq \frac{L}{2-2L}.$$

This implies that the inequality (8.1.6) holds.

Let $r = 1$ in (8.1.4). By (8.1.4),

$$\begin{aligned}
&\mu_{2^n(f(\frac{x}{2^n})+f(\frac{y}{2^n})+f(\frac{-x-y}{2^n}))}(2^n t) \\
&\geq \min\left\{\mu_{2^n f(0)}(2^{n-1}t), \frac{t}{t+\varphi(\frac{x}{2^n}, \frac{y}{2^n}, \frac{-x-y}{2^n})}\right\} \\
&= \frac{t}{t+\varphi(\frac{x}{2^n}, \frac{y}{2^n}, \frac{-x-y}{2^n})}
\end{aligned}$$

for all $x, y \in X$, $t > 0$ and $n \geq 1$ and so

$$\mu_{2^n(f(\frac{x}{2^n})+f(\frac{y}{2^n})+f(\frac{-x-y}{2^n}))}(t) \geq \frac{\frac{t}{2^n}}{\frac{t}{2^n}+\frac{L^n}{2^n}\varphi(x,y,-x-y)}$$

for all $x, y \in X$, $t > 0$ and $n \geq 1$. Since $\lim_{n\to\infty} \frac{\frac{t}{2^n}}{\frac{t}{2^n}+\frac{L^n}{2^n}\varphi(x,y,-x-y)} = 1$ for all $x, y \in X$ and $t > 0$, it follows that

$$\mu_{H(x)+H(y)+H(-x-y)}(t) \geq 1$$

for all $x, y \in X$ and $t > 0$. So the mapping $H : X \to Y$ is Cauchy additive.

Let $y = -x$ and $z = 0$ in (8.1.4). By (8.1.4), we have

$$\mu_{2^n f(\frac{rx}{2^n})-2^n rf(\frac{x}{2^n})}(2^n t) \geq \frac{t}{t+\varphi(\frac{x}{2^n}, \frac{-x}{2^n}, 0)}$$

for all $r \in \mathbb{R}$, $x \in X$, $t > 0$ and $n \geq 1$. So we have

$$\mu_{2^n f(\frac{rx}{2^n})-2^n rf(\frac{x}{2^n})}(t) \geq \frac{\frac{t}{2^n}}{\frac{t}{2^n}+\frac{L^n}{2^n}\varphi(x, -x, 0)}$$

for all $r \in \mathbb{R}$, $x \in X$, $t > 0$ and $n \geq 1$. Since $\lim_{n\to\infty} \frac{\frac{t}{2^n}}{\frac{t}{2^n}+\frac{L^n}{2^n}\varphi(x,-x,0)} = 1$ for all $x \in X$ and $t > 0$,

$$\mu_{H(rx)-rH(x)}(t) = 1$$

for all $r \in \mathbb{R}$, $x \in X$ and $t > 0$. Thus, the additive mapping $H : X \to Y$ is $\mathbb{R}$-linear. By (8.1.5), we have

$$\mu_{4^n f(\frac{x}{2^n}\cdot\frac{y}{2^n})-2^n f(\frac{x}{2^n})\cdot 2^n f(\frac{y}{2^n})}(4^n t) \geq \frac{t}{t+\varphi(\frac{x}{2^n}, \frac{y}{2^n}, 0)}$$

for all $x, y \in X$, $t > 0$ and $n \geq 1$. So we have

$$\mu_{4^n f(\frac{x}{2^n}\cdot\frac{y}{2^n})-2^n f(\frac{x}{2^n})\cdot 2^n f(\frac{y}{2^n})}(t) \geq \frac{\frac{t}{4^n}}{\frac{t}{4^n}+\frac{L^n}{2^n}\varphi(x, y, 0)}$$

for all $x, y \in X$, $t > 0$ and $n \geq 1$. Since $\lim_{n\to\infty} \frac{\frac{t}{4^n}}{\frac{t}{4^n}+\frac{L^n}{2^n}\varphi(x,y,0)} = 1$ for all $x, y \in X$ and $t > 0$,

$$\mu_{H(xy)-H(x)H(y)}(t) = 1$$

for all $x, y \in X$ and all $t > 0$. Thus, the mapping $H : X \to Y$ is multiplicative. Therefore, there exists a unique random homomorphism $H : X \to Y$ satisfying (8.1.6). This completes the proof. □

Theorem 8.1.2 *Let $\varphi : X^3 \to [0, \infty)$ be a function such that there exists $L < 1$ with*

$$\varphi(x, y, z) \leq 2L\varphi\left(\frac{x}{2}, \frac{y}{2}, \frac{z}{2}\right)$$

for all $x, y, z \in X$. Let $f : X \to Y$ be an odd mapping satisfying (8.1.4) *and* (8.1.5). *Then $H(x) := \lim_{n\to\infty} \frac{1}{2^n} f(2^n x)$ exists for any $x \in X$ and defines a random homomorphism $H : X \to Y$ such that*

$$\mu_{f(x)-H(x)}(t) \geq \frac{(2-2L)t}{(2-2L)t + \varphi(x, x, -2x)} \tag{8.1.9}$$

for all $x \in X$ and $t > 0$.

Proof Let (S, d) be the generalized metric space defined in the proof of Theorem 8.1.1. Consider the linear mapping $J : S \to S$ such that

$$Jg(x) := \frac{1}{2} g(2x)$$

for all $x \in X$. It follows from (8.1.7) that

$$\mu_{f(x)-\frac{1}{2}f(2x)}\left(\frac{1}{2}t\right) \geq \frac{t}{t + \varphi(x, x, -2x)}$$

for all $x \in X$ and all $t > 0$. So $d(f, Jf) \leq \frac{1}{2}$.

Now, there exists a mapping $H : X \to Y$ satisfying the following:

(1) H is a fixed point of J, that is,

$$H(2x) = 2H(x) \tag{8.1.10}$$

for all $x \in X$. Since $f : X \to Y$ is odd, $H : X \to Y$ is an odd mapping. The mapping H is a unique fixed point of J in the set

$$M = \{g \in S : d(f, g) < \infty\}.$$

This implies that H is a unique mapping satisfying (8.1.10) such that there exists a $\nu \in (0, \infty)$ satisfying

$$\mu_{f(x)-H(x)}(\nu t) \geq \frac{t}{t + \varphi(x, x, -2x)}$$

for all $x \in X$;

(2) $d(J^n f, H) \to 0$ as $n \to \infty$. This implies the equality

$$\lim_{n\to\infty} \frac{1}{2^n} f(2^n x) = H(x)$$

for all $x \in X$;

(3) $d(f, H) \leq \frac{1}{1-L} d(f, Jf)$, which implies the inequality

$$d(f, H) \leq \frac{1}{2-2L}.$$

This implies that the inequality (8.1.9) holds.

The rest of the proof is similar to the proof of Theorem 8.1.1. This completes the proof. □

8.1.2 The Stability of Random Derivations on Random Normed Algebras

In this section, using the fixed point method, we prove the Hyers–Ulam stability of *random derivations* on complete random normed algebras associated with the Cauchy–Jensen additive functional inequality (8.1.3).

Theorem 8.1.3 *Let $\varphi : Y^3 \to [0, \infty)$ be a function such that there exists $L < \frac{1}{2}$ with*

$$\varphi(x, y, z) \leq \frac{L}{2}\varphi(2x, 2y, 2z)$$

for all $x, y, z \in Y$. Let $f : Y \to Y$ be an odd mapping satisfying

$$\mu_{rf(x)+f(ry)+rf(2z)}(t) \geq \min\left\{\mu_{2f(\frac{rx+ry}{2}+rz)}\left(\frac{2t}{3}\right), \frac{t}{t+\varphi(x, y, z)}\right\} \tag{8.1.11}$$

and

$$\mu_{f(xy)-f(x)y-xf(y)}(t) \geq \frac{t}{t+\varphi(x, y, 0)} \tag{8.1.12}$$

for all $r \in \mathbb{R}$, $x, y, z \in Y$ and $t > 0$. Then $D(x) := \lim_{n\to\infty} 2^n f(\frac{x}{2^n})$ exists for any $x \in Y$ and defines a random derivation $D : Y \to Y$ such that

$$\mu_{f(x)-D(x)}(t) \geq \frac{(2-2L)t}{(2-2L)t + L\varphi(x, x, -x)} \tag{8.1.13}$$

for all $x \in Y$ and $t > 0$.

Proof Note that $\mu_{f(0)}(\frac{2t}{3}) = 1$. Letting $y = x = -z$ in (8.1.11), we get

$$\mu_{f(2x)-2f(x)}(t) \geq \frac{t}{t+\varphi(x, x, -x)} \tag{8.1.14}$$

for all $x \in Y$. Consider the set

$$S := \{g : Y \to Y\}$$

and introduce the generalized metric on S:

$$d(g, h) = \inf\left\{\nu \in \mathbb{R}_+ : \mu_{g(x)-h(x)}(\nu t) \geq \frac{t}{t+\varphi(x, x, -x)}, \ \forall x \in Y,\ t > 0\right\},$$

where, as usual, $\inf \phi = +\infty$. It is easy to show that (S, d) is complete (see the proof of [161, Lemma 2.1]).

Now, we consider the linear mapping $J : S \to S$ such that

$$Jg(x) := 2g\left(\frac{x}{2}\right)$$

for all $x \in Y$. It follows from (8.1.14) that

$$\mu_{f(x)-2f(\frac{x}{2})}\left(\frac{L}{2}t\right) \geq \frac{t}{t+\varphi(x,x,-x)}$$

for all $x \in Y$ and all $t > 0$. So $d(f, Jf) \leq \frac{L}{2}$.

Now, there exists a mapping $D : Y \to Y$ satisfying the following:

(1) D is a fixed point of J, that is,

$$D\left(\frac{x}{2}\right) = \frac{1}{2}D(x) \tag{8.1.15}$$

for all $x \in Y$. Since $f : Y \to Y$ is odd, $D : Y \to Y$ is an odd mapping. The mapping D is a unique fixed point of J in the set

$$M = \{g \in S : d(f, g) < \infty\}.$$

This implies that D is a unique mapping satisfying (8.1.15) such that there exists a $\nu \in (0, \infty)$ satisfying

$$\mu_{f(x)-D(x)}(\nu t) \geq \frac{t}{t+\varphi(x,x,-x)}$$

for all $x \in Y$;

(2) $d(J^n f, D) \to 0$ as $n \to \infty$. This implies the following equality:

$$\lim_{n\to\infty} 2^n f\left(\frac{x}{2^n}\right) = D(x)$$

for all $x \in Y$;

(3) $d(f, D) \leq \frac{1}{1-L}d(f, Jf)$, which implies the following inequality:

$$d(f, D) \leq \frac{L}{2-2L}.$$

This implies that the inequality (8.1.13) holds.

Let $r = 1$ in (8.1.11). By (8.1.11), we have

$$\mu_{2^n(f(\frac{x}{2^n})+f(\frac{y}{2^n})+f(\frac{-x-y}{2^n}))}(2^n t)$$
$$\geq \min\left\{\mu_{2^{n+1}f(0)}\left(\frac{2^n t}{3}\right), \frac{t}{t+\varphi(\frac{x}{2^n}, \frac{y}{2^n}, \frac{-x-y}{2^{n+1}})}\right\}$$

$$= \frac{t}{t + \varphi(\frac{x}{2^n}, \frac{y}{2^n}, \frac{-x-y}{2^{n+1}})}$$

for all $x, y \in Y$, $t > 0$ and $n \geq 1$ and so

$$\mu_{2^n(f(\frac{x}{2^n})+f(\frac{y}{2^n})+f(\frac{-x-y}{2^n}))}(t) \geq \frac{\frac{t}{2^n}}{\frac{t}{2^n} + \frac{L^n}{2^n}\varphi(x, y, \frac{-x-y}{2})}$$

for all $x, y \in Y$, $t > 0$ and $n \geq 1$. Since $\lim_{n\to\infty} \frac{\frac{t}{2^n}}{\frac{t}{2^n} + \frac{L^n}{2^n}\varphi(x,y,\frac{-x-y}{2})} = 1$ for all $x, y \in Y$ and $t > 0$,

$$\mu_{D(x)+D(y)+D(-x-y)}(t) \geq 1$$

for all $x, y \in Y$ and $t > 0$. So the mapping $D : Y \to Y$ is Cauchy additive.

Let $r = 1$, $z = 0$ and $y = -x$ in (8.1.11). By (8.1.11), we have

$$\mu_{2^n f(\frac{rx}{2^n})-2^n rf(\frac{x}{2^n})}(2^n t) \geq \frac{t}{t + \varphi(\frac{x}{2^n}, \frac{-x}{2^n}, 0)}$$

for all $r \in \mathbb{R}$, $x \in Y$, $t > 0$ and $n \geq 1$. So we have

$$\mu_{2^n f(\frac{rx}{2^n})-2^n rf(\frac{x}{2^n})}(t) \geq \frac{\frac{t}{2^n}}{\frac{t}{2^n} + \frac{L^n}{2^n}\varphi(x, -x, 0)}$$

for all $r \in \mathbb{R}$, $x \in Y$, $t > 0$ and $n \geq 1$. Since $\lim_{n\to\infty} \frac{\frac{t}{2^n}}{\frac{t}{2^n} + \frac{L^n}{2^n}\varphi(x,-x,0)} = 1$ for all $x \in Y$ and $t > 0$,

$$\mu_{D(rx)-rD(x)}(t) = 1$$

for all $r \in \mathbb{R}$, $x \in Y$ and $t > 0$. Thus, the additive mapping $D : Y \to Y$ is $\mathbb{R}$-linear. By (8.1.12), we have

$$\mu_{4^n f(\frac{x}{2^n}\cdot\frac{y}{2^n})-2^n f(\frac{x}{2^n})\cdot y - x\cdot 2^n f(\frac{y}{2^n})}(4^n t) \geq \frac{t}{t + \varphi(\frac{x}{2^n}, \frac{y}{2^n}, 0)}$$

for all $x, y \in Y$, $t > 0$ and $n \geq 1$. So we have

$$\mu_{4^n f(\frac{x}{2^n}\cdot\frac{y}{2^n})-2^n f(\frac{x}{2^n})\cdot y - x\cdot 2^n f(\frac{y}{2^n})}(t) \geq \frac{\frac{t}{4^n}}{\frac{t}{4^n} + \frac{L^n}{2^n}\varphi(x, y, 0)}$$

for all $x, y \in Y$, $t > 0$ and $n \geq 1$. Since $\lim_{n\to\infty} \frac{\frac{t}{4^n}}{\frac{t}{4^n} + \frac{L^n}{2^n}\varphi(x,y)} = 1$ for all $x, y \in Y$ and $t > 0$,

$$\mu_{D(xy)-D(x)y-xD(y)}(t) = 1$$

for all $x, y \in Y$ and $t > 0$. Thus, the mapping $D : Y \to Y$ satisfies $D(xy) = D(x)y + xD(y)$ for all $x, y \in Y$. Therefore, there exists a unique random derivation $D : Y \to Y$ satisfying (8.1.13). This completes the proof. □

Theorem 8.1.4 *Let $\varphi : Y^3 \to [0, \infty)$ be a function such that there exists $L < 1$ with*

$$\varphi(x, y, z) \le 2L\varphi\left(\frac{x}{2}, \frac{y}{2}, \frac{z}{2}\right)$$

for all $x, y, z \in Y$. Let $f : Y \to Y$ be an odd mapping satisfying (8.1.11) *and* (8.1.12). *Then $D(x) := \lim_{n\to\infty} \frac{1}{2^n} f(2^n x)$ exists for each $x \in Y$ and defines a random derivation $D : Y \to Y$ such that*

$$\mu_{f(x)-D(x)}(t) \ge \frac{(2-2L)t}{(2-2L)t + \varphi(x, x, -x)} \tag{8.1.16}$$

for all $x \in Y$ and $t > 0$.

Proof Let (S, d) be the generalized metric space defined in the proof of Theorem 8.1.3. Consider the linear mapping $J : S \to S$ such that

$$Jg(x) := \frac{1}{2} g(2x)$$

for all $x \in Y$.

The rest of the proof is similar to the proofs of Theorems 8.1.1 and 8.1.3. □

8.2 Cauchy–Jensen Functional Equations in Banach ∗-Algebras

In this section, we prove the Hyers–Ulam stability of Cauchy–Jensen functional equations in random Banach ∗-algebras.

Definition 8.2.1 Let X be a ∗-algebra and (X, μ, T_M) be a random normed space.

(1) The RN-space (X, μ, T_M) is called a *random normed ∗-algebra* if

$$\mu_{xy}(st) \ge \mu_x(s) \cdot \mu_y(t), \qquad \mu_{x^*}(t) = \mu_x(t)$$

for all $x, y \in X$ and $s, t > 0$.

(2) A complete random normed ∗-algebra is called a *random Banach ∗-algebra*.

Example 8.2.2 Let $(X, \|\cdot\|)$ be a normed ∗-algebra. Let

$$\mu_x(t) = \begin{cases} \frac{t}{t+\|x\|}, & t > 0,\ x \in X, \\ 0, & t \le 0,\ x \in X. \end{cases}$$

Then $\mu_x(t)$ is a random norm on X and (X, μ, T_M) is a random normed ∗-algebra.

Definition 8.2.3 Let $(X, \|\cdot\|)$ be a C^*-algebra and μ be a random norm on X.

(1) The random normed $*$-algebra (X, μ, T_M) is called an *induced random normed $*$-algebra.*
(2) The random Banach $*$-algebra (X, μ, T_M) is called an *induced random C^*-algebra.*

Definition 8.2.4 Let (X, μ, T_M) and (X, μ', T_M) be random normed $*$-algebras.

(1) A multiplicative $\mathbb{C}$-linear mapping $H : (X, \mu, T_M) \to (X, \mu', T_M)$ is called a *random $*$-homomorphism* if

$$H(x^*) = H(x)^*$$

for all $x \in X$.
(2) A $\mathbb{C}$-linear mapping $D : (X, \mu, T_M) \to (X, \mu', T_M)$ is called a *random $*$-derivation* if

$$D(xy) = D(x)y + xD(y), \qquad D(x^*) = D(x)^*$$

for all $x, y \in X$.

In this section, using the fixed point method, we prove the Hyers–Ulam stability of the Cauchy–Jensen functional equation in random Banach $*$-algebras.

Theorem 8.2.5 *Let $\varphi : X^3 \to [0, \infty)$ be a function such that there exists $L < \frac{1}{2}$ with*

$$\varphi(x, y, z) \le \frac{L}{2}\varphi(2x, 2y, 2z)$$

for all $x, y, z \in X$. Let $f : X \to Y$ be a mapping satisfying

$$\mu_{2f(\frac{rx+ry}{2}+rz)-rf(x)-rf(y)-2rf(z)}(t) \ge \frac{t}{t+\varphi(x, y, z)}, \tag{8.2.1}$$

$$\mu_{f(xy)-f(x)f(y)}(t) \ge \frac{t}{t+\varphi(x, y, 0)}, \tag{8.2.2}$$

$$\mu_{f(x^*)-f(x)^*}(t) \ge \frac{t}{t+\varphi(x, 0, 0)} \tag{8.2.3}$$

for all $x, y, z \in X$, $t > 0$ and $r \in \mathbb{C}$. Then $H(x) := \lim_{n\to\infty} 2^n f(\frac{x}{2^n})$ exists for any $x \in X$ and defines a random $*$-homomorphism $H : X \to Y$ such that

$$\mu_{f(x)-H(x)}(t) \ge \frac{(1-L)t}{(1-L)t+\varphi(x, 0, 0)} \tag{8.2.4}$$

for all $x \in X$ and $t > 0$.

Proof Letting $r = 1$ and $y = z = 0$ in (8.2.1), we get

$$\mu_{2f(\frac{x}{2})-f(x)}(t) \geq \frac{t}{t+\varphi(x,0,0)} \tag{8.2.5}$$

for all $x \in X$. Consider the following set

$$S := \{g : X \to Y\}$$

and introduce the generalized metric on S:

$$d(g,h) = \inf\left\{\lambda \in \mathbb{R}_+ : \mu_{g(x)-h(x)}(\lambda t) \geq \frac{t}{t+\varphi(x,0,0)}, \ \forall x \in X, \ t > 0\right\},$$

where, as usual, $\inf \phi = +\infty$. It is easy to show that (S,d) is complete (see the proof of [161, Lemma 2.1]).

Now, we consider the linear mapping $J : S \to S$ such that

$$Jg(x) := 2g\left(\frac{x}{2}\right)$$

for all $x \in X$. Let $g, h \in S$ be given such that $d(g,h) = \varepsilon$. Then we have

$$\mu_{g(x)-h(x)}(\varepsilon t) \geq \frac{t}{t+\varphi(x,0,0)}$$

for all $x \in X$ and $t > 0$ and hence

$$\begin{aligned}
\mu_{Jg(x)-Jh(x)}(L\varepsilon t) &= \mu_{2g(\frac{x}{2})-2h(\frac{x}{2})}(L\varepsilon t) \\
&= \mu_{g(\frac{x}{2})-h(\frac{x}{2})}\left(\frac{L}{2}\varepsilon t\right) \\
&\geq \frac{\frac{Lt}{2}}{\frac{Lt}{2}+\varphi(\frac{x}{2},0,0)} \\
&\geq \frac{\frac{Lt}{2}}{\frac{Lt}{2}+\frac{L}{2}\varphi(x,0,0)} \\
&= \frac{t}{t+\varphi(x,0,0)}
\end{aligned}$$

for all $x \in X$ and $t > 0$. So $d(g,h) = \varepsilon$ implies that $d(Jg, Jh) \leq L\varepsilon$. This means that

$$d(Jg, Jh) \leq Ld(g,h)$$

for all $g, h \in S$. It follows from (8.2.5) that $d(f, Jf) \leq 1$.

Now, there exists a mapping $H : X \to Y$ satisfying the following:

(1) H is a fixed point of J, that is,

$$H\left(\frac{x}{2}\right)=\frac{1}{2}H(x) \tag{8.2.6}$$

for all $x\in X$. The mapping H is a unique fixed point of J in the set

$$M=\{g\in S:d(f,g)<\infty\}.$$

This implies that H is a unique mapping satisfying (8.2.6) such that there exists a $\lambda\in(0,\infty)$ satisfying

$$\mu_{f(x)-H(x)}(\lambda t)\geq\frac{t}{t+\varphi(x,0,0)}$$

for all $x\in X$;

(2) $d(J^nf,H)\to 0$ as $n\to\infty$. This implies the following equality:

$$\lim_{n\to\infty}2^nf\left(\frac{x}{2^n}\right)=H(x)$$

for all $x\in X$;

(3) $d(f,H)\leq\frac{1}{1-L}d(f,Jf)$, which implies the following inequality:

$$d(f,H)\leq\frac{1}{1-L}.$$

This implies that the inequality (8.2.4) holds.

By (8.2.1), it follows that

$$\begin{aligned}&\mu_{2^{k+1}f(\frac{rx+ry}{2^{k+1}}+\frac{rz}{2^k})-2^krf(\frac{x}{2^k})-2^krf(\frac{y}{2^k})-2^{k+1}rf(\frac{z}{2^k})}(2^ktt)\\&\quad\geq\frac{t}{t+\varphi(\frac{x}{2^k},\frac{y}{2^k},\frac{z}{2^k})}\end{aligned}$$

for all $x,y,z\in X$, $t>0$ and $r\in\mathbb{C}$. Then we have

$$\begin{aligned}&\mu_{2^{k+1}f(\frac{rx+ry}{2^{k+1}}+\frac{rz}{2^k})-2^krf(\frac{x}{2^k})-2^krf(\frac{y}{2^k})-2^{k+1}rf(\frac{z}{2^k})}(t)\\&\quad\geq\frac{\frac{t}{2^k}}{\frac{t}{2^k}+\frac{L^k}{2^k}\varphi(x,y,z)}\end{aligned}$$

for all $x,y,z\in X$, $t>0$ and $r\in\mathbb{C}$. Since $\lim_{k\to\infty}\frac{\frac{t}{2^k}}{\frac{t}{2^k}+\frac{L^k}{2^k}\varphi(x,y,z)}=1$ for all $x,y,z\in X$, $t>0$ and $r\in\mathbb{C}$,

$$\mu_{2H(\frac{rx+ry}{2}+rz)-rH(x)-rH(y)-2rH(z)}(t)=1$$

for all $x, y, z \in X$, $t > 0$ and $r \in \mathbb{C}$. Thus, we have

$$2H\left(\frac{rx+ry}{2}+rz\right) - rH(x) - rH(y) - 2rH(z) = 0.$$

So the mapping $H : X \to Y$ is additive and $\mathbb{C}$-linear. By (8.2.2), we have

$$\mu_{4^k f(\frac{xy}{4^k}) - 2^k f(\frac{x}{2^k}) \cdot 2^k f(\frac{y}{2^k})}(4^k t) \geq \frac{t}{t + \varphi(\frac{x}{2^k}, \frac{y}{2^k}, 0)}$$

for all $x, y \in X$ and $t > 0$ and so

$$\mu_{4^k f(\frac{xy}{4^k}) - 2^k f(\frac{x}{2^k}) \cdot 2^k f(\frac{y}{2^k})}(t) \geq \frac{\frac{t}{4^k}}{\frac{t}{4^k} + \frac{L^k}{2^k}\varphi(x, y, 0)}$$

for all $x, y \in X$ and $t > 0$. Since $\lim_{k\to\infty} \frac{\frac{t}{4^k}}{\frac{t}{4^k} + \frac{L^k}{2^k}\varphi(x,y,0)} = 1$ for all $x, y \in X$ and $t > 0$, it follows that

$$\mu_{H(xy) - H(x)H(y)}(t) = 1$$

for all $x, y \in X$ and $t > 0$. Thus, $H(xy) - H(x)H(y) = 0$. By (8.2.3), we have

$$\mu_{2^k f(\frac{x^*}{2^k}) - 2^k f(\frac{x}{2^k})^*}(2^k t) \geq \frac{t}{t + \varphi(\frac{x}{2^k}, 0, 0)}$$

for all $x, y \in X$, $t > 0$ and $r \in \mathbb{C}$ and so

$$\mu_{2^k f(\frac{x^*}{2^k}) - 2^k f(\frac{x}{2^k})^*}(t) \geq \frac{\frac{t}{2^k}}{\frac{t}{2^k} + \frac{L^k}{2^k}\varphi(x, 0, 0)}$$

for all $x, y \in X$ and $t > 0$. Since $\lim_{k\to\infty} \frac{\frac{t}{2^k}}{\frac{t}{2^k} + \frac{L^k}{2^k}\varphi(x,y,0)} = 1$ for all $x, y \in X$ and $t > 0$,

$$\mu_{H(x^*) - H(x)^*}(t) = 1$$

for all $x, y \in X$ and $t > 0$. Thus, $H(x^*) - H(x)^* = 0$. Therefore, the mapping $H : X \to Y$ is a random $*$-homomorphism, as desired. This completes the proof. □

Theorem 8.2.6 *Let $\varphi : X^3 \to [0, \infty)$ be a function such that there exists $L < 1$ with*

$$\varphi(x, y, z) \leq 2L\varphi\left(\frac{x}{2}, \frac{y}{2}, \frac{z}{2}\right)$$

for all $x, y, z \in X$. Let $f : X \to Y$ be a mapping satisfying (8.2.1), (8.2.2) *and* (8.2.3). *Then $H(x) := \lim_{k\to\infty} \frac{1}{2^k} f(2^k x)$ exists for each $x \in X$ and defines a random $*$-homomorphism $H : X \to Y$ such that*

$$\mu_{f(x)-H(x)}(t) \geq \frac{(1-L)t}{(1-L)t + L\varphi(x,0,0)} \tag{8.2.7}$$

for all $x \in X$ and $t > 0$.

Proof Let (S, d) be the generalized metric space defined in the proof of Theorem 8.2.5. Consider the linear mapping $J : S \to S$ such that

$$Jg(x) := \frac{1}{2} g(2x)$$

for all $x \in X$. It follows from (8.2.5) that

$$N\left(f(x) - \frac{1}{2} f(2x), \frac{1}{2} t\right) \geq \frac{t}{t + \varphi(2x,0,0)} \geq \frac{t}{t + 2L\varphi(x,0,0)}$$

for all $x \in X$ and $t > 0$. So $d(f, Jf) \leq L$. Hence

$$d(f, H) \leq \frac{L}{1-L},$$

which implies that the inequality (8.2.7) holds.

The rest of the proof is similar to the proof of Theorem 8.2.5. This completes the proof. □

Theorem 8.2.7 *Let $\varphi : X^3 \to [0, \infty)$ be a function such that there exists $L < \frac{1}{2}$ with*

$$\varphi(x, y, z) \leq \frac{L}{2} \varphi(2x, 2y, 2z)$$

for all $x, y, z \in X$. Let $f : X \to Y$ be a mapping satisfying (8.2.2), (8.2.3) *and*

$$\mu_{f(\frac{rx+ry}{2} + rz) - \frac{r}{2} f(x) - \frac{r}{2} f(y) - r f(z)}(t) \geq \frac{t}{t + \varphi(x, y, z)} \tag{8.2.8}$$

for all $x, y, z \in X$, $t > 0$ and $r \in \mathbb{C}$. Then $H(x) := \lim_{n\to\infty} 2^n f(\frac{x}{2^n})$ exists for any $x \in X$ and defines a random $*$-homomorphism $H : X \to Y$ such that

$$\mu_{f(x)-H(x)}(t) \geq \frac{(2-2L)t}{(2-2L)t + L\varphi(x,x,x)} \tag{8.2.9}$$

for all $x \in X$ and $t > 0$.

Proof Letting $r = 1$ and $y = z = x$ in (8.2.8), we get

$$\mu_{f(2x)-2f(x)}(t) \geq \frac{t}{t + \varphi(x,x,x)} \tag{8.2.10}$$

for all $x \in X$. Consider the following set

$$S := \{g : X \to Y\}$$

and introduce the generalized metric on S:

$$d(g,h) = \inf\left\{\lambda \in \mathbb{R}_+ : \mu_{g(x)-h(x)}(\lambda t) \geq \frac{t}{t+\varphi(x,x,x)},\ \forall x \in X,\ t > 0\right\},$$

where, as usual, $\inf \phi = +\infty$. It is easy to show that (S, d) is complete (see the proof of [161, Lemma 2.1]).

Now, we consider the linear mapping $J : S \to S$ such that

$$Jg(x) := 2g\left(\frac{x}{2}\right)$$

for all $x \in X$. It follows from (8.2.10) that

$$\mu_{f(x)-2f(\frac{x}{2})}\left(\frac{L}{2}t\right) \geq \frac{t}{t+\varphi(x,x,x)}$$

for all $x \in X$ and $t > 0$. So $d(f, Jf) \leq \frac{L}{2}$. Hence

$$d(f,H) \leq \frac{L}{2-2L},$$

which implies that the inequality (8.2.9) holds.

The rest of the proof is similar to the proof of Theorem 8.2.5. This completes the proof. □

Theorem 8.2.8 *Let $\varphi : X^3 \to [0,\infty)$ be a function such that there exists $L < 1$ with*

$$\varphi(x,y,z) \leq 2L\varphi\left(\frac{x}{2},\frac{y}{2},\frac{z}{2}\right)$$

for all $x, y, z \in X$. Let $f : X \to Y$ be a mapping satisfying (8.2.2), (8.2.3) *and* (8.2.8). *Then $H(x) := \lim_{k\to\infty} \frac{1}{2^k} f(2^k x)$ exists for any $x \in X$ and defines a random $*$-homomorphism $H : X \to Y$ such that*

$$\mu_{f(x)-H(x)}(t) \geq \frac{(2-2L)t}{(2-2L)t+\varphi(x,x,x)} \tag{8.2.11}$$

for all $x \in X$ and $t > 0$.

Proof Let (S, d) be the generalized metric space defined in the proof of Theorem 8.2.7. Consider the linear mapping $J : S \to S$ such that

$$Jg(x) := \frac{1}{2}g(2x)$$

for all $x \in X$. It follows from (8.2.10) that

$$\mu_{f(x)-\frac{1}{2}f(2x)}\left(\frac{1}{2}t\right) \geq \frac{t}{t+\varphi(x,x,x)}$$

for all $x \in X$ and $t > 0$. So $d(f, Jf) \leq \frac{1}{2}$.

The rest of the proof is similar to the proof of Theorem 8.2.7. This completes the proof. □

8.2.1 The Hyers–Ulam Stability of Cauchy–Jensen Functional Equations in Induced Random C^*-algebras

Throughout this section, we assume that X is a unital C^*-algebra with the unit e and the unitary group $U(X) := \{u \in X : u^*u = uu^* = e\}$ and Y is a unital C^*-algebra.

Using the fixed point method, we prove the Hyers–Ulam stability of the Cauchy–Jensen functional equation in induced random C^*-algebras.

Theorem 8.2.9 *Let $\varphi : X^3 \to [0, \infty)$ be a function such that there exists $L < \frac{1}{2}$ with*

$$\varphi(x,y,z) \leq \frac{L}{2}\varphi(2x,2y,2z)$$

for all $x, y, z \in X$. Let $f : X \to Y$ be a mapping satisfying (8.2.1) *and*

$$\mu_{f(uv)-f(u)f(v)}(t) \geq \frac{t}{t+\varphi(u,v,0)} \tag{8.2.12}$$

and

$$\mu_{f(u^*)-f(u)^*}(t) \geq \frac{t}{t+\varphi(u,0,0)} \tag{8.2.13}$$

for all $u, v \in U(X)$ and $t > 0$. Then there exists a random $*$-homomorphism $H : X \to Y$ satisfying (8.2.4).

Proof By the same reasoning as in the proof of Theorem 8.2.5, there exists a $\mathbb{C}$-linear mapping $H : X \to Y$ satisfying (8.2.4). The mapping $H : X \to Y$ is given by

$$\lim_{n\to\infty} 2^n f\left(\frac{x}{2^n}\right) = H(x)$$

for all $x \in X$. By (8.2.12),

$$\mu_{4^k f(\frac{uv}{4^k})-2^k f(\frac{u}{2^k})\cdot 2^k f(\frac{v}{2^k})}(4^k t) \geq \frac{t}{t+\varphi(\frac{u}{2^k},\frac{v}{2^k},0)}$$

for all $u, v \in U(X)$ and $t > 0$. So we have

$$\mu_{4^k f(\frac{uv}{4^k})-2^k f(\frac{u}{2^k})\cdot 2^k f(\frac{v}{2^k})}(t) \geq \frac{\frac{t}{4^k}}{\frac{t}{4^k} + \frac{L^k}{2^k}\varphi(u, v, 0)}$$

for all $u, v \in U(X)$ and $t > 0$. Since $\lim_{k\to\infty} \frac{\frac{t}{4^k}}{\frac{t}{4^k}+\frac{L^k}{2^k}\varphi(u,y,0)} = 1$ for all $u, v \in U(X)$ and $t > 0$,

$$\mu_{H(uv)-H(u)H(v)}(t) = 1$$

for all $u, v \in U(X)$ and $t > 0$. Thus,

$$H(uv) = H(u)H(v) \tag{8.2.14}$$

for all $u, v \in U(X)$. Since H is $\mathbb{C}$-linear and each $x \in X$ is a finite linear combination of unitary elements (see [131, Theorem 4.1.7]), that is, $x = \sum_{j=1}^{m} \lambda_j u_j$ for all $\lambda_j \in \mathbb{C}$ and $u_j \in U(X)$, it follows from (8.2.14) that

$$\begin{aligned} H(xv) &= H\left(\sum_{j=1}^{m} \lambda_j u_j v\right) = \sum_{j=1}^{m} \lambda_j H(u_j v) = \sum_{j=1}^{m} \lambda_j H(u_j) H(v) \\ &= H\left(\sum_{j=1}^{m} \lambda_j u_j\right) H(v) = H(x)H(v) \end{aligned}$$

for all $v \in U(X)$.

Similarly, one can obtain

$$H(xy) = H(x)H(y)$$

for all $x, y \in X$. By (8.2.13), we have

$$\mu_{2^k f(\frac{u^*}{2^k})-2^k f(\frac{u}{2^k})^*}(2^k t) \geq \frac{t}{t + \varphi(\frac{u}{2^k}, 0, 0)}$$

for all $u \in U(X)$ and $t > 0$. So we have

$$\mu_{2^k f(\frac{u^*}{2^k})-2^k f(\frac{u}{2^k})^*}(t) \geq \frac{\frac{t}{2^k}}{\frac{t}{2^k} + \frac{L^k}{2^k}\varphi(u, 0, 0)}$$

for all $u \in U(X)$ and $t > 0$. Since $\lim_{k\to\infty} \frac{\frac{t}{2^k}}{\frac{t}{2^k}+\frac{L^k}{2^k}\varphi(u,0,0)} = 1$ for all $u \in U(X)$ and $t > 0$, it follows that

$$\mu_{H(u^*)-H(u)^*}(t) = 1$$

for all $u \in U(X)$ and $t > 0$. Thus, we have

$$H(u^*) = H(u)^* \tag{8.2.15}$$

for all $u \in U(X)$. Since H is $\mathbb{C}$-linear and each $x \in X$ is a finite linear combination of unitary elements (see [131, Theorem 4.1.7]), that is, $x = \sum_{j=1}^{m} \lambda_j u_j$ for all $\lambda_j \in \mathbb{C}$ and $u_j \in U(X)$, it follows from (8.2.15) that

$$\begin{aligned} H(x^*) &= H\left(\sum_{j=1}^{m} \overline{\lambda_j} u_j^*\right) = \sum_{j=1}^{m} \overline{\lambda_j} H(u_j^*) = \sum_{j=1}^{m} \overline{\lambda_j} H(u_j)^* \\ &= H\left(\sum_{j=1}^{m} \lambda_j u_j\right)^* = H(x)^* \end{aligned}$$

and so $H(x^*) = H(x)^*$ for all $x \in X$. Therefore, the mapping $H : X \to Y$ is a random $*$-homomorphism. This completes the proof. □

Similarly, we can obtain the following, but we omit the proofs.

Theorem 8.2.10 *Let $\varphi : X^3 \to [0, \infty)$ be a function such that there exists $L < 1$ with*

$$\varphi(x, y, z) \le 2L\varphi\left(\frac{x}{2}, \frac{y}{2}, \frac{z}{2}\right)$$

for all $x, y, z \in X$. Let $f : X \to Y$ be a mapping satisfying (8.2.1), (8.2.12) *and* (8.2.13). *Then there exists a random $*$-homomorphism $H : X \to Y$ satisfying* (8.2.7).

Theorem 8.2.11 *Let $\varphi : X^3 \to [0, \infty)$ be a function such that there exists $L < \frac{1}{2}$ with*

$$\varphi(x, y, z) \le \frac{L}{2}\varphi(2x, 2y, 2z)$$

for all $x, y, z \in X$. Let $f : X \to Y$ be a mapping satisfying (8.2.8), (8.2.12) *and* (8.2.2). *Then there exists a random $*$-homomorphism $H : X \to Y$ satisfying* (8.2.9).

Theorem 8.2.12 *Let $\varphi : X^3 \to [0, \infty)$ be a function such that there exists $L < 1$ with*

$$\varphi(x, y, z) \le 2L\varphi\left(\frac{x}{2}, \frac{y}{2}, \frac{z}{2}\right)$$

for all $x, y, z \in X$. Let $f : X \to Y$ be a mapping satisfying (8.2.8), (8.2.12) *and* (8.2.2). *Then there exists a random $*$-homomorphism $H : X \to Y$ satisfying* (8.2.11).

8.2.2 The Hyers–Ulam Stability of Cauchy–Jensen Functional Inequalities in Random Banach ∗-Algebras and Induced Random C*-Algebras

We need the following lemma for the proofs of the main results in this section.

Lemma 8.2.13 [36, 194] *Let (Y, N) be a random normed vector spaces. Let $f : X \to Y$ be a mapping such that*

$$N\big(f(x) + f(y) + 2f(z), t\big) \geq N\left(2f\left(\frac{x+y}{2} + z\right), \frac{2t}{3}\right)$$

for all $x, y, z \in X$ and $t > 0$. Then f is Cauchy additive, that is., $f(x+y) = f(x) + f(y)$ for all $x, y \in X$.

Using the fixed point method, we prove the Hyers–Ulam stability of the Cauchy–Jensen functional inequality in random Banach ∗-algebras.

Theorem 8.2.14 *Let $\varphi : X^3 \to [0, \infty)$ be a function such that there exists $L < \frac{1}{2}$ with*

$$\varphi(x, y, z) \leq \frac{L}{2}\varphi(2x, 2y, 2z)$$

for all $x, y, z \in X$. Let $f : X \to Y$ be an odd mapping satisfying (8.2.2), (8.2.3) *and*

$$\mu_{rf(x)+rf(y)+f(2rz)}(t) \geq \min\left\{\mu_{2rf(\frac{x+y}{2}+z)}\left(\frac{2t}{3}\right), \frac{t}{t+\varphi(x, y, z)}\right\} \tag{8.2.16}$$

for all $x, y, z \in X$, $t > 0$ and $r \in \mathbb{C}$. Then $F(x) := \lim_{n\to\infty} 2^n f(\frac{x}{2^n})$ exists for any $x \in X$ and defines a random ∗-homomorphism $F : X \to Y$ such that

$$\mu_{f(x)-F(x)}(t) \geq \frac{(2-2L)t}{(2-2L)t + L\varphi(x, x, -x)} \tag{8.2.17}$$

for all $x \in X$ and $t > 0$.

Proof Letting $r = 1$ and $y = x = -z$ in (8.2.16), we get

$$\mu_{f(2x)-2f(x)}(t) \geq \frac{t}{t+\varphi(x, x, -x)} \tag{8.2.18}$$

for all $x \in X$. Consider the following set

$$S := \{g : X \to Y\}$$

and introduce the generalized metric on S:

$$d(g,h)=\inf\left\{\lambda\in\mathbb{R}_+:\mu_{g(x)-h(x)}(\lambda t)\geq\frac{t}{t+\varphi(x,x,-x)},\ \forall x\in X,\ t>0\right\},$$

where, as usual, $\inf\phi=+\infty$. It is easy to show that (S,d) is complete (see [161, Lemma 2.1]).

Now, we consider the linear mapping $J:S\to S$ such that

$$Jg(x):=2g\left(\frac{x}{2}\right)$$

for all $x\in X$. Let $g,h\in S$ be given such that $d(g,h)=\varepsilon$. Then we have

$$\mu_{g(x)-h(x)}(t)\geq\frac{t}{t+\varphi(x,x,-x)}$$

for all $x\in X$ and $t>0$. Hence, we have

$$\begin{aligned}\mu_{Jg(x)-Jh(x)}(L\varepsilon t)&=\mu_{2g(\frac{x}{2})-2h(\frac{x}{2})}(L\varepsilon t)\\&=\mu_{g(\frac{x}{2})-h(\frac{x}{2})}\left(\frac{L}{2}\varepsilon t\right)\\&\geq\frac{\frac{Lt}{2}}{\frac{Lt}{2}+\varphi(\frac{x}{2},\frac{x}{2},-\frac{x}{2})}\\&\geq\frac{\frac{Lt}{2}}{\frac{Lt}{2}+\frac{L}{2}\varphi(x,x,-x)}\\&=\frac{t}{t+\varphi(x,x,-x)}\end{aligned}$$

for all $x\in X$ and $t>0$. So $d(g,h)=\varepsilon$ implies that $d(Jg,Jh)\leq L\varepsilon$. This means that

$$d(Jg,Jh)\leq Ld(g,h)$$

for all $g,h\in S$. It follows from (8.2.18) that

$$\mu_{f(x)-2f(\frac{x}{2})}\left(\frac{L}{2}t\right)\geq\frac{t}{t+\varphi(x,x,-x)}$$

for all $x\in X$ and $t>0$. So $d(f,Jf)\leq\frac{L}{2}$.

Now, there exists a mapping $F:X\to Y$ satisfying the following:

(1) F is a fixed point of J, that is,

$$F\left(\frac{x}{2}\right)=\frac{1}{2}F(x)\tag{8.2.19}$$

for all $x \in X$. Since $f : X \to Y$ is odd, $F : X \to Y$ is an odd mapping. The mapping F is a unique fixed point of J in the set

$$M = \{g \in S : d(f, g) < \infty\}.$$

This implies that F is a unique mapping satisfying (8.2.19) such that there exists a $\lambda \in (0, \infty)$ satisfying

$$\mu_{f(x)-F(x)}(\lambda t) \geq \frac{t}{t + \varphi(x, x, -x)}$$

for all $x \in X$;

(2) $d(J^n f, F) \to 0$ as $n \to \infty$. This implies the following equality:

$$\lim_{n\to\infty} 2^n f\left(\frac{x}{2^n}\right) = F(x)$$

for all $x \in X$;

(3) $d(f, F) \leq \frac{1}{1-L} d(f, Jf)$, which implies the following inequality:

$$d(f, F) \leq \frac{L}{2 - 2L}.$$

This implies that the inequality (8.2.17) holds. By (8.2.16), we have

$$\begin{aligned}&\mu_{2^n(rf(\frac{x}{2^n})+rf(\frac{y}{2^n})+f(\frac{rz}{2^{n-1}}))}(2^n t)\\ &\quad\geq \min\left\{\mu_{2^{n+1}rf(\frac{x+y}{2^{n+1}}+\frac{z}{2^n})}\left(\frac{2^{n+1}}{3}t\right), \frac{t}{t + \varphi(\frac{x}{2^n}, \frac{y}{2^n}, \frac{z}{2^n})}\right\}\end{aligned}$$

for all $x, y, z \in X$, $t > 0$, $r \in \mathbb{C}$ and $n \geq 1$. So we have

$$\begin{aligned}&\mu_{2^n(rf(\frac{x}{2^n})+rf(\frac{y}{2^n})+f(\frac{rz}{2^{n-1}}))}(t)\\ &\quad\geq \min\left\{\mu_{2^{n+1}rf(\frac{x+y}{2^{n+1}}+\frac{z}{2^n})}\left(\frac{2t}{3}\right), \frac{\frac{t}{2^n}}{\frac{t}{2^n} + \frac{L^n}{2^n}\varphi(x, y, z)}\right\}\end{aligned}$$

for all $x, y, z \in X$, $t > 0$, $r \in \mathbb{C}$ and $n \geq 1$. Since $\lim_{n\to\infty} \frac{\frac{t}{2^n}}{\frac{t}{2^n}+\frac{L^n}{2^n}\varphi(x,y,z)} = 1$ for all $x, y, z \in X$ and $t > 0$,

$$\mu_{rF(x)+rF(y)+F(2rz)}(t) \geq \mu_{2rF(\frac{x+y}{2}+z)}\left(\frac{2t}{3}\right) \tag{8.2.20}$$

for all $x, y, z \in X$, $t > 0$ and $r \in \mathbb{C}$. Let $r = 1$ in (8.2.20). By Lemma 8.2.13, the mapping $F : X \to Y$ is Cauchy additive. Letting $y = x = -z$ in (4.5), we get

$$\mu_{2rF(x)+F(-2rx)}(t) \geq N\left(0, \frac{2t}{3}\right) = 1$$

for all $x \in X$, $t > 0$ and $r \in \mathbb{C}$. So $2rF(x) = -F(-2rx) = 2F(rx)$ for all $x \in X$ and $r \in \mathbb{C}$. Thus F is $\mathbb{C}$-linear.

The rest of the proof is similar to the proof of Theorem 8.2.5. This completes the proof. □

Theorem 8.2.15 *Let $\varphi : X^3 \to [0,\infty)$ be a function such that there exists $L < 1$ with*

$$\varphi(x,y,z) \le 2L\varphi\left(\frac{x}{2},\frac{y}{2},\frac{z}{2}\right)$$

for all $x,y,z \in X$. Let $f : X \to Y$ be an odd mapping satisfying (8.2.1), (8.2.3) *and* (8.2.16). *Then $F(x) := \lim_{n\to\infty} \frac{1}{2^n} f(2^n x)$ exists for each $x \in X$ and defines a random $*$-homomorphism $F : X \to Y$ such that*

$$\mu_{f(x)-F(x)}(t) \ge \frac{(2-2L)t}{(2-2L)t + \varphi(x,x,-x)} \tag{8.2.21}$$

for all $x \in X$ and $t > 0$.

Proof Let (S,d) be the generalized metric space defined in the proof of Theorem 8.2.14. It follows from (8.2.18) that

$$\mu_{f(x)-\frac{1}{2}f(2x)}\left(\frac{1}{2}t\right) \ge \frac{t}{t+\varphi(x,x,-x)}$$

for all $x \in X$ and $t > 0$. So $d(f,Jf) \le \frac{1}{2}$. Hence

$$d(f,F) \le \frac{1}{2-2L},$$

which implies that the inequality (8.2.21) holds.

The rest of the proof is similar to the proofs of Theorems 8.2.5 and 8.2.14. This completes the proof. □

From now on, assume that X is a unital C^*-algebra with the unit e and unitary group $U(X) := \{u \in X : u^*u = uu^* = e\}$ and that Y is a unital C^*-algebra.

Using the fixed point method, we prove the Hyers–Ulam stability of the Cauchy–Jensen functional inequality in the induced random C^*-algebras.

Theorem 8.2.16 *Let $\varphi : X^3 \to [0,\infty)$ be a function such that there exists $L < \frac{1}{2}$ with*

$$\varphi(x,y,z) \le \frac{L}{2}\varphi(2x,2y,2z)$$

for all $x,y,z \in X$. Let $f : X \to Y$ be an odd mapping satisfying (8.2.12), (8.2.13) *and* (8.2.16). *Then $F(x) := \lim_{n\to\infty} 2^n f(\frac{x}{2^n})$ exists for each $x \in X$ and defines a random $*$-homomorphism $F : X \to Y$ satisfying* (8.2.17).

Proof The proof is similar to the proofs of Theorems 8.2.5, 8.2.9 and 8.2.14. □

Similarly, we can obtain the following, but we omit the proof.

Theorem 8.2.17 *Let $\varphi : X^3 \to [0, \infty)$ be a function such that there exists $L < 1$ with*

$$\varphi(x, y, z) \le 2L\varphi\left(\frac{x}{2}, \frac{y}{2}, \frac{z}{2}\right)$$

for all $x, y, z \in X$. Let $f : X \to Y$ be an odd mapping satisfying (8.2.12), (8.2.13) *and* (8.2.16). *Then $F(x) := N\text{-}\lim_{n\to\infty} \frac{1}{2^n} f(2^n x)$ exists for any $x \in X$ and defines a random $*$-homomorphism $F : X \to Y$ satisfying* (8.2.21).

8.3 Random ∗-Derivations in Banach ∗-Algebras

In this section, assume that $(Y, \mu, T_M) = (X, \mu, T_M)$ in the previous sections, and that (X, μ, T_M) is a random Banach $*$-algebra.

Using the fixed point method, we prove the Hyers–Ulam stability of random $*$-derivations in random Banach $*$-algebras.

Theorem 8.3.1 *Let $\varphi : X^3 \to [0, \infty)$ be a function such that there exists an $L < \frac{1}{2}$ with*

$$\varphi(x, y, z) \le \frac{L}{2}\varphi(2x, 2y, 2z)$$

for all $x, y, z \in X$. Let $f : X \to X$ be a mapping satisfying (8.2.1), (8.2.3) *and*

$$\mu_{f(xy)-f(x)y-xf(y)}(t) \ge \frac{t}{t + \varphi(x, y, 0)} \tag{8.3.1}$$

for all $x, y \in X$ and $t > 0$. Then $D(x) := \lim_{n\to\infty} 2^n f(\frac{x}{2^n})$ exists for any $x \in X$ and defines a random $*$-derivation $D : X \to X$ such that

$$\mu_{f(x)-D(x)}(t) \ge \frac{(1 - L)t}{(1 - L)t + \varphi(x, 0, 0)} \tag{8.3.2}$$

for all $x \in X$ and $t > 0$.

Proof The proof is similar to the proof of Theorem 8.2.5. □

Similarly, we can obtain the following, but we omit the proof here.

Theorem 8.3.2 *Let $\varphi : X^3 \to [0, \infty)$ be a function such that there exists $L < 1$ with*

$$\varphi(x, y, z) \le 2L\varphi\left(\frac{x}{2}, \frac{y}{2}, \frac{z}{2}\right)$$

for all $x, y, z \in X$. Let $f : X \to X$ be a mapping satisfying (8.2.1), (8.2.3) *and* (8.3.1). *Then $D(x) := \lim_{n\to\infty} \frac{1}{2^n} f(2^n x)$ exists for any $x \in X$ and defines a random $*$-derivation $D : X \to X$ such that*

$$\mu_{f(x)-D(x)}(t) \geq \frac{(1-L)t}{(1-L)t + L\varphi(x,0,0)} \tag{8.3.3}$$

for all $x \in X$ and $t > 0$.

From now on, assume that X is a unital C^*-algebra with the unit e and unitary group $U(X) := \{u \in X : u^*u = uu^* = e\}$.

Using the fixed point method, we prove the Hyers–Ulam stability of random $*$-derivations in the induced random C^*-algebras.

Theorem 8.3.3 *Let $\varphi : X^3 \to [0,\infty)$ be a function such that there exists $L < \frac{1}{2}$ with*

$$\varphi(x,y,z) \leq \frac{L}{2}\varphi(2x,2y,2z)$$

for all $x, y, z \in X$. Let $f : X \to X$ be an odd mapping satisfying (8.2.12), (8.2.14) *and* (8.3.1). *Then $D(x) := \lim_{n\to\infty} 2^n f(\frac{x}{2^n})$ exists for any $x \in X$ and defines a random $*$-derivation $D : X \to X$ satisfying* (8.3.2).

Proof The proof is similar to the proofs of Theorems 8.2.5 and 8.2.9. □

Similarly, we can obtain the following, but we omit the proof here.

Theorem 8.3.4 *Let $\varphi : X^3 \to [0,\infty)$ be a function such that there exists $L < 1$ with*

$$\varphi(x,y,z) \leq 2L\varphi\left(\frac{x}{2},\frac{y}{2},\frac{z}{2}\right)$$

for all $x, y, z \in X$. Let $f : X \to X$ be an odd mapping satisfying (8.2.12), (8.2.14) *and* (8.3.1). *Then $D(x) := \lim_{n\to\infty} \frac{1}{2^n} f(2^n x)$ exists for each $x \in X$ and defines a random $*$-derivation $D : X \to X$ satisfying* (8.3.3).

8.3.1 The Stability of Homomorphisms and Derivations in Non-Archimedean Random C-Algebras*

Definition 8.3.5 Let $\mathcal{U}$ be a non-Archimedean random Banach algebra. An *involution* on $\mathcal{U}$ is a mapping $u \to u^*$ from $\mathcal{U}$ into $\mathcal{U}$ satisfying the following conditions:

(1) $u^{**} = u$ for any $u \in \mathcal{U}$;
(2) $(\alpha u + \beta v)^* = \overline{\alpha}u^* + \overline{\beta}v^*$;
(3) $(uv)^* = v^*u^*$ for any $u, v \in \mathcal{U}$.

If, in addition, $\mu_{u^*u}(t) = \mu_u(t)^2$ for any $u \in \mathcal{U}$ and $t > 0$, then $\mathcal{U}$ is called a *non-Archimedean random C^*-algebra.*

Throughout this section, assume that $\mathcal{A}$ is a non-Archimedean random C^*-algebra with the norm $\mu^{\mathcal{A}}_{\cdot}$ and $\mathcal{B}$ is a non-Archimedean random C^*-algebra with the norm $\mu^{\mathcal{B}}_{\cdot}$.

For any mapping $f : \mathcal{A} \to \mathcal{B}$, we define

$$D_\lambda f(x_1, \dots, x_m) := \sum_{i=1}^{m} \lambda f\left(mx_i + \sum_{j=1,\, j\neq i}^{m} x_j\right) + f\left(\lambda \sum_{i=1}^{m} x_i\right) - 2f\left(\lambda \sum_{i=1}^{m} mx_i\right)$$

for all $\lambda \in \mathbb{T}^1 := \{\nu \in \mathbb{C} : |\nu| = 1\}$ and $x_1, \dots, x_m \in \mathcal{A}$.

Note that a $\mathbb{C}$-linear mapping $H : \mathcal{A} \to \mathcal{B}$ is called a *homomorphism* in non-Archimedean random C^*-algebras if H satisfies $H(xy) = H(x)H(y)$ and $H(x^*) = H(x)^*$ for all $x, y \in \mathcal{A}$.

We prove the generalized Hyers–Ulam stability of homomorphisms in non-Archimedean random C^*-algebras for the following functional equation:

$$D_\lambda f(x_1, \dots, x_m) = 0.$$

Theorem 8.3.6 *Let $f : \mathcal{A} \to \mathcal{B}$ be a mapping for which there exist the functions $\varphi : \mathcal{A}^m \to D^+$, $\psi : \mathcal{A}^2 \to D^+$ and $\eta : \mathcal{A} \to D^+$ such that $|m| < 1$ is far from zero and*

$$\mu^{\mathcal{B}}_{D_\lambda f(x_1,\dots,x_m)}(t) \geq \varphi_{x_1,\dots,x_m}(t), \tag{8.3.4}$$

$$\mu^{\mathcal{B}}_{f(xy)-f(x)f(y)}(t) \geq \psi_{x,y}(t), \tag{8.3.5}$$

$$\mu^{\mathcal{B}}_{f(x^*)-f(x)^*}(t) \geq \eta_x(t), \tag{8.3.6}$$

for all $\lambda \in \mathbb{T}^1$, $x_1, \dots, x_m, x, y \in A$ and $t > 0$. If there exists $L < 1$ such that

$$\varphi_{mx_1,\dots,mx_m}\left(|m|Lt\right) \geq \varphi_{x_1,\dots,x_m}(t), \tag{8.3.7}$$

$$\psi_{mx,my}\left(|m|^2Lt\right) \geq \psi_{x,y}(t), \tag{8.3.8}$$

$$\eta_{mx}\left(|m|Lt\right) \geq \eta_x(t), \tag{8.3.9}$$

for all $x, y, x_1, \dots, x_m \in \mathcal{A}$ and $t > 0$, then there exists a unique random homomorphism $H : \mathcal{A} \to \mathcal{B}$ such that

$$\mu^{\mathcal{B}}_{f(x)-H(x)}(t) \geq \varphi_{x,0,\dots,0}\left(\left(|m| - |m|L\right)t\right) \tag{8.3.10}$$

for all $x \in \mathcal{A}$ and $t > 0$.

Proof It follows from (8.3.7), (8.3.8), (8.3.9) and $L < 1$ that

$$\lim_{n\to\infty} \varphi_{m^n x_1,\dots,m^n x_m}\left(|m|^n t\right) = 1, \tag{8.3.11}$$

$$\lim_{n\to\infty} \psi_{m^n x,m^n y}\left(|m|^{2n} t\right) = 1, \tag{8.3.12}$$

$$\lim_{n\to\infty} \eta_{m^n x}\left(|m|^n t\right) = 1 \tag{8.3.13}$$

for all $x, y, x_1, \dots, x_m \in \mathcal{A}$ and $t > 0$.

Now, let Ω be the set of all mappings $g : \mathcal{A} \to \mathcal{B}$ and introduce a generalized metric on Ω as follows:

$$d(g,h) = \inf\left\{k \in (0,\infty) : \mu^{\mathcal{B}}_{g(x)-h(x)}(kt) > \phi_{x,0,\dots,0}(t),\ \forall x \in \mathcal{A},\ t > 0\right\}.$$

It is easy to show that (Ω, d) is a generalized complete metric space (see [26]).

Now, we consider the function $J : \Omega \to \Omega$ defined by $Jg(x) = \frac{1}{m}g(mx)$ for all $x \in \mathcal{A}$ and $g \in \Omega$. Note that, for all $g, h \in \Omega$,

$$\begin{aligned} d(g,h) < k \quad &\Longrightarrow \quad \mu^{\mathcal{B}}_{g(x)-h(x)}(kt) > \phi_{x,0,\dots,0}(t) \\ &\Longrightarrow \quad \mu^{\mathcal{B}}_{\frac{1}{m}g(mx)-\frac{1}{m}h(mx)}(kt) > |m|\phi_{mx,0,\dots,0}\left(|m|t\right) \\ &\Longrightarrow \quad \mu^{\mathcal{B}}_{\frac{1}{m}g(mx)-\frac{1}{m}h(mx)}(kLt) > \phi_{mx,0,\dots,0}(t) \\ &\Longrightarrow \quad d(Jg, Jh) < kL. \end{aligned}$$

From this, it is easy to see that $d(Jg, Jk) \le Ld(g,h)$ for all $g, h \in \Omega$, that is, J is a self-function of Ω with the Lipschitz constant L.

Putting $\mu = 1$, $x = x_1$ and $x_2 = x_3 = \cdots = x_m = 0$ in (8.3.4), respectively, we have

$$\mu^{\mathcal{B}}_{f(mx)-mf(x)}(t) \ge \phi_{x,0,\dots,0}(t)$$

for all $x \in \mathcal{A}$ and $t > 0$. Then we have

$$\mu^{\mathcal{B}}_{f(x)-\frac{1}{m}f(mx)}(t) \ge \phi_{x,0,\dots,0}\left(|m|t\right)$$

for all $x \in \mathcal{A}$ and $t > 0$, that is, $d(Jf, f) \le \frac{1}{|m|} < \infty$. Now, from the fixed point alternative, it follows that there exists a fixed point H of J in Ω such that

$$H(x) = \lim_{n\to\infty} \frac{1}{|m|^n} f\left(m^n x\right) \tag{8.3.14}$$

for all $x \in \mathcal{A}$ since $\lim_{n\to\infty} d(J^n f, H) = 0$.

On the other hand, it follows from (8.3.4), (8.3.11) and (8.3.14) that

$$\mu^{\mathcal{B}}_{D_\lambda H(x_1,\dots,x_m)}(t) = \lim_{n\to\infty} \mu^{\mathcal{B}}_{\frac{1}{m^n}Df(m^n x_1,\dots,m^n x_m)}(t)$$

$$\geq \lim_{n\to\infty} \phi_{m^n x_1,\ldots,m^n x_m}\left(|m|^n t\right)$$
$$= 1.$$

By a similar method to the above, we get $\lambda H(mx) = H(m\lambda x)$ for all $\lambda \in \mathbb{T}^1$ and $x \in \mathcal{A}$. Thus, one can show that the mapping $H : \mathcal{A} \to \mathcal{B}$ is $\mathbb{C}$-linear.

Further, it follows from (8.3.5), (8.3.12) and (8.3.14) that

$$\begin{aligned}\mu^{\mathcal{B}}_{H(xy)-H(x)H(y)}(t) &= \lim_{n\to\infty} \mu^{\mathcal{B}}_{f(m^{2n}xy)-f(m^n x)f(m^n y)}\left(|m|^{2n} t\right)\\ &\geq \lim_{n\to\infty} \psi_{m^n x, m^n y}\left(|m|^{2n} t\right)\\ &= 1\end{aligned}$$

for all $x, y \in \mathcal{A}$. So $H(xy) = H(x)H(y)$ for all $x, y \in \mathcal{A}$. Thus $H : \mathcal{A} \to \mathcal{B}$ is a homomorphism satisfying (8.3.10), as desired.

Also, by (8.3.6), (8.3.13) and (8.3.14), we have $H(x^*) = H(x)^*$ by a similar method. This completes the proof. □

Corollary 8.3.7 *Let $r > 1$ and θ be nonnegative real numbers and $f : \mathcal{A} \to \mathcal{B}$ be a mapping such that*

$$\mu^{\mathcal{B}}_{D_\lambda f(x_1,\ldots,x_m)}(t) \geq \frac{t}{t + \theta \cdot (\|x_1\|^r_{\mathcal{A}} + \|x_2\|^r_{\mathcal{A}} + \cdots + \|x_m\|^r_{\mathcal{A}})},$$

$$\mu^{\mathcal{B}}_{f(xy)-f(x)f(y)}(t) \geq \frac{t}{t + \theta \cdot (\|x\|^r_{\mathcal{A}} \cdot \|y\|^r_{\mathcal{A}})},$$

$$\mu^{\mathcal{B}}_{f(x^*)-f(x)^*}(t) \geq \frac{t}{t + \theta \cdot \|x\|^r_{\mathcal{A}}}$$

for all $\lambda \in \mathbb{T}^1, x_1, \ldots, x_m, x, y \in \mathcal{A}$ and $t > 0$. Then there exists a unique homomorphism $H : \mathcal{A} \to \mathcal{B}$ such that

$$\mu^{\mathcal{B}}_{f(x)-H(x)}(t) \geq \frac{(|m| - |m|^r)t}{(|m| - |m|^r)t + \theta|m| - |m|^r \|x\|^r_{\mathcal{A}}}$$

for all $x \in \mathcal{A}$ and $t > 0$.

Proof The proof follows from Theorem 8.3.6 by taking

$$\varphi_{x_1,\ldots,x_m}(t) = \frac{t}{t + \theta \cdot (\|x_1\|^r_{\mathcal{A}} + \|x_2\|^r_{\mathcal{A}} + \cdots + \|x_m\|^r_{\mathcal{A}})},$$

$$\psi_{x,y}(t) := \frac{t}{t + \theta \cdot (\|x\|^r_{\mathcal{A}} \cdot \|y\|^r_{\mathcal{A}})}$$

$$\eta_x(t) = \frac{t}{t + \theta \cdot \|x\|^r_{\mathcal{A}}}$$

for all $x_1, \dots, x_m, x, y \in A$, $L = |m|^{r-1}$ and $t > 0$, we get the desired result. □

Now, we prove the generalized Hyers–Ulam stability of derivations on non-Archimedean random C^*-algebras for the functional equation $D_\lambda f(x_1, \dots, x_m) = 0$.

Theorem 8.3.8 *Let $f : \mathcal{A} \to \mathcal{A}$ be a mapping for which there exist the functions $\varphi : \mathcal{A}^m \to D^+$, $\psi : \mathcal{A}^2 \to D^+$ and $\eta : \mathcal{A} \to D^+$ such that $|m| < 1$ is far from zero and*

$$\mu^{\mathcal{A}}_{D_\lambda f(x_1,\dots,x_m)}(t) \geq \varphi_{x_1,\dots,x_m}(t),$$

$$\mu^{\mathcal{A}}_{f(xy)-f(x)y-xf(y)}(t) \geq \psi_{x,y}(t),$$

$$\mu^{\mathcal{A}}_{f(x^*)-f(x)^*}(t) \geq \eta_x(t)$$

for all $\lambda \in \mathbb{T}^1$ and $x_1, \dots, x_m, x, y \in A$ and $t > 0$. If there exists $L < 1$ such that (8.3.7), (8.3.8) *and* (8.3.9) *hold, then there exists a unique derivation $\delta : \mathcal{A} \to \mathcal{A}$ such that*

$$\mu^{\mathcal{A}}_{f(x)-\delta(x)}(t) \geq \varphi_{x,0,\dots,0}\big((|m| - |m|L)t\big)$$

for all $x \in \mathcal{A}$ and $t > 0$.

8.3.2 The Stability of Homomorphisms and Derivations in Non-Archimedean Lie C-Algebras*

A non-Archimedean random C^*-algebra $\mathcal{C}$, endowed with the Lie product

$$[x, y] := \frac{xy - yx}{2}$$

on $\mathcal{C}$, is called a *Lie non-Archimedean random C^*-algebra*.

Definition 8.3.9 Let $\mathcal{A}$ and $\mathcal{B}$ be random Lie C^*-algebras. A $\mathbb{C}$-linear mapping $H : \mathcal{A} \to \mathcal{B}$ is called a *non-Archimedean Lie C^*-algebra homomorphism* if

$$H\big([x, y]\big) = \big[H(x), H(y)\big]$$

for all $x, y \in \mathcal{A}$.

Throughout this section, assume that $\mathcal{A}$ is a non-Archimedean random Lie C^*-algebra with the norm $\mu^{\mathcal{A}}$ and $\mathcal{B}$ is a non-Archimedean random Lie C^*-algebra with the norm $\mu^{\mathcal{B}}$.

Now, we prove the generalized Hyers–Ulam stability of homomorphisms in non-Archimedean random Lie C^*-algebras for the functional equation $D_\lambda f(x_1, \dots, x_m) = 0$.

Theorem 8.3.10 *Let $f : \mathcal{A} \to \mathcal{B}$ be a mapping for which there exist the functions $\varphi : \mathcal{A}^m \to D^+$ and $\psi : \mathcal{A}^2 \to D^+$ such that* (8.3.4) *and* (8.3.6) *hold and*

$$\mu^{\mathcal{B}}_{f([x,y])-[f(x),f(y)]}(t) \geq \psi_{x,y}(t) \tag{8.3.15}$$

for all $\lambda \in \mathbb{T}^1$, $x, y \in \mathcal{A}$ and $t > 0$. If there exists $L < 1$ and (8.3.7), (8.3.8) *and* (8.3.9) *hold, then there exists a unique homomorphism $H : \mathcal{A} \to \mathcal{B}$ such that* (8.3.10) *hold.*

Proof By the same reasoning as in the proof of Theorem 8.3.6, we can find the mapping $H : \mathcal{A} \to \mathcal{B}$ defined by

$$H(x) = \lim_{n\to\infty} \frac{f(m^n x)}{|m|^n} \tag{8.3.16}$$

for all $x \in \mathcal{A}$. It follows from (8.3.8) and (8.3.16) that

$$\begin{aligned}\mu^{\mathcal{B}}_{H([x,y])-[H(x),H(y)]}(t) &= \lim_{n\to\infty} \mu^{\mathcal{B}}_{f(m^{2n}[x,y])-[f(m^n x),f(m^n y)]}\big(|m|^{2n} t\big) \\ &\geq \lim_{n\to\infty} \psi_{m^n x, m^n y}\big(|m|^{2n} t\big) \\ &= 1\end{aligned}$$

for all $x, y \in \mathcal{A}$ and $t > 0$, then we have

$$H\big([x,y]\big) = \big[H(x), H(y)\big]$$

for all $x, y \in \mathcal{A}$. Thus $H : \mathcal{A} \to \mathcal{B}$ is a Lie C^*-algebra homomorphism satisfying (8.3.10), as desired. This completed the proof. □

Corollary 8.3.11 *Let $r > 1$ and θ be nonnegative real numbers and $f : \mathcal{A} \to \mathcal{B}$ be a mapping such that*

$$\mu^{\mathcal{B}}_{D_\lambda f(x_1,\ldots,x_m)}(t) \geq \frac{t}{t + \theta(\|x_1\|^r_A + \|x_2\|^r_A + \cdots + \|x_m\|^r_A)},$$

$$\mu^{\mathcal{B}}_{f([x,y])-[f(x),f(y)]}(t) \geq \frac{t}{t + \theta \cdot \|x\|^r_A \cdot \|y\|^r_A},$$

$$\mu^{\mathcal{B}}_{f(x^*)-f(x)^*}(t) \geq \frac{t}{t + \theta \cdot \|x\|^r_A}$$

for all $\lambda \in \mathbb{T}^1, x_1, \ldots, x_m, x, y \in \mathcal{A}$ and $t > 0$. Then there exists a unique homomorphism $H : \mathcal{A} \to \mathcal{B}$ such that

$$\mu^{\mathcal{B}}_{f(x)-H(x)}(t) \geq \frac{(|m| - |m|^r)t}{(|m| - |m|^r)t + \theta \|x\|^r_A}$$

for all $x \in \mathcal{A}$ and $t > 0$.

Proof The proof follows from Theorem 8.3.10 and a method similar to Corollary 8.3.7. □

Definition 8.3.12 Let $\mathcal{A}$ be a non-Archimedean random Lie C^*-algebra. A $\mathbb{C}$-linear mapping $\delta : \mathcal{A} \to \mathcal{A}$ is called a *Lie derivation* if

$$\delta([x, y]) = [\delta(x), y] + [x, \delta(y)]$$

for all $x, y \in \mathcal{A}$.

Now, we prove the generalized Hyers–Ulam stability of derivations on non-Archimedean random Lie C^*-algebras for the following functional equation:

$$D_\lambda f(x_1, \dots, x_m) = 0.$$

Theorem 8.3.13 *Let* $f : \mathcal{A} \to \mathcal{A}$ *be a mapping for which there exist the functions* $\varphi : A^m \to D^+$ *and* $\psi : A^2 \to D^+$ *such that* (8.3.4) *and* (8.3.6) *hold and*

$$\mu^{\mathcal{A}}_{f([x,y])-[f(x),y]-[x,f(y)]}(t) \geq \psi_{x,y}(t) \tag{8.3.17}$$

for all $x, y \in \mathcal{A}$. *If there exists* $L < 1$ *and* (8.3.7), (8.3.8) *and* (8.3.9) *hold, then there exists a unique Lie derivation* $\delta : \mathcal{A} \to \mathcal{A}$ *such that such that* (8.3.10) *hold.*

Proof By the same reasoning as the proof of Theorem 8.3.10, there exists a unique $\mathbb{C}$-linear mapping $\delta : \mathcal{A} \to \mathcal{A}$ satisfying (8.3.10), where the mapping $\delta : \mathcal{A} \to \mathcal{A}$ is defined by

$$\delta(x) = \lim_{n\to\infty} \frac{f(m^n x)}{|m|^n} \tag{8.3.18}$$

for all $x \in \mathcal{A}$. It follows from (8.3.8) and (8.3.18) that

$$\begin{aligned}
&\mu^{\mathcal{A}}_{\delta([x,y])-[\delta(x),y]-[x,\delta(y)]}(t) \\
&= \lim_{n\to\infty} \mu^{\mathcal{A}}_{f(m^{2n}[x,y])-[f(m^n x),\cdot m^n y]-[m^n x, f(m^n y)]}\left(|m|^{2n} t\right) \\
&\geq \lim_{n\to\infty} \psi_{m^n x, m^n y}\left(|m|^{2n} t\right) \\
&= 1
\end{aligned}$$

for all $x, y \in \mathcal{A}$ and $t > 0$, then we have

$$\delta([x, y]) = [\delta(x), y] + [x, \delta(y)]$$

for all $x, y \in \mathcal{A}$. Thus, $\delta : \mathcal{A} \to \mathcal{A}$ is a Lie derivation satisfying (8.3.10). This completes the proof. □

Chapter 9
Related Results on Stability of Functional Inequalities and Equations

In this chapter, we prove some stability results for certain functional inequalities and functional equations in latticetic random φ-normed spaces, r-divisible groups and homogeneous probabilistic (random) modular spaces.

9.1 Latticetic Stability of the Functional Inequalities

In this section, we prove some stability results concerning the following functional inequalities:

$$\|f(x)+f(y)+f(z)\| \leq \|f(x+y+z)\| \tag{9.1.1}$$

and

$$\|f(x)+f(y)+2f(z)\| \leq \left\|2f\left(\frac{x+y}{2}+z\right)\right\| \tag{9.1.2}$$

in the setting of latticetic random φ-normed spaces. The generalized Hyers–Ulam–Rassias stability of the following functional inequality:

$$\|2f(x)+2f(y)-f(x-y)\| \leq \|f(x+y)\| \tag{9.1.3}$$

has been proved by Fechner [86] and Gilányi [94]. In [93], Gilányi showed that, if f satisfies the functional inequality (9.1.3), then f also satisfies the Jordan–von Neumann functional equation:

$$2f(x)+2f(y)=f(x-y)+f(x+y)$$

(see also [232]).

In [195], Park et al. investigated the Cauchy additive functional inequality (9.1.1) and the Cauchy–Jensen additive functional inequality: (9.1.2) and proved the generalized Hyers–Ulam–Rassias stability of the functional inequalities (9.1.1) and (9.1.2) in Banach spaces. We also mention the paper [193] in this section.

Y.J. Cho et al., *Stability of Functional Equations in Random Normed Spaces*, Springer Optimization and Its Applications 86, DOI 10.1007/978-1-4614-8477-6_9,

A *latticetic random* φ*-normed space* is a triple $(X, \mu, \mathcal{T}_\wedge)$, where X is a vector space and μ is a mapping from X into D_L^+ (for all $x \in X$, the function $\mu(x)$ is denoted by μ_x and $\mu_x(t)$ is the value μ_x at $t \in \mathbb{R}$) such that the following conditions hold:

(φ-LRN1) $\mu_x(t) = 1_\mathcal{L}$ for all $t > 0$ if and only if $x = 0$;
(φ-LRN2) $\mu_{\alpha x}(t) = \mu_x(\frac{t}{\varphi(\alpha)})$ for all $x \in X$, $\alpha \neq 0$ and $t \geq 0$;
(φ-LRN3) $\mu_{x+y}(t+s) \geq_L \mathcal{T}_\wedge(\mu_x(t), \mu_y(s))$ for all $x, y \in X$ and $t, s \geq 0$.

Now, we give some lemmas for the main results in latticetic random φ-normed spaces.

Lemma 9.1.1 *Let X be a linear space, $(Z, \mu, \mathcal{T}_\wedge)$ be a latticetic random φ-normed space and $f : X \to Z$ be a mapping such that*

$$\mu_{f(x)+f(y)+f(z)}(t) \geq_L \mu_{f(x+y+z)}\left(\frac{t}{\varphi(2)}\right) \tag{9.1.4}$$

for all $x, y, z \in X$ and $t > 0$. Then f is Cauchy additive, that is,

$$f(x+y) = f(x) + f(y)$$

for all $x, y \in X$.

Proof Putting $x = y = z = 0$ in (9.1.4), we obtain

$$\mu_{3f(0)}(t) \geq_L \mu_{f(0)}\left(\frac{t}{\varphi(3)}\right) \geq_L \mu_{f(0)}\left(\frac{t}{\varphi(2)}\right)$$

for any $t > 0$. It follows that $f(0) = 0$. Putting $y = -x$ and $z = 0$ in (9.1.4), respectively, one obtains

$$\mu_{f(x)+f(-x)}(t) \geq_L \mu_{f(0)}\left(\frac{t}{\varphi(2)}\right) = \mu_0\left(\frac{t}{\varphi(2)}\right) = 1_\mathcal{L}$$

for any $t > 0$ and hence

$$f(x) = -f(-x)$$

for all $x \in X$. Putting $z = -x - y$ in (9.1.4), we deduce that

$$\begin{aligned} \mu_{f(x)+f(y)-f(x+y)}(t) &= \mu_{f(x)+f(y)+f(-x-y)}(t) \\ &\geq_L \mu_{f(0)}\left(\frac{t}{\varphi(2)}\right) \\ &= \mu_0\left(\frac{t}{\varphi(2)}\right) \\ &= 1_\mathcal{L} \end{aligned}$$

and so

$$f(x)+f(y)=f(x+y)$$

for all $x, y \in X$. This completes the proof. □

Similarly, one can prove the following lemma.

Lemma 9.1.2 *Let X be a linear space, $(Z, \mu, \mathcal{T}_\wedge)$ be a latticetic random φ-normed space and $f : X \to Z$ be a mapping such that*

$$\mu_{f(x)+f(y)+2f(z)}(t) \geq_L \mu_{2f(\frac{x+y}{2}+z)}\left(\frac{\varphi(2)t}{\varphi(3)}\right) \tag{9.1.5}$$

for all $x, y, z \in X$ and $t > 0$. Then f is Cauchy additive.

Theorem 9.1.3 *Let X be a linear space and Φ be a mapping from X^3 to D_L^+ ($\Phi(x, y, z)(t)$ is denoted by $\Phi_{x,y,z}(t)$) such that, for some $0 < \alpha < \varphi(2)$,*

$$\Phi_{2x,2y,2z}(\alpha t) \geq_L \Phi_{x,y,z}(t) \tag{9.1.6}$$

for all $x, y, z \in X$ and $t > 0$. Let $(Y, \mu, \mathcal{T}_\wedge)$ be a complete a latticetic random φ-normed space. If $f : X \to Y$ is an odd mapping satisfying the inequality:

$$\mathcal{T}_\wedge\big(\mu_{f(x)+f(y)+f(z)}(t), \mu_{f(x+y+z)}(t)\big) \geq_L \Phi_{x,y,z}(t) \tag{9.1.7}$$

for all $x, y, z \in X$ and $t > 0$, then there exists a unique Cauchy additive mapping $A : X \to Y$ such that

$$\mu_{f(x)-A(x)}(t) \geq_L \Phi_{x,x,-2x}\big((\varphi(2)-\alpha)t\big) \tag{9.1.8}$$

for all $x \in X$ and $t > 0$.

Proof Putting $x = y$ and $z = -2x$ in (9.1.7), respectively, we get

$$\begin{aligned} \mu_{2f(x)-f(2x)}(t) &= \wedge\big(\mu_{2f(x)-f(2x)}(t), 1_{\mathcal{L}}\big) \\ &\geq_L \mathcal{T}_\wedge\big(\mu_{2f(x)-f(2x)}(t), \mu_{f(0)}(t)\big) \\ &\geq_L \Phi_{x,x,-2x}(t) \end{aligned} \tag{9.1.9}$$

for all $x \in X$ and $t > 0$. From (9.1.9), it follows that

$$\mu_{\frac{f(2x)}{2}-f(x)}\left(\frac{t}{\varphi(2)}\right) = \mu_{2f(x)-f(2x)}(t) \geq_L \Phi_{x,x,-2x}(t) \tag{9.1.10}$$

for all $x \in X$ and $t > 0$. Replacing x by $2^n x$ in (9.1.10) and using (9.1.6), we obtain

$$\mu_{\frac{f(2^{n+1}x)}{2^{n+1}}-\frac{f(2^n x)}{2^n}}\left(\frac{t}{\varphi(2^{n+1})}\right) \geq_L \Phi_{2^n x,2^n x,-2^{n+1}x}(t) \tag{9.1.11}$$

for all $x \in X$, $t > 0$ and $n \geq 1$, that is,

$$\mu_{\frac{f(2^{n+1}x)}{2^{n+1}} - \frac{f(2^n x)}{2^n}}(t) \geq_L \Phi_{2^n x, 2^n x, -2^{n+1}x}(\varphi(2^{n+1})t)$$
$$\geq_L \Phi_{x,x,-2x}\left(\frac{\varphi(2^{n+1})t}{\alpha^n}\right) \tag{9.1.12}$$

for all $x \in X$, $t > 0$ and $n \geq 1$. Since

$$\frac{f(2^n x)}{2^n} - f(x) = \sum_{k=0}^{n-1}\left(\frac{f(2^{k+1}x)}{2^{k+1}} - \frac{f(2^k x)}{2^k}\right),$$

it follows from (9.1.12) that

$$\mu_{\frac{f(2^n x)}{2^n} - f(x)}\left(t\sum_{k=0}^{n}\frac{\alpha^k}{\varphi(2^{k+1})}\right) \geq_L (\wedge)_{k=0}^{n-1}\Phi_{x,x,-2x}(t) = \Phi_{x,x,-2x}(t),$$

that is,

$$\mu_{\frac{f(2^n x)}{2^n} - f(x)}(t) \geq_L \Phi_{x,x,-2x}\left(\frac{t}{\sum_{k=0}^{n}\frac{\alpha^k}{\varphi(2^{k+1})}}\right). \tag{9.1.13}$$

By replacing x with $2^m x$ in (9.1.13), we obtain

$$\mu_{\frac{f(2^{n+m}x)}{2^{n+m}} - \frac{f(2^m x)}{2^m}}(t) \geq_L \Phi_{2^m x, 2^m x, -2^{m+1}x}\left(\frac{t}{\sum_{k=0}^{n}\frac{\alpha^k}{\varphi(2)^{m+k+1}}}\right)$$
$$\geq_L \Phi_{x,x,-2x}\left(\frac{t}{\sum_{k=0}^{n}\frac{\alpha^{m+k}}{\varphi(2)^{m+k+1}}}\right)$$
$$\geq_L \Phi_{x,x,-2x}\left(\frac{t}{\sum_{k=m}^{n+m}\frac{\alpha^k}{\varphi(2)^{k+1}}}\right). \tag{9.1.14}$$

Since $\Phi_{x,x,-2x}(\frac{t}{\sum_{k=m}^{n+m}\frac{\alpha^k}{\varphi(2)^{k+1}}})$ tends to $1_{\mathcal{L}}$ as m, n tend to ∞, we conclude that $\{\frac{f(2^n x)}{2^n}\}$ is a Cauchy sequence in $(Y, \mu, \mathcal{T}_\wedge)$. Since $(Y, \mu, \mathcal{T}_\wedge)$ is a complete latticetic random φ-normed space, this sequence converges to a point $A(x) \in Y$. Fix $x \in X$ and put $m = 0$ in (9.1.14). Then we have

$$\mu_{\frac{f(2^n x)}{2^n} - f(x)}(t) \geq_L \Phi_{x,x,-2x}\left(\frac{t}{\sum_{k=0}^{n}\frac{\alpha^k}{\varphi(2)^{k+1}}}\right), \tag{9.1.15}$$

which implies that, for all $t, \delta > 0$,

$$
\begin{aligned}
&\mu_{A(x)-f(x)}(t+\delta) \\
&\geq_L \mathcal{T}_\wedge\big(\mu_{A(x)-\frac{f(2^n x)}{2^n}}(\delta), \mu_{\frac{f(2^n x)}{2^n}-f(x)}(t)\big) \\
&\geq_L \mathcal{T}_\wedge\left(\mu_{A(x)-\frac{f(2^n x)}{2^n}}(\delta), \Phi_{x,x,-2x}\left(\frac{t}{\sum_{k=0}^{n}\frac{\alpha^k}{\varphi(2)^{k+1}}}\right)\right).
\end{aligned}
\tag{9.1.16}
$$

Taking $n \to \infty$ and using (9.1.16), we get

$$
\mu_{A(x)-f(x)}(t+\delta) \geq_L \Phi_{x,x,-2x}\big(t(\varphi(2)-\alpha)\big).
\tag{9.1.17}
$$

Since δ is arbitrary, by taking $\delta \to 0$, one obtains

$$
\mu_{A(x)-f(x)}(t) \geq_L \Phi_{x,x,-2x}\big(t(\varphi(2)-\alpha)\big).
$$

Now, we show that the mapping A is Cauchy additive. Observe that

$$
\begin{aligned}
&\mu_{A(x)+A(y)+A(z)}(t) \\
&\geq_L \mathcal{T}_\wedge\left(\mu_{A(x)-\frac{f(2^n x)}{2^n}}\left(\frac{t(1-\frac{1}{\varphi(2)})}{8}\right), \mu_{A(y)-\frac{f(2^n y)}{2^n}}\left(\frac{t(1-\frac{1}{\varphi(2)})}{8}\right),\right. \\
&\qquad \mu_{A(z)-\frac{f(2^n z)}{2^n}}\left(\frac{t(1-\frac{1}{\varphi(2)})}{8}\right), \mu_{A(x+y+z)-\frac{f(2^n (x+y+z))}{2^n}}\left(\frac{t(1-\frac{1}{\varphi(2)})}{8}\right), \\
&\qquad \mu_{\frac{f(2^n (x+y+z))}{2^n}-\frac{f(2^n x)}{2^n}-\frac{f(2^n y)}{2^n}-\frac{f(2^n z)}{2^n}}\left(\frac{t(1-\frac{1}{\varphi(2)})}{2}\right), \\
&\qquad \left.\mu_{A(x+y+z)}\left(\frac{t}{\varphi(2)}\right)\right)
\end{aligned}
\tag{9.1.18}
$$

for all $x, y, z \in X$ and $t > 0$. The first four terms on the right-hand side of the above inequality tend to $1_{\mathcal{L}}$ as $n \to \infty$. Also, from (φ-LRN3), it follows that

$$
\begin{aligned}
&\mu_{\frac{f(2^n (x+y+z))}{2^n}-\frac{f(2^n x)}{2^n}-\frac{f(2^n y)}{2^n}-\frac{f(2^n z)}{2^n}}\left(\frac{t(1-\frac{1}{\varphi(2)})}{2}\right) \\
&\geq_L \mathcal{T}_\wedge\left(\mu_{f(2^n x)+f(2^n y)+f(2^n z)}\left(\frac{\varphi(2)^n}{4}\left(1-\frac{1}{\varphi(2)}\right)t\right),\right. \\
&\qquad\qquad \left.\mu_{f(2^n (x+y+z))}\left(\frac{\varphi(2)^n}{4}\left(1-\frac{1}{\varphi(2)}\right)t\right)\right) \\
&\geq_L \Phi_{2^n x, 2^n y, 2^n z}\left(\frac{\varphi(2)^n}{4}\left(1-\frac{1}{\varphi(2)}\right)t\right) \\
&\geq_L \Phi_{x,y,z}\left(\frac{\varphi(2)^n}{4\alpha^n}\left(1-\frac{1}{\varphi(2)}\right)t\right),
\end{aligned}
$$

that is, the fifth term also tends to $1_{\mathcal{L}}$ when n tends to ∞. Therefore, we have

$$\mu_{A(x)+A(y)+A(z)}(t) \geq_L \mu_{A(x+y+z)}\left(\frac{t}{\varphi(2)}\right)$$

and so, by Lemma 9.1.1, we conclude that the mapping A is Cauchy additive.

Finally, to prove the uniqueness of the Cauchy additive function A, assume that there exists a Cauchy additive function $B : X \to Y$ which satisfies (9.1.8). Fix $x \in X$. Clearly, $A(2^n x) = 2^n A(x)$ and $B(2^n x) = 2^n B(x)$ for all $n \geq 1$. It follows from (9.1.8) that

$$\begin{aligned}
\mu_{A(x)-B(x)}(t) &= \mu_{\frac{A(2^n x)}{2^n} - \frac{B(2^n x)}{2^n}}(t) \\
&\geq_L \mathcal{T}_\wedge\left(\mu_{\frac{A(2^n x)}{2^n} - \frac{f(2^n x)}{2^n}}\left(\frac{t}{2}\right), \mu_{\frac{B(2^n x)}{2^n} - \frac{f(2^n x)}{2^n}}\left(\frac{t}{2}\right)\right) \\
&\geq_L \Phi_{2^n x, 2^n x, -2^{n+1} x}\left(\frac{\varphi(2^n)(\varphi(2) - \alpha)t}{2}\right) \\
&\geq_L \Phi_{x,x,-2x}\left(\left(\frac{\varphi(2)}{\alpha}\right)^n \frac{(\varphi(2) - \alpha)t}{2}\right).
\end{aligned}$$

Since $\alpha < \varphi(2)$, we get

$$\lim_{n\to\infty} \Phi_{x,x,-2x}\left(\left(\frac{\varphi(2)}{\alpha}\right)^n \frac{(\varphi(2) - \alpha)t}{2}\right) = 1_{\mathcal{L}}.$$

Therefore, $\mu_{A(x)-B(x)}(t) = 1_{\mathcal{L}}$ for all $t > 0$ and so $A(x) = B(x)$. This completes the proof. □

Corollary 9.1.4 *Consider Example* 2.1.10. *If $f : X \to Y$ is a mapping such that, for some $p < 1$,*

$$\mathcal{T}_\wedge\left(\mu_{f(x)+f(y)+f(z)}(t), \mu_{f(x+y+z)}(t)\right) \geq_L \left(\frac{t}{t + (\|x\|^p + \|y\|^p + \|z\|^p)}, \frac{\|x\|^p + \|y\|^p + \|z\|^p}{t + (\|x\|^p + \|y\|^p + \|z\|^p)}\right)$$

for all $x, y, z \in X$ and $t > 0$, then there exists a unique Cauchy additive mapping $A : X \to Y$ such that

$$\mu_{f(x)-A(x)}(t) \geq_L \left(\frac{(2 - 2^p)t}{(2 - 2^p)t + (2 + 2^p)\|x\|^p}, \frac{(2 + 2^p)\|x\|^p}{(2 - 2^p)t + (2 + 2^p)\|x\|^p}\right)$$

for all $x \in X$ and $t > 0$.

Proof Let $\Phi : X^3 \to D_L^+$ be a mapping defined by

$$\Phi_{x,y,z}(t) = \left(\frac{t}{t + (\|x\|^p + \|y\|^p + \|z\|^p)}, \frac{\|x\|^p + \|y\|^p + \|z\|^p}{t + (\|x\|^p + \|y\|^p + \|z\|^p)}\right).$$

Then the conclusion follows from Theorem 9.1.3 with $\alpha = 2^p$. □

Corollary 9.1.5 *Consider Example* 2.1.10. *If $f : X \to Y$ is a mapping such that*

$$\mathcal{T}_\wedge\big(\mu_{f(x)+f(y)+f(z)}(t), \mu_{f(x+y+z)}(t)\big) \geq_L \left(\frac{t}{t+\varepsilon}, \frac{\varepsilon}{t+\varepsilon}\right)$$

for all $x, y, z \in X$ and $t > 0$ and $f(0) = 0$, then there exists a unique Cauchy additive mapping $A : X \to Y$ such that

$$\mu_{f(x)-A(x)}(t) \geq_L \left(\frac{t}{t+\varepsilon}, \frac{\varepsilon}{t+\varepsilon}\right)$$

for all $x \in X$ and $t > 0$.

Proof Let $\Phi : X^3 \to D_L^+$ be a mapping defined by

$$\Phi_{x,y,z}(t) = \left(\frac{t}{t+\varepsilon}, \frac{\varepsilon}{t+\varepsilon}\right).$$

Then the conclusion follows from Theorem 9.1.3 with $\alpha = 1$. □

Theorem 9.1.6 *Let X be a linear space and Φ be a mapping from $X^3 \times [0, \infty)$ to D_L^+ such that, for some $0 < \alpha < \varphi(3)$,*

$$\Phi_{3x,3y,3z}(\alpha t) \geq_L \Phi_{x,y,z}(t) \tag{9.1.19}$$

for all $x, y, z \in X$ and $t > 0$. Let $(Y, \mu, \mathcal{T}_\wedge)$ be a complete latticetic random φ-normed space. If $f : X \to Y$ is an odd mapping such that

$$\mathcal{T}_\wedge\big(\mu_{f(x)+f(y)+2f(z)}(t), \mu_{f(\frac{x+y}{2}+z)}(t)\big) \geq_L \Phi_{x,y,z}(t) \tag{9.1.20}$$

for all $x, y, z \in X$ and $t > 0$, then there exists a unique Cauchy additive mapping $A : X \to Y$ such that

$$\mu_{f(x)-A(x)}(t) \geq_L \Phi_{x,-3x,x}\big((\varphi(3) - \alpha)t\big) \tag{9.1.21}$$

for all $x \in X$ and $t > 0$.

Proof Since the proof is similar to that of the proceeding result, we only sketch it. Putting $y = -3x$ and $z = x$ in (9.1.20), respectively, we get

$$\mu_{3f(x)-f(3x)}(t) \geq_L \Phi_{x,-3x,x}(t) \tag{9.1.22}$$

for all $x \in X$ and $t > 0$. From this relation, it follows that

$$\mu_{\frac{f(3^n x)}{3^n} - f(x)}(t) \geq_L \Phi_{x,-3x,x}\left(\frac{t}{\sum_{k=0}^{n} \frac{\alpha^k}{\varphi(3)^{k+1}}}\right) \tag{9.1.23}$$

and so, as in the proof of Theorem 9.1.3, we have

$$\mu_{\frac{f(3^{n+m}x)}{3^{n+m}}-\frac{f(3^m x)}{3^m}}(t) \geq_L \Phi_{x,-3x,x}\left(\frac{t}{\sum_{k=m}^{n+m}\frac{\alpha^k}{\varphi(3)^{k+1}}}\right),$$

which proves that, for all $x \in X$, $\{\frac{f(3^n x)}{3^n}\}$ is a Cauchy sequence in $(Y, \mu, \mathcal{T}_\wedge)$. Denote $A(x) \in Y$ by the limit. From

$$\mu_{\frac{f(3^n x)}{3^n}-f(x)}(t) \geq_L \Phi_{x,-3x,x}\left(\frac{t}{\sum_{k=0}^{n}\frac{\alpha^k}{\varphi(3)^{k+1}}}\right) \tag{9.1.24}$$

and

$$\begin{aligned}
&\mu_{A(x)-f(x)}(t+\delta) \\
&\geq_L \mathcal{T}_\wedge\big(\mu_{A(x)-\frac{f(3^n x)}{3^n}}(\delta), \mu_{\frac{f(3^n x)}{3^n}-f(x)}(t)\big) \\
&\geq_L \mathcal{T}_\wedge\left(\mu_{A(x)-\frac{f(3^n x)}{3^n}}(\delta), \Phi_{x,-3x,x}\left(\frac{t}{\sum_{k=0}^{n}\frac{\alpha^k}{\varphi(3)^{k+1}}}\right)\right)
\end{aligned} \tag{9.1.25}$$

for all $x \in X$ and $t > 0$, it follows that

$$\mu_{A(x)-f(x)}(t) \geq_L \Phi_{x,-3x,x}\big(t\big(\varphi(3)-\alpha\big)\big).$$

The additivity of A follows from

$$\begin{aligned}
&\mu_{A(x)+A(y)+2A(z)}(t) \\
&\geq_L \mathcal{T}_\wedge\left(\mu_{A(x)-\frac{f(3^n x)}{3^n}}\left(\frac{(1-\frac{\varphi(2)}{\varphi(3)})t}{12}\right), \mu_{A(y)-\frac{f(3^n y)}{3^n}}\left(\frac{(1-\frac{\varphi(2)}{\varphi(3)})t}{12}\right),\right. \\
&\quad \mu_{2A(z)-2\frac{f(3^n z)}{3^n}}\left(\frac{(1-\frac{\varphi(2)}{\varphi(3)})t}{12}\right), \mu_{2A(\frac{x+y}{2}+z)-\frac{2f(3^n(\frac{x+y}{2}+z))}{3^n}}\left(\frac{(1-\frac{\varphi(2)}{\varphi(3)})t}{12}\right), \\
&\quad \mu_{\frac{2f(3^n(\frac{x+y}{2}+z))}{3^n}-\frac{f(3^n x)}{3^n}-\frac{f(3^n y)}{3^n}-\frac{2f(3^n z)}{3^n}}\left(\frac{(1-\frac{\varphi(2)}{\varphi(3)})2t}{3}\right), \\
&\quad \left.\mu_{2A(\frac{x+y}{2}+z)}\left(\frac{\varphi(2)t}{\varphi(3)}\right)\right)
\end{aligned}$$

for all $x, y, z \in X$ and $t > 0$ and, by using Lemma 2.2,

$$\mu_{\frac{2f(3^n(\frac{x+y}{2}+z))}{3^n}-\frac{f(3^n x)}{3^n}-\frac{f(3^n y)}{3^n}-\frac{2f(3^n z)}{3^n}}\left(\frac{(1-\frac{\varphi(2)}{\varphi(3)})2t}{3}\right)$$

$$
\begin{aligned}
&\geq_L \mathcal{T}_\wedge\left(\mu_{f(3^n x)+f(3^n y)+2f(3^n z)}\left(\frac{(1-\frac{\varphi(2)}{\varphi(3)})\varphi(3)^n t}{3}\right),\right.\\
&\qquad \left.\mu_{2f(3^n(\frac{x+y}{2}+z))}\left(\frac{(1-\frac{\varphi(2)}{\varphi(3)})\varphi(3)^n t}{3}\right)\right)\\
&\geq_L \Phi_{3^n x,3^n y,3^n z}\left(\frac{(1-\frac{\varphi(2)}{\varphi(3)})\varphi(3)^n t}{3}\right)\\
&\geq_L \Phi_{x,y,z}\left(\frac{(1-\frac{\varphi(2)}{\varphi(3)})\varphi(3)^n t}{3\alpha^n}\right).
\end{aligned}
$$

Finally, the uniqueness of the Cauchy additive mapping A subject (9.1.21) follows from the following:

$$
\begin{aligned}
\mu_{A(x)-B(x)}(t) &= \mu_{\frac{A(3^n x)}{3^n}-\frac{B(3^n x)}{3^n}}(t)\\
&\geq_L \mathcal{T}_\wedge\left(\mu_{\frac{A(3^n x)}{3^n}-\frac{f(3^n x)}{3^n}}\left(\frac{t}{2}\right),\mu_{\frac{B(3^n x)}{3^n}-\frac{f(3^n x)}{3^n}}\left(\frac{t}{2}\right)\right)\\
&\geq_L \Phi_{3^n x,-3^{n+1}x,2^n x}\left(\frac{\varphi(3)^n(\varphi(3)-\alpha)}{2}t\right)\\
&\geq_L \Phi_{x,-3x,x}\left(\left(\frac{\varphi(3)}{\alpha}\right)^n\frac{(\varphi(3)-\alpha)t}{2}\right).
\end{aligned}
$$

This completes the proof. □

9.2 Systems of QC and AQC Functional Equations

In this section, using the fixed point method, we investigate the stability of the systems of quadratic–cubic and additive–quadratic–cubic functional equations with constant coefficients form r-divisible groups into Šerstnev probabilistic Banach spaces.

The functional equation

$$
f(x+y)+f(x-y)=2f(x)+2f(y) \tag{9.2.1}
$$

is related to a symmetric bi-additive function [4, 132]. It is natural that this equation is called a *quadratic functional equation*. In particular, every solution of the quadratic equation (9.2.1) is called a *quadratic function*. The Hyers–Ulam stability problem for the quadratic functional equation was solved by Skof [244]. In [45], Czerwik proved the Hyers–Ulam–Rassias stability of the equation (9.2.1).

Recently, in [68], Eshaghi Gordji and Khodaei obtained the general solution and the generalized Hyers–Ulam–Rassias stability of the following quadratic functional equation: for all $a, b \in \mathbb{Z}\backslash\{0\}$ with $a \neq \pm 1, \pm b$,

$$f(ax+by)+f(ax-by)=2a^2f(x)+2b^2f(y). \tag{9.2.2}$$

In [116], Jun and Kim introduced the following cubic functional equation:

$$f(2x+y)+f(2x-y)=2f(x+y)+2f(x-y)+12f(x) \tag{9.2.3}$$

and they established the general solution and the generalized Hyers–Ulam–Rassias stability for the functional equation (9.2.3). Also, in [117], Jun et al. investigated the solution and the Hyers–Ulam stability for the cubic functional equation

$$\begin{aligned} &f(ax+by)+f(ax-by) \\ &\quad = ab^2\big(f(x+y)+f(x-y)\big)+2a\big(a^2-b^2\big)f(x), \end{aligned} \tag{9.2.4}$$

where $a, b \in \mathbb{Z}\backslash\{0\}$ with $a \neq \pm 1, \pm b$. For other cubic functional equations, see [180].

In [151], Lee et al. considered the following functional equation:

$$\begin{aligned} &f(2x+y)+f(2x-y) \\ &\quad =4f(x+y)+4f(x-y)+24f(x)-6f(y). \end{aligned} \tag{9.2.5}$$

In fact, they proved that a function f between two real vector spaces X and Y is a solution of the equation (9.2.5) if and only if there exists a unique symmetric bi-quadratic function $B_2 : X \times X \to Y$ such that $f(x) = B_2(x, x)$ for all $x \in X$, where the bi-quadratic function B_2 is given by

$$B_2(x,y)=\frac{1}{12}\big(f(x+y)+f(x-y)-2f(x)-2f(y)\big).$$

Obviously, the function $f(x) = cx^4$ satisfies the functional equation (9.2.5), which is called the *quartic functional equation*. For other quartic functional equations, see [40].

Ebadian, Najati and Eshaghi Gordji [53], considered the generalized Hyers–Ulam stability of the following systems of the *additive-quartic* functional equations:

$$\begin{cases} f(x_1+x_2,y)=f(x_1,y)+f(x_2,y), \\ f(x,2y_1+y_2)+f(x,2y_1-y_2) \\ \quad =4f(x,y_1+y_2)+4f(x,y_1-y_2)+24f(x,y_1)-6f(x,y_2) \end{cases} \tag{9.2.6}$$

and the quadratic-cubic functional equations:

$$\begin{cases} f(x,2y_1+y_2)+f(x,2y_1-y_2) \\ \quad =2f(x,y_1+y_2)+2f(x,y_1-y_2)+12f(x,y_1), \\ f(x,y_1+y_2)+f(x,y_1-y_2)=2f(x,y_1)+2f(x,y_2). \end{cases} \tag{9.2.7}$$

For more details about the results concerning mixed type functional equations, the readers refer to [56, 60, 64] and [65].

Recently, Ghaemi, Eshaghi Gordji and Majani [92] investigated the stability of the following systems of the quadratic-cubic functional equations:

$$\begin{cases} f(ax_1+bx_2,y)+f(ax_1-bx_2,y)=2a^2f(x_1,y)+2b^2f(x_2,y),\\ f(x,ay_1+by_2)+f(x,ay_1-by_2)\\ \quad =ab^2(f(x,y_1+y_2)+f(x,y_1-y_2))+2a(a^2-b^2)f(x,y_1) \end{cases} \tag{9.2.8}$$

and the *additive–quadratic-cubic* functional equations:

$$\begin{cases} f(ax_1+bx_2,y,z)+f(ax_1-bx_2,y,z)=2af(x_1,y,z),\\ f(x,ay_1+by_2,z)+f(x,ay_1-by_2,z)\\ \quad =2a^2f(x,y_1,z)+2b^2f(x,y_2,z),\\ f(x,y,az_1+bz_2)+f(x,y,az_1-bz_2)\\ \quad =ab^2(f(x,y,z_1+z_2)+f(x,y,z_1-z_2))+2a(a^2-b^2)f(x,y,z_1) \end{cases} \tag{9.2.9}$$

in probabilistic (random) normed spaces, where $a,b\in\mathbb{Z}\backslash\{0\}$ with $a\neq\pm1,\pm b$. The function $f:\mathbb{R}\times\mathbb{R}\to\mathbb{R}$ given by $f(x,y)=cx^2y^3$ is a solution of the system (9.2.8). In particular, letting $y=x$, we get a quintic function $g:\mathbb{R}\to\mathbb{R}$ in one variable given by $g(x):=f(x,x)=cx^5$. Also, it is easy to see that the function $f:\mathbb{R}\times\mathbb{R}\times\mathbb{R}\to\mathbb{R}$ defined by $f(x,y,z)=cxy^2z^3$ is a solution of the system (9.2.9). In particular, letting $y=z=x$, we get a sextic function $h:\mathbb{R}\to\mathbb{R}$ in one variable given by $h(x):=f(x,x,x)=cx^6$.

The proof of the following propositions are evident.

Proposition 9.2.1 *Let X and Y be real linear spaces. If a function*

$$f:X\times X\to Y$$

satisfies the system (9.2.8), *then*

$$f(\lambda x,\mu y)=\lambda^2\mu^3f(x,y)$$

for all $x,y\in X$ and rational numbers λ,μ.

Proposition 9.2.2 *Let X and Y be real linear spaces. If a function*

$$f:X\times X\times X\to Y$$

satisfies the system (9.2.9), *then*

$$f(\lambda x,\mu y,\eta z)=\lambda\mu^2\eta^3f(x,y,z)$$

for all $x,y,z\in X$ and rational numbers λ,μ,η.

Now, we investigate the stability problem for the system of the functional equations (9.2.8) form r-divisible groups into random Banach spaces by using fixed point methods.

Recall that an Abelian group G is *divisible* if and only if, for all positive integer n and $g \in G$, there exists $y \in G$ such that $ny = g$. An equivalent condition is as follows:

for any positive integer $n \geq 1$, $nG = G$ since the existence of y for all $n \geq 1$ and g implies that $nG \supseteq G$, and in the other direction $nG \subseteq G$ is true for all group. An Abelian group is said to be *p-divisible* for a prime p if, for all positive integer $n \geq 1$ and $g \in G$, there exists $y \in G$ such that $p^n y = g$. Equivalently, an Abelian group is p-divisible if and only if $pG = G$.

It is well known that an Abelian group is divisible if and only if it is p-divisible for all prime p.

For the main results in this section, we introduce the typical continuous triangular function $\varPi_T : \Delta^+ \times \Delta^+ \to \Delta^+$ defined by

$$\varPi_T(F, G)(x) = T\big(F(x), G(x)\big)$$

for all $F, G \in \Delta^+ \times \Delta^+$, where T is a continuous t-norm, that is, a continuous binary operation on [0, 1] that is commutative, associative, non-decreasing in each variable and has 1 as the identity. For example, the t-norm M on $[0, 1] \times [0, 1]$ defined by

$$M(x, y) = \min\{x, y\}$$

for all $x, y \in [0, 1]$ is a continuous and maximal t-norm, namely, for any t-norm T, $M \geq T$. Also, note that $\varPi_M$ is a maximal triangle function, that is, for all triangle function τ, $\varPi_M \geq \tau$.

Theorem 9.2.3 *Let $s \in \{-1, 1\}$ be fixed. Let G be a r-divisible group and $(Y, \nu, \varPi_T)$ be a random Banach space. Let $\phi, \psi : G \times G \times G \to D^+$ be two functions such that*

$$\begin{aligned}\varPhi(x, y)(t) &\\ := \varPi_T\Big(\phi\big(a^{\frac{s-1}{2}}x, 0, a^{\frac{s-1}{2}}y\big)\big(2a^{\frac{5s-1}{2}}t\big), \psi\big(a^{\frac{s+1}{2}}x, a^{\frac{s-1}{2}}y, 0\big)\big(2a^{\frac{5s+5}{2}}t\big)\Big) \quad (9.2.10)\end{aligned}$$

for all $x, y \in G$ and, for some $0 < k < a^{10s}$,

$$\varPhi\big(a^s x, a^s y\big)\big(ka^{-2s}t\big) \geq \varPhi(x, y)(t) \quad (9.2.11)$$

and

$$\begin{aligned}&\lim_{n\to\infty} \phi\big(a^{sn}x_1, a^{sn}x_2, a^{sn}y\big)\big(a^{-5sn}t\big)\\ &= \lim_{n\to\infty} \psi\big(a^{sn}x, a^{sn}y_1, a^{sn}y_2\big)\big(a^{-5sn}t\big)\\ &= 1 \quad (9.2.12)\end{aligned}$$

for all $x, y, x_1, x_2, y_1, y_2 \in G$ *and* $t > 0$. *If* $f : G \times G \to Y$ *is a function such that* $f(0, y) = 0$ *for all* $y \in G$,

$$\nu_{Df(x_1,x_2,y)}(t) \geq \phi(x_1, x_2, y) \tag{9.2.13}$$

and

$$\nu_{Df(x,y_1,y_2)}(t) \geq \psi(x, y_1, y_2) \tag{9.2.14}$$

for all $x, y, x_1, x_2, y_1, y_2 \in G$, *where*

$$\begin{aligned} &Df(x_1, x_2, y) \\ &\quad = f(ax_1 + bx_2, y) + f(ax_1 - bx_2, y) - 2a^2 f(x_1, y) - 2b^2 f(x_2, y) \end{aligned}$$

and

$$\begin{aligned} Df(x, y_1, y_2) = {} & f(x, ay_1 + by_2) + f(x, ay_1 - by_2) - ab^2 f(x, y_1 + y_2) \\ & - ab^2 f(x, y_1 - y_2) - 2a\big(a^2 - b^2\big) f(x, y_1), \end{aligned}$$

then there exists a unique quintic function $T : G \times G \to Y$ *satisfying the system* (9.2.8) *and*

$$\nu_{f(x,y)-T(x,y)}(t) \geq \Phi(x, y)\big(\big(1 - ka^{-10s}\big)t\big) \tag{9.2.15}$$

for all $x, y \in G$.

Proof Putting $x_1 = 2x$ and $x_2 = 0$ and replacing y by $2y$ in (9.2.13), we have

$$\nu_{f(2ax,2y)-a^2 f(2x,2y)}(t) \geq \phi(2x, 0, 2y)(2t) \tag{9.2.16}$$

for all $x, y \in G$. Putting $y_1 = 2y$ and $y_2 = 0$ and replacing x by $2ax$ in (9.2.14), we have

$$\nu_{f(2ax,2ay)-a^3 f(2ax,2y)}(t) \geq \psi(2ax, 2y, 0)(2t) \tag{9.2.17}$$

for all $x, y \in G$. Thus, we have

$$\nu_{f(2ax,2ay)-a^5 f(2x,2y)}(t) \geq \Pi_T\big(\phi(2x, 0, 2y)\big(2a^{-3}t\big), \psi(2ax, 2y, 0)(2t)\big) \tag{9.2.18}$$

for all $x, y \in G$. Replacing x, y by $\frac{x}{2}, \frac{y}{2}$ in (9.2.18), we have

$$\nu_{f(ax,ay)-a^5 f(x,y)}(t) \geq \Pi_T\big(\phi(x, 0, y)\big(2a^{-3}t\big), \psi(ax, y, 0)(2t)\big) \tag{9.2.19}$$

for all $x, y \in G$. It follows from (9.2.19) that

$$\nu_{a^{-5} f(ax,ay)-f(x,y)}(t) \geq \Pi_T\big(\phi(x, 0, y)\big(2a^2 t\big), \psi(ax, y, 0)\big(2a^5 t\big)\big) \tag{9.2.20}$$

and

$$\nu\big(a^5 f\big(a^{-1}x, a^{-1}y\big) - f(x,y)\big)(t) \\ \geq \Pi_T\big(\phi\big(a^{-1}x, 0, a^{-1}y\big)\big(2a^{-3}t\big), \psi\big(ax, a^{-1}y, 0\big)(2t)\big) \tag{9.2.21}$$

for all $x, y \in G$. So we have

$$\nu_{a^{-5s} f(a^s x, a^s y) - f(x,y)}(t) \geq \Phi(x,y)(t) \tag{9.2.22}$$

for all $x, y \in G$.

Let S be the set of all mappings $h : G \times G \to Y$ with $h(0,x) = 0$ for all $x \in G$ and define a generalized metric on S as follows:

$$d(h,k) = \inf\big\{u \in \mathbb{R}^+ : \nu\big(h(x,y) - k(x,y)\big)(ut) \geq \Phi(x,y)(t), \ \forall x, y \in G, \ t > 0\big\},$$

where, as usual, $\inf \emptyset = +\infty$. The proof of the fact that (S,d) is a complete generalized metric space follows from [27, 103].

Now, we consider the mapping $J : S \to S$ defined by

$$Jh(x,y) := a^{-5s} h\big(a^s x, a^s y\big)$$

for all $h \in S$ and $x, y \in G$. Let $f, g \in S$ such that $d(f,g) < \varepsilon$. Then we have

$$\begin{aligned} &\nu_{Jg(x,y) - Jf(x,y)}\big(kua^{-10s}t\big) \\ &\quad = \nu_{a^{-5s} g(a^s x, a^s y) - a^{-5s} f(a^s x, a^s y)}\big(kua^{-10s}t\big) \\ &\quad = \nu_{g(a^s x, a^s y) - f(a^s x, a^s y)}\big(kua^{-2s}t\big) \\ &\quad \geq \Phi\big(a^s x, a^s y\big)\big(ka^{-2s}t\big) \\ &\quad \geq \Phi(x,y)(t), \end{aligned} \tag{9.2.23}$$

that is, if $d(f,g) < \varepsilon$, then we have $d(Jf, Jg) < ka^{-10s}\varepsilon$. This means that

$$d(Jf, Jg) \leq ka^{-10s} d(f,g)$$

for all $f, g \in S$, that is, J is a strictly contractive self-mapping on S with the Lipschitz constant ka^{-10s}. It follows from (9.2.22) that

$$\nu_{Jf(x,y) - f(x,y)}(t) \geq \Phi(x,y)(t)$$

for all $x, y \in G$ and $t > 0$, which implies that $d(Jf, f) \leq 1$. From Luxemburg–Jung's theorem, it follows that there exists a unique mapping $T : G \times G \to Y$ such that T is a fixed point of J, that is, $T(a^s x, a^s y) = a^{5s} T(x,y)$ for all $x, y \in G$. Also, we have $d(J^m g, T) \to 0$ as $m \to \infty$, which implies the equality

$$\lim_{m \to \infty} a^{-5sm} f\big(a^{sm} x, a^{sm} y\big) = T(x)$$

for all $x, y \in G$. It follows from (9.2.13) that

$$\begin{aligned}
&\nu_{T(ax_1+bx_2,y)+T(ax_1-bx_2,y)-2a^2T(x_1,y)-2b^2T(x_2,y)}(t)\\
&\quad=\lim_{n\to\infty}\nu_{Df(x_1,x_2,y)}(t)\\
&\quad\geq\lim_{n\to\infty}\phi\big(a^{sn}x_1,a^{sn}x_2,a^{sn}y\big)\big(a^{-5sn}t\big)\\
&\quad=1
\end{aligned}$$

for all $x_1, x_2, y \in G$, where

$$\begin{aligned}
&Df(x_1,x_2,y)\\
&\quad=a^{-5sn}f\big(a^{sn}(ax_1+bx_2),a^{sn}y\big)+a^{-5sn}f\big(a^{sn}(ax_1-bx_2),a^{sn}y\big)\\
&\qquad-2a^{-5sn}a^2f\big(a^{sn}x_1,a^{sn}y\big)-2a^{-5sn}b^2f\big(a^{sn}x_2,a^{sn}y\big)
\end{aligned}$$

Also, it follows from (9.2.14) that

$$\begin{aligned}
\nu_{D_1f(x,y_1,y_2)}(t)&=\lim_{n\to\infty}\nu_{D_2f(x,y_1,y_2)}(t)\\
&\geq\lim_{n\to\infty}a^{-5sn}\psi\big(a^{sn}x,a^{sn}y_1,a^{sn}y_2\big)(t)\\
&=1
\end{aligned}$$

for all $x, y_1, y_2 \in G$, where

$$\begin{aligned}
D_1f(x,y_1,y_2)=T(x,ay_1+by_2)+T(x,ay_1-by_2)-ab^2\big(T(x,y_1+y_2)\\
-T(x,y_1-y_2)\big)-2a\big(a^2-b^2\big)T(x,y_1)
\end{aligned}$$

and

$$\begin{aligned}
&D_2f(x,y_1,y_2)\\
&\quad=a^{-5sn}f\big(a^{sn}x,a^{sn}(ay_1+by_2)\big)+a^{-5sn}f\big(a^{sn}x,a^{sn}(ay_1-by_2)\big)\\
&\qquad-a^{-5sn}ab^2f\big(a^{sn}x,a^{sn}(y_1+y_2)\big)-a^{-5sn}ab^2f\big(a^{sn}x,a^{sn}(y_1-y_2)\big)\\
&\qquad-2a^{-5sn}a\big(a^2-b^2\big)f\big(a^{sn}x,a^{sn}y_1\big).
\end{aligned}$$

This means that T satisfies (9.2.8), that is, T is quintic. According to Luxemburg–Jung's theorem, since T is the unique fixed point of J in the set $\Omega = \{g \in S : d(f, g) < \infty\}$, T is the unique mapping such that

$$\nu_{f(x,y)-T(x,y)}(ut)\geq\Phi(x,y)(t)$$

for all $x, y \in G$ and $t > 0$. Again, using Luxemburg–Jung's theorem, we obtain

$$d(f,T)\leq\frac{1}{1-L}d(f,Jf)\leq\frac{1}{1-ka^{-10s}},$$

which implies that

$$\nu_{f(x,y)-T(x,y)}\left(\frac{t}{1-ka^{-10s}}\right) \geq \Phi(x,y)(t)$$

for all $x, y \in G$ and $t > 0$. Therefore, we have

$$\nu_{f(x,y)-T(x,y)}(t) \geq \Phi(x,y)\big(\big(1-ka^{-10s}\big)t\big)$$

for all $x, y \in G$ and $t > 0$. This completes the proof. □

Next, we investigate the stability problem for the system of the functional equations (9.2.9) form r-divisible groups into random Banach spaces by using Luxemburg–Jung's theorem.

Theorem 9.2.4 *Let $s \in \{-1, 1\}$ be fixed. Let G be a r-divisible group and (Y, ν, Π_T) be a random Banach space. Let $\Phi, \Psi, \Upsilon : G \times G \times G \times G \to D^+$ be the functions such that*

$$\begin{aligned}\Theta(x,y,z)(t) := \Pi_T\big(&\Upsilon\big(a^{\frac{s+1}{2}}x, a^{\frac{s+1}{2}}y, a^{\frac{s+1}{2}}z, 0\big)\big(2a^{3s+3}t\big),\\ &\Pi_T\big(\Psi\big(a^{\frac{s+1}{2}}x, a^{\frac{s+1}{2}}y, 0, a^{\frac{s-1}{2}}z\big)\big(2a^{3s+6}t\big),\\ &\Phi\big(a^{\frac{s-1}{2}}x, 0, a^{\frac{s-1}{2}}y, a^{\frac{s-1}{2}}z\big)\big(2a^{3s+8}t\big)\big)\big)\end{aligned} \tag{9.2.24}$$

for all $x, y, z \in G$ and $t > 0$ and, for some $0 < k < a^{6s}$,

$$\Phi\big(a^s x, a^s y, a^s z\big)(kt) \geq \Phi(x,y,z)(t) \tag{9.2.25}$$

and

$$\begin{aligned}&\lim_{n\to\infty} \Phi\big(a^{sn}x_1, a^{sn}x_2, a^{sn}y, a^{sn}z\big)\big(a^{-6sn}t\big)\\ &\quad= \lim_{n\to\infty} \Psi\big(a^{sn}x, a^{sn}y_1, a^{sn}y_2, a^{sn}z\big)\big(a^{-6sn}t\big)\\ &\quad= \lim_{n\to\infty} \Upsilon\big(a^{sn}x, a^{sn}y, a^{sn}z_1, a^{sn}z_2\big)\big(a^{-6sn}t\big)\\ &\quad= 1\end{aligned} \tag{9.2.26}$$

for all $x, y, x_1, x_2, y_1, y_2, z_1, z_2 \in G$ and $t > 0$. If $f : G \times G \times G \to Y$ is a function such that $f(x, 0, z) = 0$ for all $x, z \in G$ and

$$\nu_{f(ax_1+bx_2,y,z)+f(ax_1-bx_2,y,z)-2af(x_1,y,z)}(t) \geq \Phi(x_1,x_2,y,z)(t), \tag{9.2.27}$$

$$\nu_{Df(x,y_1,y_2,z)}(t) \geq \Psi(x,y_1,y_2,z)(t), \tag{9.2.28}$$

$$\nu_{Df(x,y,z_1,z_2)}(t) \geq \Upsilon(x,y,z_1,z_2)(t) \tag{9.2.29}$$

for all $x, y, x_1, x_2, y_1, y_2, z_1, z_2 \in G$ *and* $t > 0$, *where*

$$\begin{aligned}Df(x, y, z_1, z_2) &= f(x, ay_1 + by_2, z) + f(x, ay_1 - by_2, z)\\ &\quad - 2a^2 f(x, y_1, z) - 2b^2 f(x, y_2, z)\end{aligned}$$

and

$$\begin{aligned}&Df(x, y, z_1, z_2)\\ &\quad = f(x, y, az_1 + bz_2) + f(x, y, az_1 - bz_2)\\ &\qquad - ab^2\big(f(x, y, z_1 + z_2) + f(x, y, z_1 - z_2)\big) - 2a\big(a^2 - b^2\big) f(x, y, z_1),\end{aligned}$$

then there exists a unique quintic function $T : G \times G \times G \to Y$ *satisfying* (9.2.9) *and*

$$\nu_{f(x,y,z)-T(x,y,z)}(t) \geq \Theta(x, y, z)\big(\big(1 - ka^{-6s}\big)t\big) \tag{9.2.30}$$

for all $x, y, z \in G$ *and* $t > 0$.

Proof Putting $x_1 = 2x$ and $x_2 = 0$ and replacing y, z by $2y, 2z$ in (9.2.27), respectively, we have

$$\nu_{f(2ax,2y,2z)-af(2x,2y,2z)}\left(\frac{1}{2}t\right) \geq \Phi(2x, 0, 2y, 2z)(t) \tag{9.2.31}$$

for all $x, y, z \in G$ and $t > 0$. Putting $y_1 = 2y$ and $y_2 = 0$ and replacing x, z by $2ax, 2z$ in (9.2.28), respectively, we have

$$\nu_{f(2ax,2ay,2z)-a^2 f(2ax,2y,2z)}\left(\frac{1}{2}t\right) \geq \Psi(2ax, 2y, 0, 2z)(t) \tag{9.2.32}$$

for all $x, y, z \in G$ and $t > 0$. Putting $z_1 = 2z$ and $z_2 = 0$ and replacing x, y by $2ax, 2ay$ in (9.2.29), respectively, we have

$$\nu_{f(2ax,2ay,2az)-a^3 f(2ax,2ay,2z)}\left(\frac{1}{2}t\right) \geq \Upsilon(2ax, 2ay, 2z, 0)(t) \tag{9.2.33}$$

for all $x, y, z \in G$ and $t > 0$. Thus, it follows that

$$\begin{aligned}&\nu_{f(2ax,2ay,2az)-a^6 f(2x,2y,2z)}(t)\\ &\quad \geq \Pi_T\big(\Upsilon(2ax, 2ay, 2z, 0)(2t), \Pi_T\big(\Psi(2ax, 2y, 0, 2z)\big(2a^3t\big),\\ &\qquad \Phi(2x, 0, 2y, 2z)\big(2a^5t\big)\big)\big)\end{aligned} \tag{9.2.34}$$

for all $x, y, z \in G$ and $t > 0$. Replacing x, y and z by $\frac{x}{2}$, $\frac{y}{2}$ and $\frac{z}{2}$ in (9.2.34), respectively, we have

$$\begin{aligned}
&\nu_{f(ax,ay,az)-a^6 f(x,y,z)}(t)\\
&\quad\geq \Pi_T\big(\Upsilon(ax,ay,z,0)(2t), \Pi_T\big(\Psi(ax,y,0,z)\big(2a^3 t\big),\\
&\qquad \Phi(x,0,y,z)\big(2a^5 t\big)\big)\big)
\end{aligned}\tag{9.2.35}$$

for all $x, y, z \in G$ and $t > 0$. It follows from (9.2.35) that

$$\begin{aligned}
&\nu_{a^{-6} f(ax,ay,az)-f(x,y,z)}(t)\\
&\quad\geq \Pi_T\big(\Upsilon(ax,ay,z,0)\big(2a^6 t\big), \Pi_T\big(\Psi(ax,y,0,z)\big(2a^9 t\big),\\
&\qquad \Phi(x,0,y,z)\big(2a^{11} t\big)\big)\big)
\end{aligned}\tag{9.2.36}$$

and

$$\begin{aligned}
&\nu_{a^6 f(a^{-1}x,a^{-1}y,a^{-1}z)-f(x,y,z)}(t)\\
&\quad\geq \Pi_T\big(\Upsilon\big(x,y,a^{-1}z,0\big)(2at), \Pi_T\big(\Psi\big(x,a^{-1}y,0,a^{-1}z\big)\big(2a^3 t\big),\\
&\qquad \Phi\big(a^{-1}x,0,a^{-1}y,a^{-1}z\big)\big(2a^5 t\big)\big)\big)
\end{aligned}\tag{9.2.37}$$

for all $x, y, z \in G$ and $t > 0$. Thus, we have

$$\nu_{a^{-6s} f(a^s x,a^s y,a^s z)-f(x,y,z)}(t) \geq \Theta(x,y,z)(t) \tag{9.2.38}$$

for all $x, y, z \in G$ and $t > 0$.

Let S be the set of all mappings $h : X \times X \times X \to Y$ with $h(x,0,z) = 0$ for all $x, z \in G$ and define a generalized metric on S as follows:

$$\begin{aligned}
d(h,k) = \inf\big\{&u \in \mathbb{R}^+ : \nu\big(h(x,y,z) - k(x,y,z)\big)(ut) \geq \Theta(x,y,z)(t),\\
&\forall x,y,z \in G,\ t > 0\big\},
\end{aligned}$$

where, as usual, $\inf\emptyset = +\infty$. For the proof of the fact that (S,d) is a complete generalized metric space, see [27, 103].

Now, we consider the mapping $J : S \to S$ defined by

$$Jh(x,y,z) := a^{-6s} h\big(a^s x, a^s y, a^s z\big)$$

for all $h \in S$ and $x, y, z \in G$. Let $f, g \in S$ be such that $d(f,g) < \varepsilon$. Then we have

$$\begin{aligned}
&\nu_{Jg(x,y,z)-Jf(x,y,z)}\big(kua^{-6s} t\big)\\
&\quad= \nu_{a^{-6s} g(a^s x,a^s y,a^s z)-a^{-6s} f(a^s x,a^s y,a^s z)}\big(kua^{-6s} t\big)\\
&\quad= \nu_{g(a^s x,a^s y,a^s z)-f(a^s x,a^s y,a^s z)}(kut)\\
&\quad\geq \Theta\big(a^s x, a^s y, a^s z\big)(kt)\\
&\quad\geq \Theta(x,y,z)(t),
\end{aligned}\tag{9.2.39}$$

that is, if $d(f, g) < \varepsilon$, then we have $d(Jf, Jg) < ka^{-6s}\varepsilon$. This means that

$$d(Jf, Jg) \leq ka^{-6s} d(f, g)$$

for all $f, g \in S$, that is, J is a strictly contractive self-mapping on S with the Lipschitz constant ka^{-6s}. It follows from (9.2.38) that

$$\nu_{Jf(x,y,z)-f(x,y,z)}(t) \geq \Theta(x, y, z)(t)$$

for all $x, y.z \in G$ and $t > 0$, which implies that $d(Jf, f) \leq 1$. From Luxemburg–Jung's theorem, it follows that there exists a unique mapping $T : G \times G \times G \to Y$ such that T is a fixed point of J, that is, $T(a^s x, a^s y, a^s z) = a^{6s} T(x, y, z)$ for all $x, y, z \in G$. Also, $d(J^m g, T) \to 0$ as $m \to \infty$, which implies that

$$\lim_{m\to\infty} a^{-6sm} f\left(a^{sm} x, a^{sm} y, a^{sm} z\right) = T(x)$$

for all $x \in X$. It follows from (9.2.27), (9.2.28) and (9.2.29) that

$$\begin{aligned}
&\nu_{T(ax_1+bx_2,y,z)+T(ax_1-bx_2,y,z)-2aT(x_1,y,z)}(t) \\
&\quad = \lim_{n\to\infty} \nu_{Df(x_1,x_2,y,z)}(t) \\
&\quad \geq \lim_{n\to\infty} \Phi\left(a^{sn}x_1, a^{sn}x_2, a^{sn}y, a^{sn}z\right)\left(a^{-6sn}t\right) \\
&\quad = 1,
\end{aligned} \tag{9.2.40}$$

$$\begin{aligned}
\nu_{D_1 f(x,y_1,y_2,z)}(t) &= \lim_{n\to\infty} \nu_{D_2 f(x,y_1,y_2,z)}(t) \\
&\geq \lim_{n\to\infty} \Psi\left(a^{sn}x, a^{sn}y_1, a^{sn}y_2, a^{sn}z\right)\left(a^{-6sn}t\right) \\
&= 1
\end{aligned} \tag{9.2.41}$$

and

$$\begin{aligned}
\nu_{D_1 f(x,y,z_1,z_2)}(t) &= \lim_{n\to\infty} \nu_{D_2 f(x,y,z_1,z_2)}(t) \\
&\geq \lim_{n\to\infty} \Upsilon\left(a^{sn}x, a^{sn}y, a^{sn}z_1, a^{sn}z_2\right)\left(a^{-6sn}t\right) \\
&= 1
\end{aligned} \tag{9.2.42}$$

for all $x, y, x_1, x_2, y_1, y_2, z_1, z_2 \in G$ and $t > 0$, where

$$\begin{aligned}
Df(x_1, x_2, y, z) = a^{-6sn} & f\left(a^{sn}(ax_1 + bx_2), a^{sn}y, a^{sn}z\right) \\
& + a^{-6sn} f\left(a^{sn}(ax_1 - bx_2), a^{sn}y, a^{sn}z\right) \\
& - 2aa^{-6sn} f\left(a^{sn}x_1, a^{sn}y, a^{sn}z\right),
\end{aligned} \tag{9.2.43}$$

$$\begin{aligned} D_1 f(x, y_1, y_2, z) &= T(x, ay_1 + by_2, z) + T(x, ay_1 - by_2, z) \\ &\quad - 2a^2 T(x, y_1, z) - 2b^2 T(x, y_2, z), \end{aligned} \tag{9.2.44}$$

$$\begin{aligned} &D_2 f(x, y_1, y_2, z) \\ &= a^{-6sn} f\big(a^{sn}x, a^{sn}(ay_1 + by_2), a^{sn}z\big) \\ &\quad + f\big(a^{sn}x, a^{sn}ay_1 - a^{sn}by_2, a^{sn}z\big) - 2a^2a^{-6sn} f\big(a^{sn}x, a^{sn}y_1, a^{sn}z\big) \\ &\quad + 2b^2a^{-6sn} f\big(a^{sn}x, a^{sn}y_2, a^{sn}z\big), \end{aligned} \tag{9.2.45}$$

$$\begin{aligned} &D_1 f(x, y, z_1, z_2) \\ &= T(x, y, az_1 + bz_2) + T(x, y, az_1 - bz_2) - ab^2\big(T(x, y, z_1 + z_2) \\ &\quad - T(x, y, z_1 - z_2)\big) - 2a\big(a^2 - b^2\big)T(x, y, z_1) \end{aligned} \tag{9.2.46}$$

and

$$\begin{aligned} D_2 f(x, y, z_1, z_2) &= a^{-6sn} f\big(a^{sn}x, a^{sn}y, a^{sn}(az_1 + bz_2)\big) \\ &\quad + a^{-6sn} f\big(a^{sn}x, a^{sn}y, a^{sn}(az_1 - bz_2)\big) \\ &\quad - ab^2a^{-6sn}\big(f\big(a^{sn}x, a^{sn}y, a^{sn}(z_1 + z_2)\big) \\ &\quad + f\big(a^{sn}x, a^{sn}y, a^{sn}(z_1 - z_2)\big)\big) \\ &\quad - 2aa^{-6sn}\big(a^2 - b^2\big) f\big(a^{sn}x, a^{sn}y, a^{sn}z_1\big). \end{aligned} \tag{9.2.47}$$

This means that T satisfies (9.2.9), that is, T is sextic.

According to Luxemburg–Jung's theorem, since T is the unique fixed point of J in the set $\Omega = \{g \in S : d(f, g) < \infty\}$, T is the unique mapping such that

$$\nu_{f(x,y,z)-T(x,y,z)}(ut) \geq \Theta(x, y, z)(t)$$

for all $x, y, z \in G$ and $t > 0$. Again, using Luxemburg–Jung's theorem, we obtain

$$d(f, T) \leq \frac{1}{1 - L} d(f, Jf) \leq \frac{1}{1 - ka^{-6s}},$$

which implies that

$$\nu_{f(x,y,z)-T(x,y,z)}\left(\frac{t}{1 - ka^{-6s}}\right) \geq \Theta(x, y, z)(t)$$

for all $x, y, z \in G$ and $t > 0$. So, it follows that

$$\nu_{f(x,y,z)-T(x,y,z)}(t) \geq \Theta(x, y, z)\big(\big(1 - ka^{-6s}\big)t\big)$$

for all $x, y, z \in G$ and $t > 0$. This completes the proof. □

9.3 Couple Functional Equations

Solutions of the functional equation

$$f(x+y) = F\big[f(x), f(y)\big] \tag{9.3.1}$$

were investigated by Aczél [2] and the stability problem for a general equation of the form

$$f\big[G(x,y)\big] = F\big[f(x), f(y)\big] \tag{9.3.2}$$

was investigated by Cholewa [39] (see also [88]). Indeed, Cholewa proved the superstability of the equation under some additional conditions on the functions and spaces involved.

Recently, Jung and Min [127] applied the fixed point method to the investigate the stability of the functional equation (9.3.1) as follows:

Theorem 9.3.1 *Let X be a vector space over $\mathbb{K}$, $(Y, \|\cdot\|)$ be a Banach space over $\mathbb{K}$ and $(Y\times Y, \|\cdot\|_2)$ be a Banach space over $\mathbb{K}$. Assume that $F: Y\times Y \to Y$ is a bounded linear transformation, whose norm is denoted by $\|F\|$, satisfying*

$$F\big[F(u,u), F(v,v)\big] = F\big[F(u,v), F(u,v)\big]$$

for all $u, v \in Y$ and there exists a number $K > 0$ such that

$$\big\|(u,u)-(v,v)\big\|_2 \le K\|u-v\|$$

for all $u, v \in Y$. Moreover, assume that $\varphi: X\times X \to [0,\infty)$ is a function satisfying

$$\varphi\left(\frac{x}{2}, \frac{y}{2}\right) \le \varphi(x,y)$$

for all $x, y \in X$. If $K\|F\| < 1$ and a mapping $f: X \to Y$ satisfies the following inequality:

$$\big\|f(x,y) - F\big[f(x), f(y)\big]\big\| \le \varphi(x,y)$$

for all $x, y \in X$, then there exists a unique solution $f^*: X \to Y$ of (9.3.1) such that

$$\big\|f(x) - f^*(x)\big\| \le \frac{1}{1-K\|F\|}\varphi(x,x)$$

for all $x \in X$.

Obviously, for some nonnegative constants θ and p, $\varphi(x,y) = \theta(\|x\|^p + \|y\|^p)$ satisfies the following:

$$\varphi\left(\frac{x}{2}, \frac{y}{2}\right) \le \varphi(x,y)$$

for all $x, y \in X$.

Corollary 9.3.2 *Let X be a vector space over $\mathbb{K}$, $(Y, \|\cdot\|)$ be a Banach space over $\mathbb{K}$ and $(Y \times Y, \|\cdot\|_2)$ be a Banach space over $\mathbb{K}$. Assume that $F : Y \times Y \to Y$ is a bounded linear transformation, whose norm is denoted by $\|F\|$, satisfying*

$$F\big[F(u,u), F(v,v)\big] = F\big[F(u,v), F(u,v)\big]$$

for all $u, v \in Y$ and there exists a number $K > 0$ such that

$$\big\|(u,u) - (v,v)\big\|_2 \leq K\|u - v\|$$

for all $u, v \in Y$. If $K\|F\| < 1$ and a mapping $f : X \to Y$ satisfies the following inequality:

$$\big\|f(x,y) - F\big[f(x), f(y)\big]\big\| \leq \theta\big(\|x\|^p + \|y\|^p\big)$$

for all $x, y \in X$ and for some nonnegative constants θ and p, then there exists a unique solution $f^* : X \to Y$ of (9.3.1) such that

$$\big\|f(x) - f^*(x)\big\| \leq \frac{2\theta}{1 - K\|F\|}\|x\|^p$$

for all $x \in X$.

To illustrate Corollary 9.3.2, Jung and Min gave one example as follows:

Assume that $X = Y = \mathbb{C}$ and consider Banach spaces $(\mathbb{C}, |\cdot|)$ and $(\mathbb{C} \times \mathbb{C}, |\cdot|_2)$, where we define $|(u,v)|_2 = \sqrt{|u|^2 + |v|^2}$ for all $u, v \in \mathbb{C}$. Let A, B be fixed complex numbers with $|A| + |B| < \frac{1}{\sqrt{2}}$ and $F : \mathbb{C} \times \mathbb{C} \to \mathbb{C}$ be a linear transformation defined by

$$F(u,v) + Au + Bv$$

for all $u, v \in \mathbb{C}$. Then it is easy to show that F satisfies the following condition:

$$F\big[F(u,u), F(v,v)\big] = F\big[F(u,v), F(u,v)\big]$$

for all $u, v \in \mathbb{C}$. If u and v are complex numbers with $|(u,v)|_2 \leq 1$, then we have

$$\big|F(u,v)\big| \leq |A||u| + |B||v| \leq |A| + |B|$$

and so it follows that

$$\|F\| = \sup\big\{\big|F(u,v)\big| : \big|(u,v)\big|_2 \leq 1,\ u, v \in \mathbb{C}\big\} \leq |A| + |B|,$$

which implies that F is bounded. On the other hand, we have

$$\big\|(u,u) - (v,v)\big\|_2 = \big|(u - v, u - v)\big|_2 \leq 2\|u - v\|$$

for all $u, v \in \mathbb{C}$, that is, we can choose $\sqrt{2}$ for the value of K and so

$$K\|F\| \leq \sqrt{2}\big(|A| + |B|\big) < 1.$$

If a function $f:\mathbb{C}\to\mathbb{C}$ satisfies the following inequality:

$$\big\|f(x+y)-F\big[f(x),f(y)\big]\big\|\le\epsilon$$

for all $x, y\in\mathbb{C}$ and for some $\epsilon>0$, then, by Corollary 9.3.2 with $\theta=\frac{\epsilon}{2}$ and $p=0$, there exists a unique function $f^*:\mathbb{C}\to\mathbb{C}$ such that

$$f^*(x+y)=F\big[f^*(x),f^*(y)\big]$$

for all $x, y\in\mathbb{C}$ and

$$\big\|f(x)-f^*(x)\big\|\le\frac{\epsilon}{1-\sqrt{2}(|A|+|B|)}$$

for all $x\in\mathbb{C}$.

In this section, we generalize Jun and Min's result and use fixed point approach to obtain the stability of the functional equation

$$f\big(x+\sigma(y)\big)=F\big[f(x),f(y)\big] \tag{9.3.3}$$

for a class of functions of a vector space into a RN-space, where σ is an involution.

Let (U,μ,T) and (V,μ,T) be two RN-spaces over $\mathbb{K}$. The Cartesian product $U\times V$ of two sets U, V may acquire the structure of a linear vector space if the addition and scalar multiplication, which are the ordered pairs (u,v) for any $u\in U$ and $v\in V$, are defined by

$$(u_1,v_1)+(u_2,v_2)=(u_1+u_2,v_1+v_2)$$

and

$$a(u,v)=(au,av).$$

These operations are specified by only using the known operations on two vector spaces U and V. We can easily verify that the function

$$\Omega_{(u,v)}(t)=T\big(\mu_u(t),\mu_v(t)\big)$$

for all $u\in U$ and $v\in V$ is a random norm on $U\times V$. If U and V are Banach spaces, then it is quietly simple to prove that $U\times V$ is also a Banach space relative to the norm just defined.

Theorem 9.3.3 *Let X be a vector space over $\mathbb{K}$ and (Y,μ,T) be a complete RN-space over $\mathbb{K}$. Let $(X\times X,\Omega,T)$ be a complete RN-space over $\mathbb{K}$. Assume that $F:X\times X\to Y$ is a bounded linear transformation, that is, for any $M_0>0$*

$$\mu_{F(u,v)}(t)\ge\Omega_{(u,v)}\left(\frac{t}{M_0}\right) \tag{9.3.4}$$

for all $u, v \in X$, satisfying

$$F\big[F(u,u), F(v,v)\big] = F\big[F(u,v), F(u,v)\big] \tag{9.3.5}$$

for all $u, v \in X$ and there exists a number $K > 0$ such that

$$\Omega_{(u(x),u(\sigma(x)))-(v(x),v(\sigma(x)))}(t) \geq \mu_{u(x)-v(x)}(Kt) \tag{9.3.6}$$

for all $u, v : X \to X$ and $t > 0$. Moreover, assume that $\varphi : X \times X \times [0,\infty) \to [0,1]$ is a function satisfying

$$\varphi_{(x,\sigma(y))}(t) \geq \varphi_{(2x,2y)}(t) \tag{9.3.7}$$

for all $x, y \in X$ and $t > 0$. If $M_0 < K$ and a mapping $f : X \to Y$ satisfies the following inequality:

$$\mu_{f(x+\sigma(y))-F[f(x),f(y)]}(t) \geq \varphi_{(x,y)}(t) \tag{9.3.8}$$

for any $x, y \in X$ and $t > 0$, then there exists a unique solution $f^* : X \to Y$ of (9.3.3) *such that*

$$\mu_{f(x)-f^*(x)}(t) \geq \varphi_{(x,x)}\big((K - M_0)t\big) \tag{9.3.9}$$

for all $x, y \in X$ and $t > 0$.

Proof First, we denote by $\mathcal{X}$ the set of all mappings $h : X \to Y$ and by d the generalized metric on X defined as follows:

$$d(g,h) = \inf\big\{C \in [0,\infty) : \mu_{g(x)-h(x)}(Ct) \geq \varphi_{(x,x)}(t),\ \forall x \in X\big\}. \tag{9.3.10}$$

By a similar method used at the proof of Theorem 3.1 in [125], we can show that $(\mathcal{X}, d)$ is a generalized complete metric space. Now, define an operator $J : \mathcal{X} \to \mathcal{X}$ by

$$(Jh)(x) = F\left[h\left(\frac{x}{2}\right), h\left(\sigma\left(\frac{x}{2}\right)\right)\right] \tag{9.3.11}$$

for all $x \in X$.

Now, we prove that J is strictly contractive in $\mathcal{X}$. For any $g, h \in \mathcal{X}$, let $C \in [0,\infty]$ be an arbitrary constant with $d(g,h) \leq C$, that is,

$$\mu_{g(x)-h(x)}(Ct) \geq \varphi_{(x,x)}(t) \tag{9.3.12}$$

for each $x \in X$ and $t > 0$. By (9.3.6), (9.3.7), (9.3.11) and (9.3.12), we have

$$\begin{aligned}\mu_{Jg(x)-Jh(x)}(t) &\geq \mu_{F[g(\frac{x}{2}),g(\sigma(\frac{x}{2}))]-F[h(\frac{x}{2}),h(\sigma(\frac{x}{2}))]}(t)\\ &\geq \Omega_{((g(\frac{x}{2}),g(\sigma(\frac{x}{2})))-(h(\frac{x}{2}),h(\sigma(\frac{x}{2})))}\left(\frac{t}{M_0}\right)\end{aligned}$$

$$\geq \mu_{g(\frac{x}{2})-h(\frac{x}{2})}\left(\frac{\kappa t}{M_0}\right)$$

$$\geq \varphi_{(\frac{x}{2},\frac{x}{2})}\left(\frac{Kt}{CM_0}\right)$$

$$\geq \varphi_{(x,x)}\left(\frac{Kt}{CM_0}\right) \tag{9.3.13}$$

for all $x \in X$ and $t > 0$. Then it follows from (9.3.10) that

$$d(Jg, Jh) \leq \frac{M_0}{K} d(g, h)$$

for any $g, h \in \mathcal{X}$, where $\frac{M_0}{K}$ is the Lipschitz constant with $0 < \frac{M_0}{K} < 1$. Thus J is strictly contractive.

Now, we claim that $d(Jf, f) \leq \infty$. Replacing $\frac{x}{2}$ by x and $\sigma(\frac{x}{2})$ by y in (9.3.8), it follows from (9.3.7) and (9.3.11) that

$$\mu_{f(\frac{x}{2}+\sigma(\sigma(\frac{x}{2})))-F[f(\frac{x}{2}),f(\sigma(\frac{x}{2}))]}(t) \geq \varphi_{(\frac{x}{2},\sigma(\frac{x}{2}))}(t)$$

for all $x \in X$ and $t > 0$ and so

$$\mu_{f(x)-(Jf)(x)}(t) \geq \varphi_{(\frac{x}{2},\sigma(\frac{x}{2}))}(t) \geq \varphi_{(x,x)}(t) \tag{9.3.14}$$

for all $x \in X$ and $t > 0$. Then we have

$$d(Jf, f) \leq 1 \leq \infty. \tag{9.3.15}$$

Now, it follows from Luxemburg–Jung's theorem (b) (Chap. 5) that there exists a mapping $f^* : E_1 \to E_2$, which is a fixed point of J, such that

$$\lim_{n\to\infty} d\left(J^n f, f^*\right) = 0. \tag{9.3.16}$$

From Luxemburg–Jung's theorem (c), we have

$$d\left(J^n f, f^*\right) \leq \frac{1}{1-\frac{M_0}{K}} d(Jf, f) \leq \frac{1}{1-\frac{M_0}{K}}, \tag{9.3.17}$$

which implies that the validity of (9.3.9). According to Luxemburg–Jung's theorem (a), f^* is the unique fixed point of J with $d(f, f^*) < \infty$.

Now, we prove that

$$\mu_{(J^n f)(x+\sigma(y))-F[(J^n f)(x),(J^n f)(y)]}(t) \geq \varphi_{(x,x)}\left(\frac{t}{(\frac{M_0}{K})^n}\right) \tag{9.3.18}$$

for all $n \in \mathbb{N}$, $x, y \in X$ and $t > 0$. Indeed, it follows from (9.3.5), (9.3.6), (9.3.7), (9.3.8) and (9.3.11) that

$$
\begin{aligned}
&\mu_{(Jf)(x+\sigma(y))-F[(Jf)(x),(Jf)(y)]}(t)\\
&\quad = \mu_{F[f(\frac{x+\sigma(y)}{2}),f(\sigma(\frac{x+\sigma(y)}{2}))]-F[F[f(\frac{x}{2}),f(\sigma(\frac{x}{2}))],F[f(\frac{y}{2}),f(\sigma(\frac{y}{2}))]]}(t)\\
&\quad \geq \Omega_{[f(\frac{x+\sigma(y)}{2}),f(\sigma(\frac{x+\sigma(y)}{2}))]-[F[f(\frac{x}{2}),f(\sigma(\frac{x}{2}))],F[f(\frac{y}{2}),f(\sigma(\frac{y}{2}))]]}\left(\frac{t}{M_0}\right)\\
&\quad \geq \mu_{f(\frac{x+\sigma(y)}{2})-[F[f(\frac{x}{2}),f(\frac{y}{2})]]}\left(\frac{Kt}{M_0}\right)\\
&\quad \geq \varphi_{(\frac{x}{2},\frac{y}{2})}\left(\frac{Kt}{M_0}\right)\\
&\quad \geq \varphi_{(x,y)}\left(\frac{Kt}{M_0}\right)
\end{aligned} \tag{9.3.19}
$$

for all $x, y \in X$ and $t > 0$. Then it follows from (9.3.5), (9.3.6), (9.3.7), (9.3.8) and (9.3.18) that

$$
\begin{aligned}
&\mu_{(J^{n+1}f)(x+\sigma(y))-F[(J^{n+1}f)(x),(J^{n+1}f)(y)]}(t)\\
&\quad = \Omega_{F[J^n f(\frac{x+\sigma(y)}{2}),J^n f(\sigma(\frac{x+\sigma(y)}{2}))]-F[F[J^n f(\frac{x}{2}),J^n f(\sigma(\frac{x}{2}))],F[J^n f(\frac{y}{2}),J^n f(\sigma(\frac{y}{2}))]]}(t)\\
&\quad \geq \mu_{J^n f(\frac{x+\sigma(y)}{2})-[F[J^n f(\frac{x}{2}),J^n f(\frac{y}{2})]]}\left(\frac{Kt}{M_0}\right)\\
&\quad \geq \varphi_{(\frac{x}{2},\frac{y}{2})}\left(\frac{t}{(\frac{M_0}{K})^{n+1}}\right)\\
&\quad \geq \varphi_{(x,x)}\left(\frac{t}{(\frac{M_0}{K})^{n+1}}\right)
\end{aligned} \tag{9.3.20}
$$

for all $n \in \mathbb{N}$, which proves the validity of (9.3.18).

Finally, we prove that

$$f^*\big(x+\sigma(y)\big) = F\big[f^*(x), f^*(y)\big]$$

for all $x, y \in X$. Since F is a continuous and bounded linear transformation, it follows from (9.3.16) and (9.3.18) that

$$
\begin{aligned}
&\mu_{f^*(x+\sigma(y))-F[f^*(x),f^*(y)]}(t)\\
&\quad = \lim_{n\to\infty} \mu_{J^n f(\frac{x+\sigma(y)}{2})-[F[J^n f(\frac{x}{2}),J^n f(\frac{y}{2})]]}(t)\\
&\quad \geq \lim_{n\to\infty} \varphi_{(x,x)}\left(\frac{t}{(\frac{M_0}{K})^n}\right) = 1
\end{aligned} \tag{9.3.21}
$$

for all $x, y \in X$ and $t > 0$, which implies that f^* is a solution of (9.3.8). This completes the proof. □

Corollary 9.3.4 *Let X be a vector space over $\mathbb{K}$ and (Y, μ, T) be a complete RN-space over $\mathbb{K}$. Let $(X \times X, \Omega, T)$ be a complete RN-space over $\mathbb{K}$. Assume that $F : X \times X \to Y$ is a bounded linear transformation satisfying the condition* (9.3.5) *and that there exists a number $K > 0$ satisfying the condition* (9.3.6). *If $M_0 < K$ and a mapping $f : X \to Y$ satisfies the following inequality*:

$$\mu_{f(x+y)-F[f(x),f(y)]}(t) \geq \frac{t}{t+\theta(\|x\|^p + \|y\|^p)}$$

for all $x, y \in X$, $t > 0$ and for some nonnegative real constants θ and p, then there exists a unique solution $f^* : X \to Y$ of (9.3.3) *such that*

$$\mu_{f(x)-f^*(x)}(t) \geq \frac{(K - M_0)t}{(\kappa - M_0)t + 2K\theta\|x\|^p}$$

for all $x \in X$ and $t > 0$.

References

1. J. Aczél, The general solution of two functional equations by reduction to functions additive in two variables and with aid of Hamel-bases. Glas. Mat.-Fiz. Astronom. Drus. Mat. Fiz. Hrvat. **20**, 65–73 (1965)
2. J. Aczél, *Lectures on Functional Equations and Their Applications* (Academic Press, New York, 1966)
3. J. Aczél, *A Short Course on Functional Equations* (Reidel, Dordrecht, 1987)
4. J. Aczél, J. Dhombres, *Functional Equations in Several Variables* (Cambridge University Press, Cambridge, 1989)
5. R.P. Agarwal, Y.J. Cho, R. Saadati, On random topological structures. Abstr. Appl. Anal. **2011**, 762361 (2011), 41 pp.
6. J. Alonso, C. Benítez, Orthogonality in normed linear spaces: a survey. I. Main properties. Extr. Math. **3**, 1–15 (1988)
7. J. Alonso, C. Benítez, Orthogonality in normed linear spaces: a survey. II. Relations between main orthogonalities. Extr. Math. **4**, 121–131 (1989)
8. C. Alsina, On the stability of a functional equation arising in probabilistic normed spaces, in *General Inequalities*, vol. 5, Oberwolfach, 1986 (Birkhäuser, Basel, 1987), pp. 263–271
9. C. Alsina, B. Schweizer, A. Sklar, On the definition of a probabilistic normed space. Aequ. Math. **46**, 91–98 (1993)
10. C. Alsina, B. Schweizer, A. Sklar, Continuity properties of probabilistic norms. J. Math. Anal. Appl. **208**, 446–452 (1997)
11. A.M. Ampere, Recherche sur quelques points de la theorie des functions. J. E. Polytech. Cah. **6**, 148–181 (1806)
12. M. Amyari, Stability of C^*-inner products. J. Math. Anal. Appl. **322**, 214–218 (2006)
13. L.C. Andrews, *Special Functions for Engineers and Applied Mathematicians* (MacMillan, London, 1985)
14. T. Aoki, On the stability of the linear transformation in Banach spaces. J. Math. Soc. Jpn. **2**, 64–66 (1950)
15. T.M. Apostol, *Mathematical Analysis*, 2nd edn. (Addison-Wesley, Reading, 1975)
16. K.T. Atanassov, Intuitionistic fuzzy sets. Fuzzy Sets Syst. **20**, 87–96 (1986)
17. E. Baktash, Y.J. Cho, M. Jalili, R. Saadati, S.M. Vaezpour, On the stability of cubic mappings and quadratic mappings in random normed spaces. J. Inequal. Appl. **2008**, 902187 (2008), 11 pp.
18. M. Bidkham, H.A. Soleiman Mezerji, M. Eshaghi Gordji, Hyers–Ulam stability of polynomial equations. Abstr. Appl. Anal. **2010**, 754120 (2010), 7 pp.
19. N. Brillouët-Belluot, J. Brzdek, K. Ciepliński, On some developments in Ulam's type stability. Abstr. Appl. Anal. **2012**, 716936 (2012), 41 pp.
20. G. Birkhoff, Orthogonality in linear metric spaces. Duke Math. J. **1**, 169–172 (1935)

Y.J. Cho et al., *Stability of Functional Equations in Random Normed Spaces*, Springer Optimization and Its Applications 86, DOI 10.1007/978-1-4614-8477-6,

21. J. Brzdek, K. Ciepliński, A fixed point theorem and the Hyers–Ulam stability in non-Archimedean spaces. J. Math. Anal. Appl. **400**, 68–75 (2013)
22. J. Brzdek, A. Pietrzyk, A note on stability of the general linear equation. Aequ. Math. **75**, 267–270 (2008)
23. J. Brzdek, D. Popa, B. Xu, Selections of set-valued maps satisfying a linear inclusion in a single variable. Nonlinear Anal. **74**, 324–330 (2011)
24. J. Brzdek, D. Popa, B. Xu, On approximate solutions of the linear functional equation of higher order. J. Math. Anal. Appl. **373**, 680–689 (2011)
25. S.O. Carlsson, Orthogonality in normed linear spaces. Ark. Mat. **4**, 297–318 (1962)
26. L. Cădariu, V. Radu, Fixed points and the stability of Jensen's functional equation. J. Inequal. Pure Appl. Math. **4**(1), 4 (2003)
27. L. Cădariu, V. Radu, On the stability of the Cauchy functional equation: a fixed point approach. Grazer Math. Ber. **346**, 43–52 (2004)
28. L. Cădariu, V. Radu, Fixed point methods for the generalized stability of functional equations in a single variable. Fixed Point Theory Appl. **2008**, 749392 (2008)
29. L.P. Castro, A. Ramos, Hyers–Ulam–Rassias stability for a class of nonlinear Volterra integral equations. Banach J. Math. Anal. **3**, 47–56 (2009)
30. A.L. Cauchy, *Analyse Algebrique*, Cours d'Analyse de l'Ecole Polytechnique, vol. 1 (Debure, Paris, 1821)
31. S.S. Chang, Y.J. Cho, S.M. Kang, *Nonlinear Operator Theory in Probabilistic Metric Spaces* (Nova Science Publishers, New York, 2001)
32. Y.J. Cho, C.R. Diminnie, R.W. Freese, E.Z. Andalafte, Isosceles orthogonal triples in linear 2-normed spaces. Math. Nachr. **157**, 225–234 (1992)
33. Y.J. Cho, M. Eshaghi Gordji, S. Zolfaghari, Solutions and stability of generalized mixed type QC functional equations in random normed spaces. J. Inequal. Appl. **2010**, 403101 (2010), 16 pp.
34. Y.J. Cho, J.I. Kang, R. Saadati, Fixed points and stability of additive functional equations on the Banach algebras. J. Comput. Anal. Appl. **14**, 1103–1111 (2012)
35. Y.J. Cho, C. Park, Th.M. Rassias, R. Saadati, Inner product spaces and functional equations. J. Comput. Anal. Appl. **13**, 296–304 (2011)
36. Y.J. Cho, C. Park, R. Saadati, Fuzzy functional inequalities. J. Comput. Anal. Appl. **13**, 305–320 (2011)
37. Y.J. Cho, C. Park, R. Saadati, Functional inequalities in non-Archimedean Banach spaces. Appl. Math. Lett. **60**, 1994–2002 (2010)
38. Y.J. Cho, R. Saadati, Lattice non-Archimedean random stability of ACQ functional equations. Adv. Differ. Equ. **2011**, 31 (2011)
39. P.W. Cholewa, The stability problem for a generalized Cauchy type functional equation. Rev. Roum. Math. Pures Appl. **29**, 457–460 (1984)
40. J.K. Chung, P.K. Sahoo, On the general solution of a quartic functional equation. Bull. Korean Math. Soc. **40**, 565–576 (2003)
41. K. Ciepliński, Applications of fixed point theorems to the Hyers–Ulam stability of functional equations—a survey. Ann. Funct. Anal. **3**, 151–164 (2012)
42. K. Ciepliński, Generalized stability of multi-additive mappings. Appl. Math. Lett. **23**, 1291–1294 (2010)
43. K. Ciepliński, On the generalized Hyers–Ulam stability of multi-quadratic mappings. Comput. Math. Appl. **62**, 3418–3426 (2011)
44. K. Ciepliński, Stability of multi-additive mappings in β-Banach spaces. Nonlinear Anal. **75**, 4205–4212 (2012)
45. S. Czerwik, On the stability of the quadratic mapping in normed spaces. Abh. Math. Semin. Univ. Hamb. **62**, 59–64 (1992)
46. S. Czerwik, *Functional Equations and Inequalities in Several Variables* (World Scientific, New Jersey, 2002)
47. S. Czerwik, *Stability of Functional Equations of Ulam–Hyers–Rassias Type* (Hadronic Press, Palm Harbor, 2003)

48. S. Czerwik, The stability of the quadratic functional equation, in *Stability of Mappings of Hyers–Ulam Type*, ed. by Th.M. Rassias, J. Tabor (Hadronic Press, Palm Harbor, 1994), pp. 81–91
49. G. Deschrijver, E.E. Kerre, On the relationship between some extensions of fuzzy set theory. Fuzzy Sets Syst. **23**, 227–235 (2003)
50. J. Diaz, B. Margolis, A fixed point theorem of the alternative for contractions on a generalized complete metric space. Bull. Am. Math. Soc. **74**, 305–309 (1968)
51. C.R. Diminnie, A new orthogonality relation for normed linear spaces. Math. Nachr. **114**, 197–203 (1983)
52. F. Drljević, On a functional which is quadratic on A-orthogonal vectors. Publ. Inst. Math. (Belgr.) **54**, 63–71 (1986)
53. A. Ebadian, A. Najati, M. Eshaghi Gordji, On approximate additive–quartic and quadratic–cubic functional equations in two variables on Abelian groups. Results Math. (2010). doi:10.1007/s00025-010-0018-4
54. A. Ebadian, N. Ghobadipour, M. Eshaghi Gordji, A fixed point method for perturbation of bimultipliers and Jordan bimultipliers in C^*-ternary algebras. J. Math. Phys. **51**, 103508 (2010). doi:10.1063/1.3496391
55. W. Eichhorn, *Functional Equations in Economics* (Addison-Wesley, Reading, 1978)
56. M. Eshaghi Gordji, Stability of a functional equation deriving from quartic and additive functions. Bull. Korean Math. Soc. **47**, 491–502 (2010)
57. M. Eshaghi Gordji, Nearly ring homomorphisms and nearly ring derivations on non-Archimedean Banach algebras. Abstr. Appl. Anal. **2010**, 393247 (2010), 12 pp.
58. M. Eshaghi Gordji, B. Alizadeh, Y.W. Lee, G.H. Kim, Nearly quadratic n-derivations on non-Archimedean Banach algebras. Discrete Dyn. Nat. Soc. **2012**, 961642 (2012), 10 pp.
59. M. Eshaghi Gordji, M.B. Ghaemi, Y.J. Cho, H. Majani, A general system of Euler–Lagrange-type quadratic functional equations in Menger probabilistic non-Archimedean 2-normed spaces. Abstr. Appl. Anal. **2011**, 208163 (2011), 21 pp.
60. M. Eshaghi Gordji, M.B. Ghaemi, S.K. Gharetapeh, S. Shams, A. Ebadian, On the stability of J^*-derivations. J. Geom. Phys. **60**, 454–459 (2010)
61. M. Eshaghi Gordji, M.B. Ghaemi, H. Majani, Generalized Hyers–Ulam–Rassias theorem in Menger probabilistic normed spaces. Discrete Dyn. Nat. Soc. **2010**, 162371 (2010), 11 pp.
62. M. Eshaghi Gordji, M.B. Ghaemi, H. Majani, C. Park, Generalized Ulam–Hyers stability of Jensen functional equation in Šrstnev PN spaces. J. Inequal. Appl. **2010**, 868193 (2010), 14 pp.
63. M. Eshaghi Gordji, M.B. Ghaemi, J.M. Rassias, B. Alizadeh, Nearly ternary quadratic higher derivations on non-Archimedean ternary Banach algebras: a fixed point approach. Abstr. Appl. Anal. **2011**, 417187 (2011), 18 pp.
64. M. Eshaghi Gordji, S.K. Gharetapeh, C. Park, S. Zolfaghri, Stability of an additive–cubic–quartic functional equation. Adv. Differ. Equ. **2009**, 395693 (2009), 20 pp.
65. M. Eshaghi Gordji, S.K. Gharetapeh, J.M. Rassias, S. Zolfaghari, Solution and stability of a mixed type additive, quadratic and cubic functional equation. Adv. Differ. Equ. **2009**, 826130 (2009), 17 pp.
66. M. Eshaghi Gordji, S. Kaboli-Gharetapeh, C. Park, S. Zolfaghri, Stability of an additive–cubic–quartic functional equation. Adv. Differ. Equ. **2009**, 395693 (2009)
67. M. Eshaghi Gordji, R. Khodabakhsh, S.M. Jung, H. Khodaei, *AQCQ*-functional equation in non-Archimedean normed spaces. Abstr. Appl. Anal. **2010**, 741942 (2010), 22 pp.
68. M. Eshaghi Gordji, H. Khodaei, On the generalized Hyers–Ulam–Rassias stability of quadratic functional equations. Abstr. Appl. Anal. **2009**, 923476 (2009), 11 pp.
69. M. Eshaghi Gordji, H. Khodaei, Solution and stability of generalized mixed type cubic, quadratic and additive functional equation in quasi-Banach spaces. Nonlinear Anal. **71**, 5629–5643 (2009)
70. M. Eshaghi Gordji, H. Khodaei, The fixed point method for fuzzy approximation of a functional equation associated with inner product spaces. Discrete Dyn. Nat. Soc. **2010**, 140767 (2010), 15 pp.

71. M. Eshaghi Gordji, H. Khodaei, *Stability of Functional Equations* (LAP Lambert Academic Publishing, Saarbrücken, 2010)
72. M. Eshaghi Gordji, H. Khodaei, The fixed point method for fuzzy approximation of a functional equation associated with inner product spaces. Discrete Dyn. Nat. Soc. **2010**, 140767 (2010), pp. 15
73. M. Eshaghi Gordji, H. Khodaei, A. Ebadian, G.H. Kim, Nearly radical quadratic functional equations in p-2-normed spaces. Abstr. Appl. Anal. **2012**, 896032 (2012), 10 pp.
74. M. Eshaghi Gordji, H. Khodaei, R. Khodabakhsh, General quartic-cubic-quadratic functional equation in non-Archimedean normed spaces. Sci. Bull."Politeh." Univ. Buchar., Ser. A, Appl. Math. Phys. **72**, 69–84 (2010)
75. M. Eshaghi Gordji, H. Khodaei, G.H. Kim, Nearly quadratic mappings over p-adic fields. Abstr. Appl. Anal. **2012**, 285807 (2012), 12 pp.
76. M. Eshaghi Gordji, H. Khodaei, H.M. Kim, Approximate quartic and quadratic mappings in quasi-Banach spaces. Int. J. Math. Math. Sci. **2011**, 734567 (2011), 18 pp.
77. M. Eshaghi Gordji, G.H. Kim, Approximate n-Lie homomorphisms and Jordan n-Lie homomorphisms on n-Lie algebras. Abstr. Appl. Anal. **2012**, 279632 (2012), 11 pp.
78. M. Eshaghi Gordji, A. Najati, Approximately J^*-homomorphisms: a fixed point approach. J. Geom. Phys. **60**, 809–814 (2010)
79. M. Eshaghi Gordji, Th.M. Rassias, Ternary homomorphisms between unital ternary C^*-algebras. Proc. Rom. Acad., Ser. A : Math. Phys. Tech. Sci. Inf. Sci. **12**, 189–196 (2011)
80. M. Eshaghi Gordji, M.B. Savadkouhi, Stability of a mixed type cubic and quartic functional equations in random normed spaces. J. Inequal. Appl. **2009**, 527462 (2009), 9 pp.
81. M. Eshaghi Gordji, M.B. Savadkouhi, Stability of a mixed type cubic-quartic functional equation in non-Archimedean spaces. Appl. Math. Lett. **23**, 1198–1202 (2010)
82. M. Eshaghi Gordji, M.B. Savadkouhi, C. Park, Quadratic–quartic functional equations in RN-spaces. J. Inequal. Appl. **2009**, 868423 (2009), 14 pp.
83. G.Z. Eskandani, On the stability of an additive functional equation in quasi-Banach spaces. J. Math. Anal. Appl. **345**, 405–409 (2008)
84. T. Evans, A note on the associative law. J. Lond. Math. Soc. **25**, 196–201 (1950)
85. V.A. Faiziev, Th.M. Rassias, P.K. Sahoo, The space of (ψ, γ)-additive mappings on semigroups. Trans. Am. Math. Soc. **354**, 4455–4472 (2002)
86. W. Fechner, Stability of a functional inequalities associated with the Jordan–von Neumann functional equation. Aequ. Math. **71**, 149–161 (2006)
87. M. Fochi, Functional equations in A-orthogonal vectors. Aequ. Math. **38**, 28–40 (1989)
88. G.L. Forti, An existence and stability theorem for a class of functional equations. Stochastica **4**, 22–30 (1980)
89. Z. Gajda, On stability of additive mappings. Int. J. Math. Math. Sci. **149**, 431–434 (1991)
90. C.F. Gauss, *Theoria Moyus Corporum Caelestium* (Perthes–Besser, Hamburg, 1809)
91. R. Ger, J. Sikorska, Stability of the orthogonal additivity. Bull. Pol. Acad. Sci., Math. **43**, 143–151 (1995)
92. M.B. Ghaemi, M. Eshaghi Gordji, H. Majani, Approximately quintic and sextic mappings on the probabilistic normed spaces. Preprint
93. A. Gilányi, Eine zur Parallelogrammgleichung äquivalente Ungleichung. Aequ. Math. **62**, 303–309 (2001)
94. A. Gilányi, On a problem by K. Nikodem. Math. Inequal. Appl. **5**, 707–710 (2002)
95. A.M. Gleason, A definition of a quadratic form. Am. Math. Mon. **73**, 1049–1056 (1966)
96. J. Goguen, $\mathcal{L}$-fuzzy sets. J. Math. Anal. Appl. **18**, 145–174 (1967)
97. I. Goleţ, Some remarks on functions with values in probabilistic normed spaces. Math. Slovaca **57**, 259–270 (2007)
98. P.M. Gruber, Stability of isometries. Trans. Am. Math. Soc. **245**, 263–277 (1978)
99. S. Gudder, D. Strawther, Orthogonally additive and orthogonally increasing functions on vector spaces. Pac. J. Math. **58**, 427–436 (1975)
100. O. Hadžić, E. Pap, *Fixed Point Theory in PM-Spaces* (Kluwer Academic, Dordrecht, 2001)

101. O. Hadžić, E. Pap, New classes of probabilistic contractions and applications to random operators, in *Fixed Point Theory and Applications*, vol. 4, ed. by Y.J. Cho, J.K. Kim, S.M. Kong (Nova Science Publishers, Hauppauge, 2003), pp. 97–119
102. O. Hadžić, E. Pap, M. Budincević, Countable extension of triangular norms and their applications to the fixed point theory in probabilistic metric spaces. Kybernetika **38**, 363–381 (2002)
103. O. Hadžić, E. Pap, V. Radu, Generalized contraction mapping principles in probabilistic metric spaces. Acta Math. Hung. **101**, 131–148 (2003)
104. P. Hajek, *Metamathematics of Fuzzy Logic* (Kluwer Academic, Dordrecht, 1998)
105. G.H. Hardy, J.E. Littlewood, G. Polya, *Inequalities*, 2nd edn. (Cambridge University Press, Cambridge, 1952)
106. K. Hensel, Über eine neue Begrundung der Theorie der algebraischen Zahlen. Jahresber. Dtsch. Math.-Ver. **6**, 83–88 (1897)
107. D.H. Hyers, On the stability of the linear functional equation. Proc. Natl. Acad. Sci. USA **27**, 222–224 (1941)
108. D.H. Hyers, G. Isac, Th.M. Rassias, *Stability of Functional Equations in Several Variables* (Birkhäuser, Basel, 1998)
109. D.H. Hyers, Th.M. Rassias, Approximate homomorphism. Aequ. Math. **44**, 125–153 (1992)
110. G. Isac, Th.M. Rassias, Stability of ψ-additive mappings: applications to nonlinear analysis. Int. J. Math. Math. Sci. **19**, 219–228 (1996)
111. R.C. James, Orthogonality in normed linear spaces. Duke Math. J. **12**, 291–302 (1945)
112. R.C. James, Orthogonality and linear functionals in normed linear spaces. Trans. Am. Math. Soc. **61**, 265–292 (1947)
113. A. Javadian, E. Sorouri, G.H. Kim, M. Eshaghi Gordji, Generalized Hyers–Ulam stability of the second-order linear differential equations. J. Appl. Math. **2011**, 813137 (2011), 10 pp.
114. B. Jesen, J. Karpf, A. Thorup, Some functional equations in groups and rings. Math. Scand. **22**, 257–265 (1968)
115. P. Jordan, J. von Neumann, On inner products in linear metric spaces. Ann. Math. **36**, 719–723 (1935)
116. K.W. Jun, H.M. Kim, The generalized Hyers–Ulam–Rassias stability of a cubic functional equation. J. Math. Anal. Appl. **274**, 867–878 (2002)
117. K.W. Jun, H.M. Kim, I.S. Chang, On the Hyers–Ulam stability of an Euler–Lagrange type cubic functional equation. J. Comput. Anal. Appl. **7**, 21–33 (2005)
118. S.M. Jung, Hyers–Ulam–Rassias stability of Jensen's equation and its application. Proc. Am. Math. Soc. **126**, 3137–3143 (1998)
119. S.M. Jung, Stability of the quadratic equation of Pexider type. Abh. Math. Semin. Univ. Hamb. **70**, 175–190 (2000)
120. S.M. Jung, *Hyers–Ulam–Rassias Stability of Functional Equations in Mathematical Analysis* (Hadronic Press, Palm Harbor, 2001)
121. S.M. Jung, Hyers–Ulam stability of a system of first order linear differential equations with constant coefficients. J. Math. Anal. Appl. **320**, 549–561 (2006)
122. S.M. Jung, A fixed point approach to the stability of a Volterra integral equation. Fixed Point Theory Appl. **2007**, 57064 (2007), 9 pp.
123. S.M. Jung, Hyers–Ulam stability of linear differential equations of first order. Appl. Math. Lett. **22**, 70–74 (2009)
124. S.M. Jung, *Hyers–Ulam–Rassias Stability of Functional Equations in Nonlinear Analysis* (Springer, New York, 2011)
125. S.M. Jung, Z.H. Lee, A fixed point approach to the stability of quadratic functional equation with involution. Fixed Point Theory Appl. **2008**, 732086 (2008), 11 pp.
126. S.M. Jung, S. Min, On approximate Euler differential equations. Abstr. Appl. Anal. **2009**, 537963 (2009), 8 pp.
127. S.M. Jung, S. Min, A fixed point approach to the stability of the functional equation $f(x+y) = F[f(x), f(y)]$. Fixed Point Theory Appl. **2009**, 912046 (2009), 8 pp.

128. S.M. Jung, Th.M. Rassias, Ulam's problem for approximate homomorphisms in connection with Bernoulli's differential equation. Appl. Math. Comput. **187**, 223–227 (2007)
129. S.M. Jung, Th.M. Rassias, Generalized Hyers–Ulam stability of Riccati differential equation. Math. Inequal. Appl. **11**, 777–782 (2008)
130. C.F.K. Jung, On generalized complete metric spaces. Bull. Am. Math. Soc. **75**, 113–116 (1969)
131. R.V. Kadison, J.R. Ringrose, *Fundamentals of the Theory of Operator Algebras* (Academic Press, New York, 1983)
132. P. Kannappan, Quadratic functional equation and inner product spaces. Results Math. **27**, 368–372 (1995)
133. P. Kannappan, *Functional Equations and Inequalities* (Springer, New York, 2009)
134. J. Kepler, *Chilias Logarithmorum ad Totidem Numerous Rotundod*. Marburg (1624)
135. H. Khodaei, M. Eshaghi Gordji, S.S. Kim, Y.J. Cho, Approximation of radical functional equations related to quadratic and quartic mappings. J. Math. Anal. Appl. **397**, 284–297 (2012)
136. H. Khodaei, Th.M. Rassias, Approximately generalized additive functions in several variables. Int. J. Nonlinear Anal. Appl. **1**, 22–41 (2010)
137. A. Khrennikov, in *Non-Archimedean Analysis: Quantum Paradoxes, Dynamical Systems and Biological Models*. Mathematics and Its Applications, vol. 427 (Kluwer Academic, Dordrecht, 1997)
138. E.P. Klement, R. Mesiar, E. Pap, *Triangular Norms* (Kluwer Academic, Dordrecht, 2000)
139. E.P. Klement, R. Mesiar, E. Pap, Triangular norms, position paper I: basic analytical and algebraic properties. Fuzzy Sets Syst. **143**, 5–26 (2004)
140. E.P. Klement, R. Mesiar, E. Pap, Triangular norms, position paper II: general constructions and parameterized families. Fuzzy Sets Syst. **145**, 411–438 (2004)
141. E.P. Klement, R. Mesiar, E. Pap, Triangular norms, position paper III: continuous t-norms. Fuzzy Sets Syst. **145**, 439–454 (2004)
142. T. Kochanek, M. Lewicki, Stability problem for number-theoretically multiplicative functions. Proc. Am. Math. Soc. **135**, 2591–2597 (2007)
143. T. Kochanek, W. Wyrobek, Measurable orthogonally additive functions modulo a discrete subgroup. Acta Math. Hung. **123**, 239–248 (2009)
144. M. Kuczma, Functional equations on restricted domains. Aequ. Math. **18**, 1–34 (1978)
145. M. Kuczma, *An Introduction to the Theory of Functional Equations and Inequalities* (Uniwersytet Slaski, Warszawa–Krakow–Katowice, 1985)
146. M. Kuczma, On measurable functions with vanishing differences. Ann. Math. Sil. **6**, 42–60 (1992)
147. S. Kurepa, A cosine functional equation in Banach algebras. Acta Sci. Math. **23**, 255–267 (1962)
148. B. Lafuerza-Guillén, A. Rodríguez-Lallena, C. Sempi, A study of boundedness in probabilistic normed spaces. J. Math. Anal. Appl. **232**, 183–196 (1999)
149. B. Lafuerza-Guillén, D-bounded sets in probabilistic normed spaces and their products. Rend. Mat., Ser. VII **21**, 17–28 (2001)
150. B. Lafuerza-Guillén, Finite products of probabilistic normed spaces. Rad. Mat. **13**, 111–117 (2004)
151. S. Lee, S. Im, I. Hwang, Quartic functional equations. J. Math. Anal. Appl. **307**, 387–394 (2005)
152. A.M. Legendre, *Elements de Geometrie*, Vol. II (Didot, Paris, 1791)
153. Y. Li, Y. Shen, Hyers–Ulam stability of nonhomogeneous linear differential equations of second order. Int. J. Math. Math. Sci. **2009**, 576852 (2009), 7 pp.
154. E.R. Lorch, On certain implications which characterize Hilbert spaces. Ann. Math. **49**, 523–532 (1948)
155. N. Lungu, D. Popa, Hyers–Ulam stability of a first order partial differential equation. J. Math. Anal. Appl. **385**, 86–91 (2012)

156. W.A.J. Luxemburg, On the convergence of successive approximations in the theory of ordinary differential equations II. Proc. K. Ned. Akad. Wet., Ser. A, Indag. Math. **20**, 540–546 (1958)
157. G. Maksa, The stability of the entropy of degree alpha. J. Math. Anal. Appl. **346**, 17–21 (2008)
158. B. Margolis, J.B. Diaz, A fixed point theorem of the alternative for contractions on the generalized complete metric space. Bull. Am. Math. Soc. **126**, 305–309 (1968)
159. R.E. Megginson, *An Introduction to Banach Space Theory* (Springer, New York, 1998)
160. K. Menger, Statistical metrics. Proc. Natl. Acad. Sci. USA **28**, 535–537 (1942)
161. D. Miheţ, V. Radu, On the stability of the additive Cauchy functional equation in random normed spaces. J. Math. Anal. Appl. **343**, 567–572 (2008)
162. D. Miheţ, The probabilistic stability for a functional equation in a single variable. Acta Math. Hung. doi:10.1007/s10474-008-8101-y
163. D. Miheţ, The fixed point method for fuzzy stability of the Jensen functional equation. Fuzzy Sets Syst. doi:10.1016/j.fss.2008.06.014
164. D. Miheţ, R. Saadati, S.M. Vaezpour, The stability of the quartic functional equation in random normed spaces. Acta Appl. Math. **110**, 797–803 (2010)
165. D. Miheţ, R. Saadati, S.M. Vaezpour, The stability of an additive functional equation in Menger probabilistic φ-normed spaces. Math. Slovaca **61**, 817–826 (2011)
166. A.K. Mirmostafaee, M.S. Moslehian, Fuzzy versions of Hyers–Ulam–Rassias theorem. Fuzzy Sets Syst. **159**, 720–729 (2008)
167. A.K. Mirmostafaee, M.S. Moslehian, Fuzzy approximately cubic mappings. Inf. Sci. **178**, 3791–3798 (2008)
168. A.K. Mirmostafaee, M.S. Moslehian, Fuzzy stability of additive mappings in non-Archimedean fuzzy normed spaces. Fuzzy Sets Syst. **160**, 1643–1652 (2009)
169. M. Mirzavaziri, M.S. Moslehian, A fixed point approach to stability of a quadratic equation. Bull. Braz. Math. Soc. **37**, 361–376 (2006)
170. T. Miura, S.M. Jung, S.E. Takahasi, Hyers–Ulam–Rassias stability of the Banach space valued linear differential equations $y' = \lambda y$. J. Korean Math. Soc. **41**, 995–1005 (2004)
171. T. Miura, S. Miyajima, S.E. Takahasi, Hyers–Ulam stability of linear differential operator with constant coefficients. Math. Nachr. **258**, 90–96 (2003)
172. T. Miura, S. Miyajima, S.E. Takahasi, A characterization of Hyers–Ulam stability of first order linear differential operators. J. Math. Anal. Appl. **286**, 136–146 (2003)
173. M. Mohammadi, Y.J. Cho, C. Park, P. Vetro, R. Saadati, Random stability of an additive–quadratic–quartic functional equation. J. Inequal. Appl. **2010**, 754210 (2010), 18 pp.
174. M.S. Moslehian, On the orthogonal stability of the Pexiderized quadratic equation. J. Differ. Equ. Appl. **11**, 999–1004 (2005)
175. M.S. Moslehian, On the stability of the orthogonal Pexiderized Cauchy equation. J. Math. Anal. Appl. **318**, 211–223 (2006)
176. M.S. Moslehian, K. Nikodem, D. Popa, Asymptotic aspect of the quadratic functional equation in multinormed spaces. J. Math. Anal. Appl. **355**, 717–724 (2009)
177. M.S. Moslehian, Th.M. Rassias, Orthogonal stability of additive type equations. Aequ. Math. **73**, 249–259 (2007)
178. M.S. Moslehian, Th.M. Rassias, Stability of functional equations in non-Archimedean spaces. Appl. Anal. Discrete Math. **1**, 325–334 (2007)
179. D.H. Mushtari, On the linearity of isometric mappings on random normed spaces. Kazan Gos. Univ. Uchen. Zap. **128**, 86–90 (1968)
180. A. Najati, Hyers–Ulam–Rassias stability of a cubic functional equation. Bull. Korean Math. Soc. **44**, 825–840 (2007)
181. A. Najati, Th.M. Rassias, Stability of homomorphisms and (ϕ, θ)-derivations. Appl. Anal. Discrete Math. **3**, 264–281 (2009)
182. A. Najati, Y.J. Cho, Generalized Hyers–Ulam stability of the pexiderized Cauchy functional equation in non-Archimedean spaces. Fixed Point Theory Appl. **2011**, 309026 (2011), 11 pp.

183. A. Najati, G.Z. Eskandani, Stability of a mixed additive and cubic functional equation in quasi-Banach spaces. J. Math. Anal. Appl. **342**, 1318–1331 (2008)
184. A. Najati, C. Park, Hyers–Ulam–Rassias stability of homomorphisms in quasi-Banach algebras associated to the Pexiderized Cauchy functional equation. J. Math. Anal. Appl. **335**, 763–776 (2007)
185. A. Najati, Th.M. Rassias, Stability of a mixed functional equation in several variables on Banach modules. Nonlinear Anal. **72**, 1755–1767 (2010)
186. D. O'Regan, R. Saadati, Nonlinear contraction theorems in probabilistic spaces. Appl. Math. Comput. **195**, 86–93 (2008)
187. L. Paganoni, J. Rätz, Conditional function equations and orthogonal additivity. Aequ. Math. **50**, 135–142 (1995)
188. Z. Pales, R.W. Craigen, The associativity equation revisited. Aequ. Math. **37**, 306–312 (1989)
189. P.M. Pardalos, Th.M. Rassias, A.A. Khan (eds.), *Nonlinear Analysis and Variational Problems. In Honor of George Isac*. Springer Optimization and Its Applications (Springer, New York, 2010)
190. C. Park, On an approximate automorphism on a C^*-algebra. Proc. Am. Math. Soc. **132**, 1739–1745 (2004)
191. C. Park, Fixed points and Hyers–Ulam–Rassias stability of Cauchy–Jensen functional equations in Banach algebras. Fixed Point Theory Appl. **2007**, 50175 (2007)
192. C. Park, Generalized Hyers–Ulam–Rassias stability of quadratic functional equations: a fixed point approach. Fixed Point Theory Appl. **2008**, 493751 (2008)
193. C. Park, Fixed points in functional inequalities. J. Inequal. Appl. **2008**, 298050 (2008), pp. 8
194. C. Park, Fuzzy stability of additive functional inequalities with the fixed point alternative. J. Inequal. Appl. **2009**, 410576 (2009)
195. C. Park, Y. Cho, M. Han, Functional inequalities associated with Jordan–von Neumann type additive functional equations. J. Inequal. Appl. **2007**, 41820 (2007)
196. C. Park, Y.J. Cho, H.A. Kenary, Orthogonal stability of a generalized quadratic functional equation in non-Archimedean spaces. J. Comput. Anal. Appl. **14**, 526–535 (2012)
197. C. Park, M. Eshaghi Gordji, Comment on "Approximate ternary Jordan derivations on Banach ternary algebras" [B. Savadkouhi et al. J. Math. Phys. **50**, 042303 (2009)]. J. Math. Phys. **51**, 044102 (2010)
198. C. Park, A. Najati, Generalized additive functional inequalities in Banach algebras. Int. J. Nonlinear Anal. Appl. **1**, 54–62 (2010)
199. C. Park, J. Park, Generalized Hyers–Ulam stability of an Euler–Lagrange type additive mapping. J. Differ. Equ. Appl. **12**, 1277–1288 (2006)
200. C. Park, J.M. Rassias, Stability of the Jensen-type functional equation in C^*-algebras: a fixed point approach. Abstr. Appl. Anal. **2009**, 360432 (2009), 17 pp.
201. C. Park, Th.M. Rassias, On a generalized Trif's mapping in Banach modules over a C^*-algebra. J. Korean Math. Soc. **43**, 1417–1428 (2006)
202. C. Park, Th.M. Rassias, Stability of homomorphisms in JC^*-algebras. Pac.-Asian J. Math. **1**, 1–17 (2007)
203. C. Park, Th.M. Rassias, Fixed points and generalized Hyers–Ulam stability of quadratic functional equations. J. Math. Inequal. **1**, 515–528 (2007)
204. C. Park, Th.M. Rassias, Homomorphisms and derivations in proper JCQ^*-triples. J. Math. Anal. Appl. **337**, 1404–1414 (2008)
205. C. Park, Th.M. Rassias, Homomorphisms between JC^*-algebras. Stud. Univ. Babeş–Bolyai, Math. **53**, 9–24 (2008)
206. C. Park, Th.M. Rassias, Fixed points and stability of the Cauchy functional equation. Aust. J. Math. Anal. Appl. **6**, 14 (2009), 9 pp.
207. C. Park, Th.M. Rassias, Isomorphisms in unital C^*-algebras. Int. J. Nonlinear Anal. Appl. **1**, 1–10 (2010)
208. A.G. Pinsker, Sur une fonctionnelle dans l'espace de Hilbert. C. R. (Dokl.) Acad. Sci. URSS **20**, 411–414 (1938)

209. D. Popa, I. Rasa, On the Hyers–Ulam stability of the linear differential equation. J. Math. Anal. Appl. **381**, 530–537 (2011)
210. D. Popa, I. Rasa, Hyers–Ulam stability of the linear differential operator with nonconstant coefficients. Appl. Math. Comput. **291**, 1562–1568 (2012)
211. D. Popa, I. Rasa, The Fréchet functional equation with application to the stability of certain operators. J. Approx. Theory **164**, 138–144 (2012)
212. W. Prager, J. Schwaiger, Multi-affine and multi-Jensen functions and their connection with generalized polynomials. Aequ. Math. **69**, 41–57 (2005)
213. V. Radu, Some remarks on quasi-normed and random normed structures, in *Seminar on Probability Theory and Applications (STPA)*, vol. 159 (West Univ. of Timişoara, Timişoara, 2003)
214. V. Radu, The fixed point alternative and the stability of functional equations. Fixed Point Theory **4**, 91–96 (2003)
215. J.M. Rassias, H.M. Kim, Generalized Hyers–Ulam stability for general additive functional equations in quasi-β-normed spaces. J. Math. Anal. Appl. **356**, 302–309 (2009)
216. Th.M. Rassias, On the stability of the linear mapping in Banach spaces. Proc. Am. Math. Soc. **72**, 297–300 (1978)
217. Th.M. Rassias, New characterization of inner product spaces. Bull. Sci. Math. **108**, 95–99 (1984)
218. Th.M. Rassias, The stability of mappings and related topics, in *Report on the 27th Internat. Symp. on Functional Equations*. Aequat. Math., vol. 39 (1990), pp. 292–293
219. Th.M. Rassias, On a modified Hyers–Ulam sequence. J. Math. Anal. Appl. **158**, 106–113 (1991)
220. Th.M. Rassias, On the stability of the quadratic functional equation and its applications. Stud. Univ. Babeş-Bolyai, Math. **43**, 89–124 (1998)
221. Th.M. Rassias, The problem of S.M. Ulam for approximately multiplicative mappings. J. Math. Anal. Appl. **246**, 352–378 (2000)
222. Th.M. Rassias, On the stability of functional equations in Banach spaces. J. Math. Anal. Appl. **251**, 264–284 (2000)
223. Th.M. Rassias, On the stability of functional equations originated by a problem of Ulam. Mathematica **44**(67), 39–45 (2002)
224. Th.M. Rassias, On the stability of minimum points. Mathematica **45**(68), 93–104 (2003)
225. Th.M. Rassias (ed.), *Functional Equations, Inequalities and Applications* (Kluwer Academic, Dordrecht, 2003)
226. Th.M. Rassias, J. Brzdek (eds.), *Functional Equations in Mathematical Analysis* (Springer, New York, 2012)
227. Th.M. Rassias, P. Šemrl, On the behavior of mappings which do not satisfy Hyers–Ulam stability. Proc. Am. Math. Soc. **114**, 989–993 (1992)
228. Th.M. Rassias, P. Šemrl, On the Hyers–Ulam stability of linear mappings. J. Math. Anal. Appl. **173**, 325–338 (1993)
229. Th.M. Rassias, J. Tabor (eds.), *Stability of Mappings of Hyers–Ulam Type* (Hadronic Press, Palm Harbor, 1994)
230. J. Rätz, On approximately additive mappings, in *General Inequalities 2*, ed. by E.F. Beckenbach. Internat. Ser. Numer. Math., vol. 47 (Birkhäuser, Basel, 1980), pp. 233–251
231. J. Rätz, On orthogonally additive mappings. Aequ. Math. **28**, 35–49 (1985)
232. J. Rätz, On inequalities associated with the Jordan–von Neumann functional equation. Aequ. Math. **66**, 191–200 (2003)
233. J. Rätz, Gy. Szabó, On orthogonally additive mappings IV. Aequ. Math. **38**, 73–85 (1989)
234. A.M. Robert, *A Course in p-Adic Analysis*. Graduate Texts in Mathematics, vol. 198 (Springer, New York, 2000)
235. I.A. Rus (ed.), *Principles and Applications of Fixed Point Theory* (Dacia, Cluj-Napoca, 1979)
236. R. Saadati, M. Amini, D-boundedness and D-compactness in finite dimensional probabilistic normed spaces. Proc. Indian Acad. Sci. Math. Sci. **115**, 483–492 (2005)

237. R. Saadati, Y.J. Cho, J. Vahidi, The stability of the quartic functional equation in various spaces. Comput. Math. Appl. **60**, 1994–2002 (2010)
238. R. Saadati, D. O'Regan, S.M. Vaezpour, J.K. Kim, Generalized distance and common fixed point theorems in Menger probabilistic metric spaces. Bull. Iran. Math. Soc. **35**, 97–117 (2009)
239. P.K. Sahoo, T. Riedel, *Mean Value Theorem and Functional Equations* (World Scientific, Singapore, 1998)
240. J. Schwaiger, Remark 12, in *Report on the 25th Internat. Symp. on Functional Equations*. Aequat. Math., vol. 35 (1988), pp. 120–121
241. B. Schweizer, A. Sklar, *Probabilistic Metric Spaces* (Elsevier, North Holland, 1983)
242. A.N. Šerstnev, On the motion of a random normed space. Dokl. Akad. Nauk SSSR **149**, 280–283 (1963) (English translation in Sov. Math. Dokl. 4, 388–390 (1963))
243. E.V. Shulman, Group representations and stability of functional equations. J. Lond. Math. Soc. **54**, 111–120 (1996)
244. F. Skof, Proprietà locali e approssimazione di operatori. Rend. Semin. Mat. Fis. Milano **53**, 113–129 (1983)
245. K. Sundaresan, Orthogonality and nonlinear functionals on Banach spaces. Proc. Am. Math. Soc. **34**, 187–190 (1972)
246. Gy. Szabó, Sesquilinear-orthogonally quadratic mappings. Aequ. Math. **40**, 190–200 (1990)
247. L. Szekelyhidi, *Convolution Type Functional Equations on Topological Abelian Groups* (World Scientific, Singapore, 1991)
248. J. Tabor, J. Tabor, Stability of the Cauchy functional equation in metric groupoids. Aequ. Math. **76**, 92–104 (2008)
249. T. Trif, On the stability of a general gamma-type functional equation. Publ. Math. (Debr.) **60**, 47–61 (2002)
250. S.M. Ulam, *A Collection of Mathematical Problems* (Interscience, New York, 1960)
251. S.M. Ulam, *Problems in Modern Mathematics* (Wiley, New York, 1964), Chap. VI, Sci. Ed.
252. S.M. Ulam, *Sets, Numbers and Universe* (MIT Press, Cambridge, 1974)
253. F. Vajzović, Über das Funktional H mit der Eigenschaft: $(x, y) = 0 \Rightarrow H(x + y) + H(x - y) = 2H(x) + 2H(y)$. Glas. Mat. Ser. III **2**(22), 73–81 (1967)
254. Z. Wang, Th.M. Rassias, Intuitionistic fuzzy stability of functional equations associated with inner product spaces. Abstr. Appl. Anal. **2011**, 456182 (2011), 19 pp.
255. G. Wang, M. Zhou, L. Sun, Hyers–Ulam stability of linear differential equations of first order. Appl. Math. Lett. **21**, 1024–1028 (2008)
256. A. Wilansky, Additive functions, in *Lectures on Calculus*, ed. by K.O. May (Holden–Day, Toronto, 1967), pp. 97–124
257. W. Wyrobek, Orthogonally additive functions modulo a discrete subgroup, in *Aequat. Math.*, vol. 78 (Springer, New York, 2009), pp. 63–69
258. T.Z. Xu, On the stability of multi-Jensen mappings in β-normed spaces. Appl. Math. Lett. **25**, 1866–1870 (2012)
259. T.Z. Xu, Stability of multi-Jensen mappings in non-Archimedean normed spaces. J. Math. Phys. **53**, 023507 (2012), 9 pp.
260. M. Youssef, E. Elqorachi, Th.M. Rassias, Hyers–Ulam stability of the Jensen functional equation in quasi-Banach spaces. Nonlinear Funct. Anal. Appl. **15**, 581–603 (2010)
261. L.A. Zadeh, Fuzzy sets. Inf. Control **8**, 338–353 (1965)
262. O. Zariski, P. Samuel, *Commutative Algebra* (Van Nostrand, Princeton, 1958)
263. E. Zeidler, *Nonlinear Functional Analysis and Its Applications I: Fixed-Point Theorems* (Springer, New York, 1986)
264. H.J. Zimmermann, *Fuzzy Set Theory and Its Application* (Kluwer, Dordrecht, 1985)

Index

Y.J. Cho et al., *Stability of Functional Equations in Random Normed Spaces*, Springer Optimization and Its Applications 86, DOI 10.1007/978-1-4614-8477-6,

Printed in the United States
By Bookmasters